ELECTRONIC COMMUNICATION IN DEVELOPING COUNTRIES

Explanatory Theory, Volume 2

Edited by
Connie Eigenmann

ELECTRONIC COMMUNICATION IN DEVELOPING COUNTRIES

Explanatory Theory, Volume 2

Edited by
Connie Eigenmann

COMMON GROUND PUBLISHING 2015

First published in 2015 in Champaign, Illinois, USA
by Common Ground Publishing LLC
as part of Global Studies book imprint

Electronic Communication in Developing Countries Explanatory Theory, volume 2 / by Connie Eigenmann

ISBN 978-1-61229-806-1 (pbk: alk. Paper) –ISBN 978-1-61229-807-8 (pdf)

Cover Photo Credit: December Gladden

Table of Contents

ACKNOWLEDGEMENTS

When students and colleagues contribute to a great undertaking, it is for a great purpose… "that we may publish with the voice of thanksgiving, and tell of all thy wondrous works" (*Psalm* 26:7). This collection of research acknowledges God's power and foresight in its publication. I would also like to acknowledge the encouragement and skills of Ian Nelk, editor, and the dedicated researchers who came alongside. There were so many helpful academics that helped gather data and obtain research permissions within each sojourned country. They are truly worthy to be thanked and acknowledged.

Zahrah Aljaber

Zahrah Aljaber is from Alhasa, Saudi Arabia. She is a graduate student of Fort Hays State University (FHSU), Hays, Kansas, studying for the Master's degree in General Communication Studies. Zahrah received a Bachelor's degree in Arabic Language from King Fasil University, and has worked as an Arabic language instructor. Zahrah speaks Arabic, English and some Farsi. She is focused with her family life and raising her son while striving to improve her English language skills, and finish the Master's degree. Zahrah likes to help other people and volunteers in the community. She enjoys cooking Arabic and international foods.

Lizette Avalos

Lizette Avalos is an undergraduate student at Fort Hays State University, Hays, Kansas, double majoring in Leadership Studies and Communication Studies with a Business Administration minor. She received her Associate of Arts Degree in Behavioral Science from Seward County Community College (2013). Lizette worked part time in the FHSU Student Government Association as the Public Relations and Community Relations Director. Lizette is involved in several organizations at FHSU including *Omicron Delta Kappa*, Mortar Board, *Tau Sigma*, Circle K International, Leadership Studies Association, Stripes for St. Jude, *Sigma Sigma Sigma*, and Wishmakers-on-Campus. During her free time, Lizette enjoys volunteering through the Center for Student Involvement and other service opportunities offered at FHSU. Lizette's hobbies include reading, traveling, and spending time with her family and friends.

Trisha Capansky

Trisha Capansky, Ph.D. has taught five years at the undergraduate and graduate level in technical and professional communication. Her courses involve a theoretical and practical assessment of language usage as it is understood through existing and emerging media, and through social contexts. Dr. Capansky's primary research area is in the application and reception of nontraditional content/medium pairings. Particularly, her interests are in how government and workforce cultures use media to shape, modify, or reform both internal and external practices; the material history of texts; and reader/artifact relationships. In her dissertation she explored how the Declaration of Independence, as a new genre in political discourse, influences global relations. Beyond the dissertation and classroom, she is active within the local community where her involvement includes an appointment to the municipal planning board. She is currently affiliated with the Department of English and Modern Foreign Languages, University of Tennessee at Martin.

Julia Dache

Julia Daitche has a B.A. in Anglophone Studies and Philosophy. She recently graduated in Anglophone and German Studies to achieve her Master's degree from the University of Duisburg-Essen, Germany in 2015. She spent a semester abroad at the Fort Hays State University and studied Communication Studies in fall semester 2013. Julia worked several years in the international office and in the German department of her university. After her graduation she started to work at the Studierendenwerk Essen-Duisburg where she takes care of international students and offers intercultural training.

Connie S. Eigenmann

Connie S. Eigenmann (B. A. Eastern Illinois University 1991, M. A. Eastern Illinois University 1992, Ph. D. University of Oklahoma 1995) has been affiliated with 15 professional communication organizations. She has presented 30 refereed papers in intercultural communication, rhetoric, electronic communication usage, and storytelling. Dr. Eigenmann has work experience in international management and consulting; public relations, university development and telemarketing; speech clinic and forensics judging; and teaching in higher education. Before joining the Communication Studies Department at Fort Hays State University, she acted as Head of Business Communications, Sultan Qaboos University, Oman; Dean of the Indianapolis College of Business and Computer Science in Lahore, Pakistan; and Professor of English Communication at Shantou University, China. She has been affiliated with Lamson & Sessions, Inc. as a technical writer for their electrical conduit and fixture warehouse in Oklahoma City, USA. Her articles have been published in *The International Journal of Technology, Knowledge and Society, The International Journal of Diverse Identities, The International Journal of the Book,*

The Encyclopedia of Energy, Biomes and Ecosystems, and the *American Communication Journal.*

Baturalp Hamit Gursoy

Baturalp Hamit Gursoy is from Istanbul, Turkey. He graduated from Vocational Tourism and Hotel Management High School (2008), and Istanbul Kultur University (Associate Degree 2010). He has worked as an accountant, bartender, receptionist and cashier at the Istanbul Airport, Ritz Carlton, and Radisson SAS Hotel. He graduated from Fort Hays State University in Business Administration (2014), working in the Geosciences and ESL Departments. Baturalp is currently a graduate student and teaching assistant at FHSU completing his MLA concentration in Global Professional English. In his spare time, Baturalp plays soccer as he has been a semi-professional for 10 years. He also cooks Italian and Turkish specialties.

XiangXiang Jura Ji

Xiangxiang Jura Ji (B.A., Ningbo Institute of Technology, Zhejiang University 2013; M.A., Fort Hays State University 2015) is a graduate student from China. She has taught public speaking classes as a GTA at Fort Hays State University (Aug., 2014- May, 2015). This experience enriched her understanding in cultural difference, and inspired her research interests. Jura is a Rotary International Scholar who enjoys giving culture-related presentations to people in the Western Kansas area (Hays, Oberlin, Ellis). When she was a college student in China, Jura published reports on the *Qianjiang Evening News* (2010), and articles by Zhejiang University Press (2010-2011). She was selected to serve as a bilingual volunteer of World Expo (2010), and was one of the three initiators of TEDxNingbo (2012). After coming to the U.S., she spent her spare time in community activities and organizing a weekly Bible study. Jura plans to pursue a second master's degree in Human Resource Management.

Jia Jenny Jia

Jia Jenny Jia is a graduate student in the Communication Studies program at Fort Hays State University. She is from China, and has graduated from Sias International University (B.A. 2011) and Fort Hays State University (B.B.A. 2011) with a major in Accounting. After her dual undergraduate degrees, she worked full time in the Student Recruitment and Examination Information Center of Henan Province in China. There she advised students who are studying for the college entrance examinations online. The three-years of academic work gave her patience, and developed her as an out-going person willing to help. Now she is studying Social Networking Management in Fort Hays State University, and adding skills to help people.

Xiaotian Marcus Gao

Xiaotian Marcus Gao is a graduate student at Fort Hays State University majoring in Communication Studies with a focus on organizational communication. He has a Bachelor's Degree of Organizational Leadership from Fort Hays State University and another Bachelor's Degree of International Finance from Shenyang Normal University in Liaoning, China. He was the recipient of the Presidential Scholarship offered by FHSU. Marcus is also very involved in non-profit organizations and community service. In April, 2015, he was chosen to be the student Rotary scholar for Rotary Kansas District 5670. He has also worked with Habitat for Humanity to volunteer in Santa Fe, New Mexico and St. Jude Children's Research Hospital in Memphis, Tennessee. Marcus has also worked for the Department of Leadership Studies at FHSU as an international student advisor to help American faculty members to adapt class materials to the reality of China.

Naomi Nutsch

Naomi Nutsch is an undergraduate student at Fort Hays State University, Hays, Kansas, majoring in General Communication, specializing in Intercultural Studies. She has an Associate of Arts degree in Business from Cloud County Community College (2013). Naomi has academic scholarships to assist in the educational expenses and works part-time using her experience and education in business. Naomi's focus is on her education, but she also strives to improve her community, including volunteer work at her local movie theatre, fire department, and the County fair. Naomi has participated in art competitions, Future Business Leaders of America, cheerleading, and theatrics. In her free time, Naomi enjoys scrapbooking, spending time with her loved ones, as well as keeping up to date with arts, theatre, and entertainment.

Lei Joe Qiao

Lei Qiao is currently a graduate student of Fort Hays State University (FHSU) majoring in Global Management. Lei earned his Bachelor's degree of Accounting in SIAS International University in China (2012). After he graduated, he worked in the Logistics Department of SIAS International University, his first job. This job opportunity developed personal skills and adaption to an intercultural society. After two years of working experience, Lei continued his education in the United States. During his free time, he likes playing saxophone. He is also a member of the FHSU Wind Ensemble.

Brandon Steinert

Brandon Steinert is a graduate student at Fort Hays State University in Hays, Kansas, majoring in communication with an emphasis in organizational leadership. He has a Bachelor of Science degree in journalism and mass communications from Kansas State University. After finishing his undergraduate degree in 2009, Brandon worked for a group of small newspapers as a managing editor for two years. He then served as Director of Public Relations and Marketing at Barton Community College. Brandon is highly active in his community as president of Barton County Young Professionals, a member of the local community theater board and is a strong proponent of health and fitness. He strives to live by the motto, "what we consume becomes us," which refers to anything from food and drink to books and movies. He intentionally uses "we" and "us" rather than "I" and "me."

Chenyang Doris Wang

Chenyang Doris Wang currently is an international graduate student in the major of Liberal Studies at Fort Hays State University. Chenyang participated in the "3+1" program of her hometown university (Northwest University of Politics and Law) in 2012. In addition to the three years of college study in China, she also received a Bachelor's degree of General Studies with a one-year learning experience in FHSU. Now she is concentrating on finishing the graduation thesis and the Human Resources Management Certification. In April 2015, Chenyang with her group members designed the half-day face-to-face workshop on Change Management for the Management Development Center at FHSU. The workshop included the change management theory delivery, video introduction, game and case study, which allowed participants to immediately apply the newly acquired knowledge in practice.

Yuchen Wang

Yuchen Wang is currently a graduate student of Fort Hays State University (FHSU) pursuing her Master's degree in Communication. Yuchen has the

Bachelor's degree of Communication and Journalism. She completed several internships in China; including the assistant editor of *China Central Television* (CCTV) Channel 2, the assistant reporter of *Xi'an Evening News Press*, and a reporter for *Xi'an TV Station*. Currently, Yuchen is working as a resident assistant in FHSU and has been nominated as the most successful programs holder in April, 2015. As an active volunteer, Yuchen also provides presentations introducing China and Chinese culture to students of local elementary schools regularly.

Shuhao Joshua Zhang

Shuhao Zhang is currently at Fort Hays State University (FHSU) in the Communication Studies program pursuing his Master's degree. He is from China, and has a double Bachelor's degree of Finance and Leadership from Shenyang Normal University (2014) and Fort Hays State University (2014). He has completed an internship as a salesman at Beijing Hyundai Motor Co., Ltd., and worked as a shop assistant at Tianwen Kadokawa Animation and Comics Co., Ltd. in Shenyang, China. He is a member of the Chinese Students & Scholars Association in Hays, Kansas.

Jane Yajuan Zhou

Jane Yajuan Zhou, born in China, was educated in both P. R. China and the U.S., where she was awarded a Bachelor's degree in Business Administration from Golden Gate University in San Francisco. While in SF, she worked several years as an independent debt consultant. Jane is a graduate student in the communication studies program at Fort Hays State University. From Spring 2014 to Fall of 2014, serving as a GTA in Fort Hays, Jane taught two Oral Communication courses as the primary instructor. Jane currently lives in Washington DC, where she is finishing the research and writing of her master's thesis.

Introduction

Connie Eigenmann

This study examines communication and technology use in developing countries as a supplement to the plethora of research on developed countries. Its authors are primarily students charged with the task of obtaining data from lesser known areas of human endeavor. They themselves are largely from developing countries. As learners open to guidance, they have tried to fit their research and observations to previously constructed theory to offer an explanation of how they obtained data and how their respondents adopted cellular telephones and the internet, in contrast to the goals of the first volume. The researchers have diligently reviewed obtainable literature on technology and communication to incorporate existing theory that attempts to make sense of the influence of technology on communication. The countries described in this extension of volume one complete the BRICS overview, and add fourteen countries worthy of observation.

Africa is represented by three more countries: Guinea, Kenya, and Uganda. Asia has preliminary research on citizens and sojourners in Indonesia, Laos, Malaysia, Nepal, and the BRICS country of Russia. South America has been represented by Argentina and Paraguay, and the Middle East by United Arab Emirates, questionable as a developing country. Island nations are represented by Jamaica, Maldives, and the Philippines. Mexico adds the Central Americas to our growing collection of data on developing countries. Some are carefully watched to see if they pass over into developed country status economically. All are experiencing infrastructure problems. Their technology in many cases is leapfrogging into usage patterns seen in the U.S., Canada, and Western Europe.

The definition of a developing country is presented by the International Statistical Institute (2015) as slightly over US\$ 11,905 per household, per anum. The www.worldbank.org classification for the current 2015 fiscal year for low-income economies is defined as those with a GNI per capita, calculated using the *World Bank Atlas* method, of US\$ 1,045 or less in 2013. Middle-income economies are those with a GNI per capita of more than US\$ 1,045 but less than US\$ 12,746; and high-income economies are those with a GNI per capita of US\$ 12,746 or more. Lower-middle-income and upper-middle-income economies are separated at a GNI per capita of US\$ 4,125. For our research conducted between

2013-2015, a developing country was determined in part by a country's status as cited by the World Bank, and in part by a consistent economic standard across all income groups. One of the primary means of judgment of a country as developing had to do with the mean annual income of the individuals in the country. This book focuses on countries that fit the developing country definition, a rapidly changing entity.

Research in this book is exploratory. Each country represents a unique blend of cultures, and adoption of electronic communication. The researchers have collected the data utilizing existing electronic communication behaviors particular at the time within each country. A wide variety of scholars argue the need for the acknowledgement of observations of multiple cultures being done and influenced by the researcher's culture of origin (Kuo and Chew 2009, Chen 2009, Chitty 2010, Kim 2010). This heavily justifies self-reported survey research commonly done in Communication Studies. None of the researchers were of their chosen country's citizenry; however, only five were of US origin, eleven were from China, and one each was from Germany and the Kingdom of Saudi Arabia. Many spoke the language of the country they investigated, and added to our database of multi-language editions of the survey.

Our research is fundamental to the social scientific approach to achieve understanding of a complex world. This work focuses on a world that includes developing countries. The purpose of this scholarship is to add more information to a community that works together to discover or describe behaviors; and possibly, determine causes for behaviors. Our question continues, does the technology change developing society and the types of communication, or is it simply providing a new means of transmission of the developing society's norms of communication behavior?

These chapters explore electronic communication behaviors in a variety of developing countries. The focus is on reporting cellular and internet use, and behaviors that surround it for a particular time and a variety of places. This exploratory research done by the authors of the various chapters meets the goal of accumulating knowledge to enhance our understanding of others. It fits into the overall goals of research as a community of scholars sharing information and building on scholarly work. It is not about proving or disproving theories; it is about expanding our knowledge of communication phenomena by discovering and describing behaviors linked to communication through electronic means.

Individually, the various chapters in this book focus on different cultures, with different concerns, different levels of technological infrastructure, and different social norms. We are a world of diverse cultures and communication behaviors. Our hope is that this text will allow you to see some of the communication behaviors that exist outside of your own experiences (Robson 2014).

REFERENCES

Chen, Guo-Ming. "Beyond the Dichotomy of Communication Studies." *Asian Journal of Communication* 19, no. 2(2009): 398-411. doi: 10.1080/01292980903293312

Chitty, Naren. "Mapping Asian International Communication." *Asian Journal of Communication* 20, no. 2(2010): 181-196. doi: 10.1080/12929810036933377

International Statistical Institute. 2015. http://www.isi-web.org/component/ content/article/5-root/root/81-developing

Kim, Young Y. "*Unum* and *Pluribus*: Ideological Underpinnings of Interethnic Communication in the United States." In L. Samovar, R. Porter, & R. McDaniel (Eds.), *Intercultural Communication: A Reader* (13th ed., 208-222), 2010. Boston, MA: Wadsworth.

Kuo, Eddie C. Y. and Han Ei Chew. "Beyond Ethnocentrism in Communication Theory: Towards a Culture-Centric Approach." *Asian Journal of Communication* 19, no. 4(2009): 422-437. doi: 10.1080/01292980903293361

Robson, Scott. Preface, *Electronic Communication in Developing Countries*. Champaign, IL: Common Ground Publishing, LLC, 2014.

World Bank. Accessed 2013-2015. http://data.worldbank.org/about/country-and-lending-groups

CHAPTER 1

Social Networking, Cellular Telephony, and Internet Usage in Argentina

Liping Yang

For a developing country, the e-communication in Argentina is developing rapidly. The purpose of this research is to employ qualitative and quantitative methods of inquiry to record current social networking, cellular telephony, and internet usage in Argentina.

Argentina is famous due to its soccer expertise, and countless internationals travel to Argentina to watch these games every year. Acculturation Theory is widely used to explain the mix of cultures and dispersion cycles of sojourners. With the progress of a world-wide, homogeneous society, electronic communication use by human beings has become pervasive and primary. Even though electronic communication lacks human contact, it can still mediate the difficulties of intercultural communication. Intercultural communication is communication occurring between people from different cultural backgrounds. This research will introduce how Acculturation Theory explains electronic communication as it is used in Argentina in social networking, cellular telephony, and internet access by sojourners and Argentinians alike.

The 21st century is host to a gregarious society. Every day we communicate with others. Communication connects us through all time zones and everywhere; therefore, communication is a tremendous lubricant in our work and recreational lives. Communication does not just exist between parents and children, husbands and wives, and between friends: it is necessary to communicate with other people who live in this world. There are many ways to communicate. One common way is face to face communication. If you cannot meet frequently, you can communicate by the telephone or text messages, and also you can communicate with each other by the social networking if you can use it.

With the progress of society, the proportion of electronic communication in human beings is becoming more and more important. Even though electronic communication lacks human contact, it can still make up the difficulty of the intercultural communication. Those of different cultural backgrounds recognize e-communication as an important feature in today's world. With the acceleration of economies, globalization, trans-nations, and intercultural exchanges, intercultural communication has become supremely important. Because people from different cultural backgrounds accelerate cross-border activities, a large multinational workforce has emerged. Various cultural activities such as soccer and sports make the trend increasingly evident within developing countries.

Intercultural communication is becoming increasingly vital for development, and Acculturation theory can aid understanding. This research will introduce how Acculturation theory can explain documented electronic communication usage in Argentina. According to Redfield, Linton and Herskovits (1936),

> Acculturation comprehends those phenomena which result when groups of individuals having different cultures come into continuous first-hand contact with subsequent changes in the original culture patterns of either or both groups. (149)

This theory can be used in schools, enterprises and the development of countries. Because people think their cultural values are correct due to indigenous cultural values and ways of foreign affairs, certain conduct will cause different views and variance of cultural landscapes.

ACCULTURATION THEORY

Acculturation theory has a wide range of applications. For instance, it can be applied in education. Greenland and Brown (2005) studied the effects of different cultural contact depending on different countries. They showed how some Japanese students interact utilizing Acculturation theory as a group in the United Kingdom. Greenland and Brown (2005) found that the effects of cultural contact depend on students' adjustment abilities. Firstly, they admitted the positive side of acculturation theory, and then pointed out the negative. Their long-term experiment shows how these students acculturate together.

The impact of cultural penetration during education is reflected in many countries, such as Argentina. Argentina's universities have exchange programs with the universities of China, which encourage Chinese students to study in Argentina. This phenomenon can be understood by Acculturation theory. The new, mixed culture is formed by these students and many people in social reality. Argentina and China, both developing countries, are experiencing an historical phenomenon of sedimentation; which is balanced by their dynamic, constantly changing cultures.

Acculturation theory also can affect consumer behavior. Peñaloza (2009) focused on acculturation and consumer behavior and how to connect them. Different religions also have different consumer behaviors. Peñaloza (2009) talked about the role of religion in Acculturation theory and researched this question in different regions with other religions. Peñaloza (2009) concentrated on different immigrants in the United States and the United Kingdom, and the differences between consumer behaviors in these two areas. She believed it is important to research how acculturation affects consumer behaviors, because the behaviors of consumption may benefit cultural connections.

This theory can be used not only in consumer behavior, but also in health behavior. Acculturation theory was tested by Landrine (2004). The information indicated that acculturation has negative influence on health behaviors among various ethnic groups. Landrine (2004) used strong evidence to prove that previous data were not convincing. In the beginning of this research, Landrine (2004) presented Learning theory and Behavior theory, which are the basis of his research in behaviorism as applied to population and culture.

After reading *Acculturation: Advances in Theory, Measurement, and Applied Research,* Levy (2004) wrote a review. This reviewed book is understood to solve the problem of international acculturation and intra-national acculturation. Levy (2004) not only noticed the importance of this book's focus, but also isolated details of this book, such as how Asian and Latino individuals acculturate to life in America. The major purpose of this review was to critique presented means of restriction toward minorities in America and Canada.

In this fast paced era, everyone is busy with their own affairs. However, one thing seems to be forgotten by most people, communication between people. In the past, face-to-face was the only form of communication, but in this network and telephone ubiquitous era, it seems to have lost its position. People are smarter than animals, because people can use language to express their thoughts and need to communicate. Not only thoughts, but sharing feelings, creative communication, and writing for extended audiences are all required for community needs.

THE APPLICATION OF ELECTRONIC COMMUNICATION

Electronic Communication, also known as e-communication, is a kind of communication which combines computer technology with electronic communications technology. It is a new form of communication with the rise of electronic information technology and development, including fax, closed circuit television, computer networks, and e-mail.

As mentioned earlier, electronic communication can be used in many places, such as schools, businesses and the development of countries. An increasing number of students are using electronic communication currently, but there are still gaps between students and their close family members. Gentzler, Oberhauser, Westerman, and Nadorff (2011) presented a survey of 211 university students, who use electronic communication tools to communicate with their family members. The result is that the more they use electronic communication, such as cellphones, electronic mail, and computers, the more life satisfaction students have. However, the increased use of social-networking for communication produces loneliness and negative attachment for students on an increasing scale. The research presents young people as using information technology to communicate in primary ways of importance. This is not only between children and parents, but also between teachers and students.

With the development of information technology, the new forms of electronic communication have become increasingly important in educational areas. Smith and Minnick (1996) focused on how electronic communication positively affects teachers when collecting and grading homework. Smith and Minnick (1996) found that electronic communication did not restrict instructor planning for class discussion, and was beneficial for instructors to create more helpful and useful educational methods. However, these methods require a trouble-free computer environment in order to be implemented. Although students communicate with both parents and teachers, parents need to communicate directly with teachers to understand their children's situations in school. The purpose of this research determines how electronic communication affects schools and homes. Kosaretskii and Chernyshova (2013) explored this question by using various information and data, like school students involved in family activities in Russia because of the increasing number of students' parents getting involved in electronic communication. Therefore, they wonder how electronic communication affects schools and homes. Kosaretskii and Chernyshova introduce the ways of communication with teachers and families respectively, and conclude the advantages and disadvantages of electronic communication in schools and homes.

Some of the advantages of electronic communication are fast delivery, high information capacity, low cost and high efficiency. When sending a letter to a foreign country, it requires several days to reach the hands of the recipient; but by e-mail or fax, it will be instantly received. Because of this, electronic communication is widely used in commercial in enterprises. Sproull and Kiesler (1986) believed that social context cues have an influence on information exchange. Furthermore, Sproull and Kiesler (1986) claimed that electronic mail is not only beneficial for exchange of old information, but also the exchange of new. They also collected useful data to figure out how electronic mail affects all levels of organizations. Sproull and Kiesler (1986) found that electronic mail is related to self-absorption, status equalization, and uninhibited behavior. The purpose of this research is to investigate and compare staff members' communication patterns within a country, and their preferences for face-to-face or electronic communication.

Sullivan (1995) chose staff members and concluded that preferences are mainly associated with the type of communication task undertaken. It depends on various situations, and either electronic or face-to-face communication styles are acceptable. Nantz and Drexel (1995) tried to assess the security, privacy and ownership of electronic communication in order to let people realize and appropriately use electronic communication. Electronic mail has become the main tool of information technology in the Americas. In order to communicate effectively, Nantz and Drexel (1995) researched how students use electronic communication for a variety of tasks. They concluded that students should appropriately choose communication options.

One nation needs to communicate with other nations, one enterprise needs to communicate with other enterprises, and also one country needs to communicate with other countries. The development of electronic communication represents a country's progress, and that development cannot be without multiple users from all levels of the population. Liikanen (2003) believed that electronic communication will become more important for both America and Europe in business areas in the future. Liikanen (2003) was also excited about the effect of electronic communication in the economies of America and Europe at that time. The partnership between America and Europe is strong and intimate throughout their economic history. Besides, the US and Europe have strong competitiveness in the global market. Liikanen (2003) presented information technologies that are applied in business areas, and how new electronic communications increase productivity in businesses.

Relative to electronic communication, those non-electronic communications, such as face to face and printing, also are beneficial to cultural exchange. In order to meet various communication needs, research on Augmentative and Alternative Communication (AAC) has been produced. The purpose is to develop non-electronic communication aids for communication needs. Iacono, Lyon, and West

(2011) found data from an Australian pilot project and solved this problem by setting the provision of an AAC system. Although there is high demand for non-electronic communication aids, Iacono, Lyon, and West emphasized the electronic.

THE APPLICATION OF ELECTRONIC COMMUNICATION IN ARGENTINA

Argentina, as a country in Latin America, has consummate internet facilities. The penetration rate of internet in Argentina was over 50 percent in 2012, and around 71 percent of internet users in Argentina browse the news on websites. This proportion was higher than the average level in Latin America. In fact, the Argentine government plays a key role in the development of its internet. With the development of technology and globalization, electronic communication has become an important part of business. As the emerging market in Latin America, Argentina is a very potential market. Waldman and Stowe (1998) studied Argentine entrepreneurs in 1996 in order to understand the situation of the corporate culture and internal communications in Argentina. They found greetings, nonverbal communication, written communication, verbal communication, and use of communication technologies to play an important role in Argentine business. Argentine women in the business arena play a very important role, too. The following research questions are of importance:

- RQ1: Which types of online activity are most used in Argentina?
- RQ2: How long have Argentineans used cellular telephones?
- RQ3: How many Argentineans have online friends they have not yet met in person?
- RQ4: How long have Argentineans used the internet?

METHOD

Examination of the research questions was conducted using a stratified random sampling approach. Researchers sent surveys to various public, government and educational locations and asked for volunteers to participate in a survey related to cellular and internet use. Each region of the country was canvassed by purposive sampling to include all geographical areas. Government and private workplaces; government, international, and private schools; and urban and rural locales were offered surveys. Previous researchers distributed surveys to 250 respondents in each country with a similar sampling frame. This information was used for comparative purposes in this study. Systematic effort to locate certain age categories; female, illiterate, or unemployed respondents was sometimes necessary. Ages 18-65 were contacted for possible participation. The online survey was posed to people living in Argentina by Facebook and online chat software.

RESULTS

There were 61 effective responses collected in one month. The greatest age cohort for electronic use in Argentina was 18-25 years old with 83.61% response rate. The next highest age cohort was 25-35 years old with 13.11% response, making age 18-35 the largest population of users at 96.72%. No respondents were found for the 55-65 age cohort. Argentina gender demographics (n=61) were 60.66% female, and 39.34% male. Well over half, 67.21% of the participants had graduated from an institute, college or university. See figures 1.1-1.3.

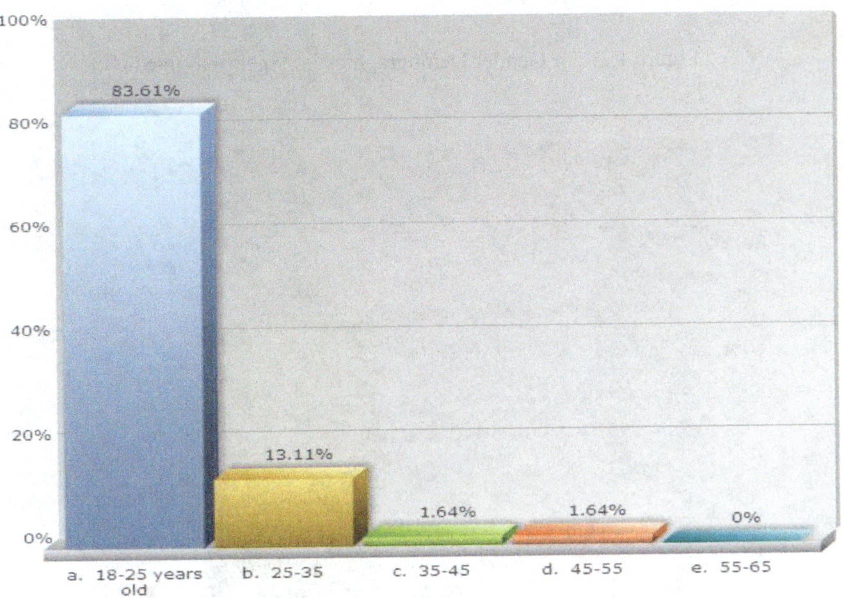

Figure 1.1:   Age Demographics in Argentina (n=61)

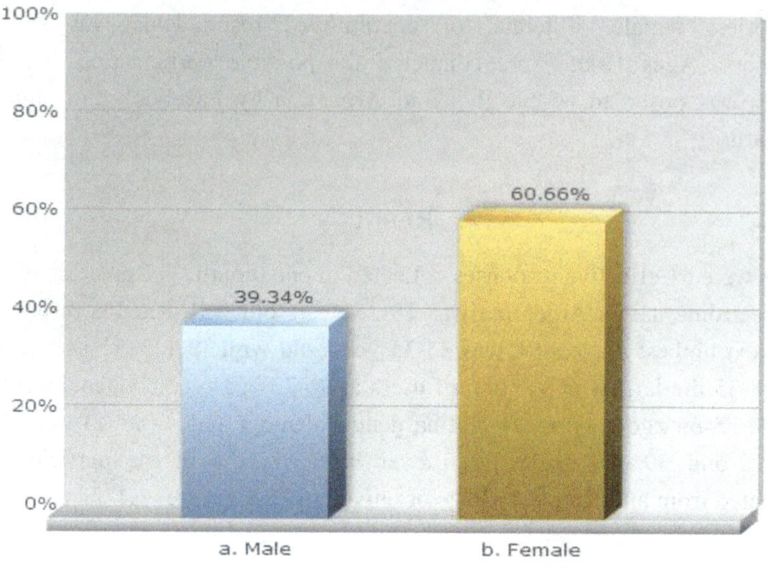

Figure 1.2:   Gender Demographics in Argentina (n=61)

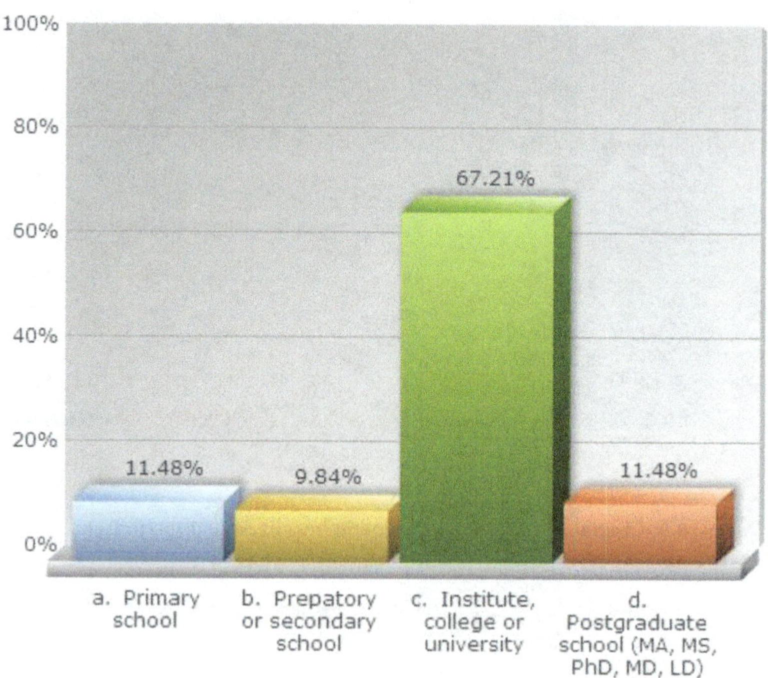

Figure 1.3:   Educational Background Demographics in Argentina (n=61)

RQ1: Which types of online activity are most used in Argentina?

The online activity most used by Argentineans is communication by cellphone, handy, or mobile phone with 86.67% response. The next highest online activity that was most reported by Argentineans is the television with 50% response. Email accounts, radio, and digital camera had 43.33% response, 36.67% response, and 36.67% response respectively. The online activity least reported as used by Argentinean is internet bulletin board with 3.33% response. See figure 1.4.

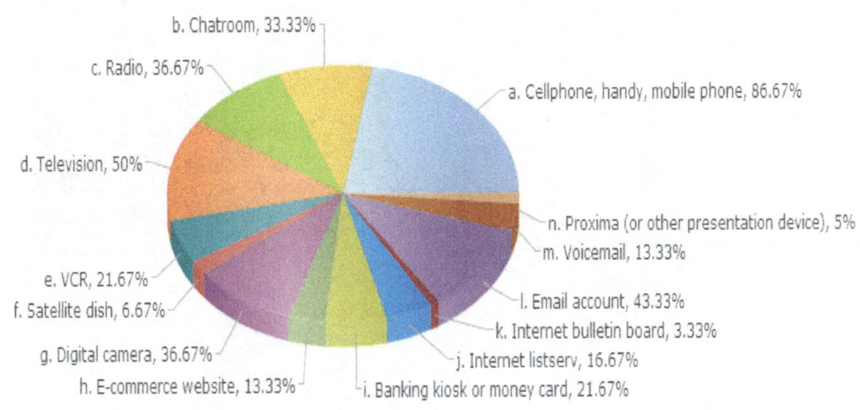

Figure 1.4: Usage Proportion of Online Activities (n=61)

RQ2: How long have Argentinean used cellular telephones?

The largest cohort of electronic usage was for cellular telephones at five years or more. Another 32.79% Argentineans had used cellular telephones for three to five years. Collapsing the cohorts, Argentineans have used cellular telephones for three to more than five years at 78.69%. Only 8.2% Argentineans surveyed had used a cellular telephone less than one year. See figure 1.5.

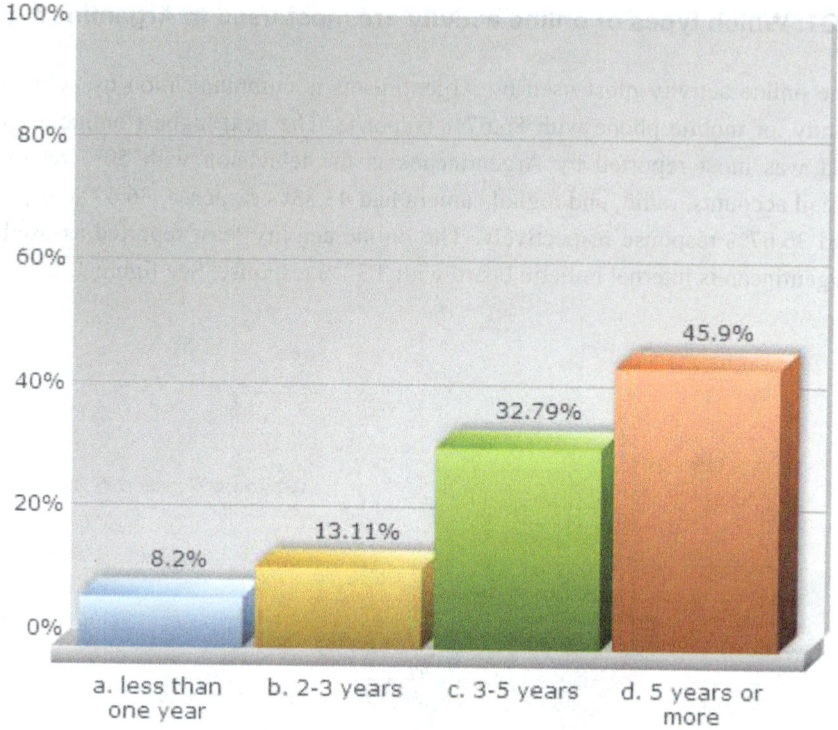

Figure 1.5: Years of Cellular Telephone Usage (n=61)

RQ3: How many Argentineans have online friends they have not yet met in person?

There were 78.69% responses of those who have online friends they have not yet met in person (yes). Only 21.31% Argentineans have met their online friends (no). See figure 1.6.

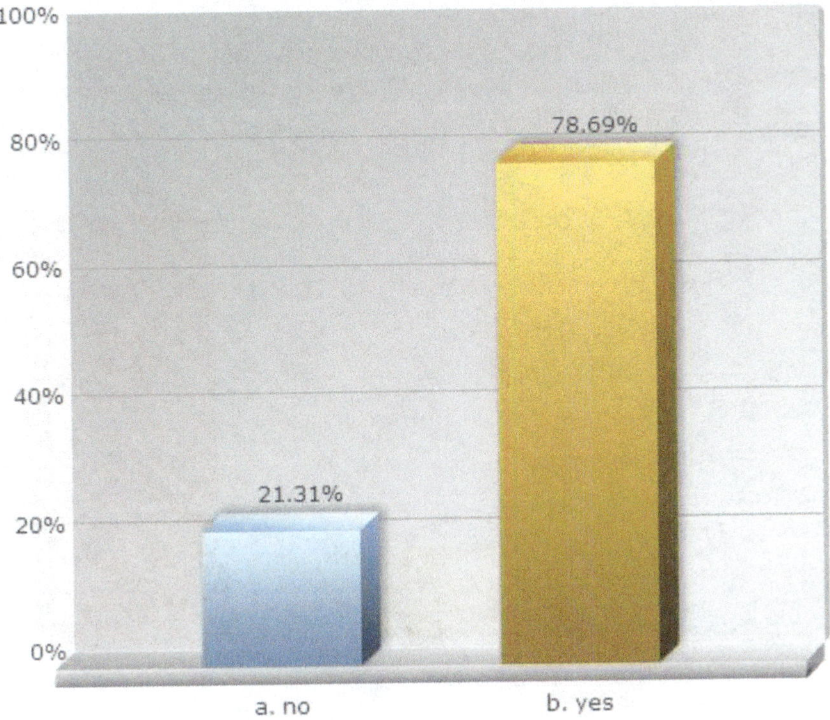

Figure 1.6: Proportion of Argentineans with Online vs Face-to-Face Friends (n=61)

RQ4: How long have Argentineans used the internet?

The highest proportion of how long Argentineans have used the internet is five or more with 68.85%. The other cohorts of duration Argentineans have used the internet are minimal by comparison at 11.48% and 13.11%. Only 6.56% Argentineans have used the internet less than one year. See figure 1.7.

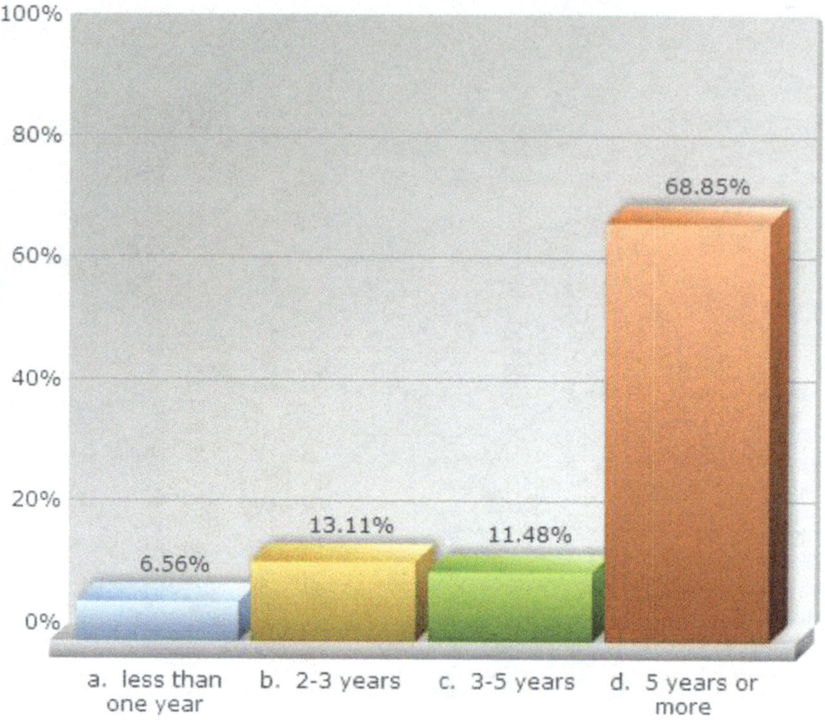

Figure 1.7: Duration of Internet Usage (n=61)

SUMMARY

Argentina is the eighth-largest country in the world by population, and it also is the second largest country in Latin America. Argentina is becoming one of the communication network's more developed countries, as it has the fastest pace of development in South America. Its mobile phones and broadband are the most active, and it also provides different types of call services: fixed telephones, 3G mobile phones and telephone booths. Because technology is convenient to Argentina's people, it has developed very quickly in recent years. Most of the users of social networking, cellular telephone, and internet are between 18 to 25 years old, and most of these are female. Their education background is higher than other users, and nearly all have a university education background.

In spite of these promising signs, there remain problems in the development of technology in Argentina. First of all, Argentina lacks the funds for technology. Investment in technology was lower for Argentina than the average Latin American country. Moreover, the investment of technology in Latin American countries is lower than the average level of the world. Secondly, Argentina's government does not actively participate in science and technology investment and technology exploration, and technological achievements cannot be effectively applied in the marketplace. Last but not least, there is an interesting phenomenon existing in Argentina called technology dependence. After World War II, Argentina began to encourage enterprises to achieve industrialization of the country by themselves because the government wanted to reduce the number of imports. Undoubtedly, this encouragement was useful and valuable; however, it was not a stable form for Argentina. Once the conditions of debt and exporting accelerate, financial crisis is inevitable. These stringent measures have had a negative, overall influence on Argentinian technology.

REFERENCES

Gentzler, Amy L., Ann M. Oberhauser, David Westerman, and Danielle K. Nadorff. "College Students' Use of Electronic Communication with Parents: Links to Loneliness, Attachment, and Relationship Quality." *Cyberpsychology, Behavior & Social Networking* 14, no.1/2(2011): 71-74.

Greenland, Katy, and Rupert Brown. "Acculturation and Contact in Japanese Students Studying in the United Kingdom." *Journal of Social Psychology* 145, no. 4(2005): 373-389.

Iacono, Teresa, Katie Lyon, and Denise West. "Non-electronic Communication Aids for People with Complex Communication Needs." *International Journal of Speech-Language Pathology* 13, no. 5(2011): 399-410.

Kosaretskii, S. G., and D. V. Chernyshova. "Electronic Communication Between the School and the Home." *Russian Education & Society* 55, no. 10(2013): 81-89.

Landrine, Hope, and Elizabeth A. Kolbert. "Culture Change and Ethnic-minority Health   Behavior: An Operant Theory of Acculturation." *Journal of Behavioral Medicine* 27, no. 6(2004): 527-555.

Levy, Jacob J. "Review of *'Acculturation: Advances in Theory, Measurement, and Applied Research'*" [Book]. *Counselling Psychology Quarterly* 17, no. 1(2004): 135-136.

Liikanen, Erkki. "Future of Electronic Communications and E-Business in Europe." *I-Ways* 26, no. 4(2003): 165-167.

Nantz, Karen S., and Cynthia L. Drexel. "Incorporating Electronic Mail into the Business Communication Course." *Business Communication Quarterly* 58, no. 3(1995): 45-51.

Peñaloza, Lisa. "Acculturation and Consumer Behavior: Building Cultural Bridges through Consumption." *Advances in Consumer Research* 36(2009): 16-19.

Smith, Douglas K., and Barbara J. Minnick. "Electronic Teacher-Students Communication." *Business Communication Quarterly* 59, no. 1(1996): 74-81.

Sproull, Lee, and Sara Kiesler. "Reducing Social Context Cues: Electronic Mail in Organizational Communication." *Management Science* 32, no. 11(1986): 1492-1512.

Sullivan, Christopher B. "Preferences for Electronic Mail in Organizational Communication Tasks." *Journal of Business Communication* 32, no. 1(1995): 49-64.

Waldman, Lila D., and Jessica Stowe. "Business Communication in Argentina." *Delta Pi Epsilon Journal* 40, no. 3(1998): 145-157.

CHAPTER 2

New Media Usage in Ghana: An Exploratory Study

Jia Jenny Jia

Ghana, officially called the Republic of Ghana, is a country in the west of Africa. It is located along the Gulf of Guinea and Atlantic Ocean. Ghana is an agricultural country and has abundant natural resources. Even so, about 28 percent still live below the poverty line. It is one of the fastest growing countries in the world in recent years.

This study focuses on the cellular telephone and internet usage in Ghana based on age, gender, and education level. Data are collected from Ghana citizens, Ghanaians who permanently live in the US, and those who had studied abroad. Five research questions were posed. Questionnaires were distributed electronically to determine the demographics, descriptors and impacts of cellular telephone and internet use.

LITERATURE REVIEW

Reception Theory

Shi (2013) gives us a detailed introduction of Wolfgang Iser, and analyzes his Reception theory deeply from seven aspects. As the originators of the Reception theory, Jauss and Iser both have been concerned with a reconstitution of literary theory, but they have different emphases on it. The research of Jauss' prefers the macroscopic research and study of the Reception theory from literary history. However, Iser prefers studying from a microscopic viewpoint, which is how readers relate to the individual text. Then Shi (2013) showed us the seven aspects of Iser's Reception theory. The main content is the implied reader. From Iser's standpoint, he places emphasis on reading and interpreting literary works. At the same time, he focuses on the reader's reaction and how literature is re-constructed. There are uncertain implications in literature. The implied reader means some person, who could understand the author's real intention, and is the key factor to realize the meaning of literature.

Hamilton and Schneider (2002) introduce two famous literary scholars, Wolfgang Iser and Mark Turner, and analyze their literary works, which are related to the Reception theory and cognitive criticism. In Iser's Reception theory, his point is to discuss how to enhance the readers' active roles. Every text places its implied readers into different positions, and motivates the readers to take action. Iser's Reception theory places too much emphasis on the implied reader, and never defines reading as natural. Cognitive criticism is a combination of cognitive science and literary criticism. The literary critics can move beyond the biographic pattern to understand and reflect the reality of the literary works in the latest scientific, cognitive way. Although Iser's and Turner's standpoints are similar, their approaches are different in some aspects. For instance, according to Iser, agency resides first with the text, then with the reader; however, Turner holds that agency is always with humans, never with objects.

By inspecting the theory of the historical meaning of literary reception and Anglo-American theorists of the "new" history of political thought, which show insights in different ways, Thompson (1993) analyzes the interpretation of historical meaning of literary reception. The two perspectives have different standpoints on the character of a text. Among the former, the reader's thinking is most important. Among the latter, what the author intends to express plays an important role. However, the writer's viewpoint is that the two perspectives are valuable, but interpreting historical meanings is one-sided.

Lv and Ning (2013) take the Reception theory as an efficient way in analyzing how to translate EST, which is worthy of reference for this survey. English for Science and Technology (EST) translation means the translators want to convey to readers the new knowledge, the new theories, and information in the field of science and technology through translation. In the light of the Reception theory, the translator should not only pay attention to transfer the original author's intention, but also consider the reception ability and the reader requirements.

EST has three characteristics, which are accuracy, objectivity, and brevity. The translator needs to follow the principles in order, so his readers can easily and fully understand his works. Reception theory influences the original author, the translator, and the readers. The translator should consider the reader's acceptability. On the other hand, the translator to the original author is just as the reader to the translator. The original author should allow the translator to make some changes without altering the style of the source text.

Electronic Communication Usage in Developing Countries

Donner (2008) reviewed the usage of cellular phones as the main content of research in developing countries. With the rapid development of communication technology and the increasing number of cellular phone users worldwide, the developing countries are in the grip of an unprecedented increase of cellular phone usage. Donner (2008) reviewed from two aspects. On the one hand, these researches are focused on the usage of cellular phones in developing countries. On the other hand, these researches were classified according to the impact on the usage of cellular phones and the interrelationships between mobile technologies and users. In the world, different cultural backgrounds and social economics, our knowledge about the usage of cellular phones has increased. The largest number of cellular phone users is in developed countries. In low-income countries, cellular phone users are concentrated in the relatively prosperous areas. In lower-middle-income countries, cellular phones are more widespread than low-income countries.

In Asia, cellphones have become the most widely used medium of communication. But for most people, they do not realize that cellphones are a potential tool to gain knowledge. Librero and Others (2007) introduced us to two major projects, which are studying the potential of cellphones and the short message service (SMS) techniques for education in the Philippines and Mongolia. People use cellphones to send and receive messages via SMS. It has become a universal phenomenon in the world. The use of cellphones is convenient and efficient for learning. Numerous cellphone tools, such as text messaging, iPod and MP3 players, can be used to help people learn, but there are a few limitations on the usage of cellphones for learning. Some advanced functions are only available by using multimedia message service (MMS). However, most people still use the lower-end cellphones in the Philippines and Mongolia, which can only receive SMS messages. On the other hand, some people do not want to spend money on the use of cell phones to send messages for educational purposes. Therefore, some SMS technical issues need to be solved.

In Malaysia, most mobile phone users live in the main cities, such as Kuala Lumpur and Malacca. A lot of researchers are investigating the usage of mobile phones but seldom pay attention on the overall mobile phone satisfaction (OMPS). Yeow and Others (2008) examined the factors, which affect people's satisfaction with daily mobile phone use in Malaysia. OMPS is influenced by six mobile phone usage factors, which are Peer Chatting, Family Coordination, Social Interruption, Public Disturbance, Radio-Frequency Radiation Health Effects, and Road Accidents. Yeow and Others (2008) developed a survey in which they asked three hundred students and working adults to answer. The result showed that Peer Chatting and Family Coordination factors have a positive impact on OMPS, and the other four factors imposed negative influences. According to this result, Yeow and Others (2008) offered some specific advice on how to increase OMPS. For example, it is unhealthy to put our phones too near our heads when we are making a mobile phone call; a habit which needs to be stopped.

In India, Kumar and Others (2011) conducted a 26-week survey where they observed rural children using cellphones every day without suggesting them to use the cellphones to study. The study took place in two stages: summer 2008, and spring and summer 2009. The study also took place in two areas: one village, which was relatively prosperous, and the other was less prosperous. Then, the cellphones were pre-loaded with some applications. The researchers and the children gathered to understand the children's mobile learning behaviors in their everyday life. Next, they allowed these children to use cellphones freely. Finally, the researchers started looking at these children's behaviors. The children did things differently because of different genders, ages, and social classes. The results showed that these rural India children all voluntarily engaged in mobile learning when they were unsupervised.

Internet Usage in Ghana

Amenyedzi and Others (2011) evaluated the usage of computers and the internet on education in the Tema Metropolis in Ghana; and researched how students use it to study, and how teachers use it to guide students. Amenyedzi and Others (2011) applied a stratified sampling method to investigate students and teachers in three senior high schools in the Tema metropolis. The results showed that a large percentage of students and teachers had basic knowledge of computers, but only few of them used the computers and internet to improve learning and work efficiency.

Computers and internet can help the students complete their studies, increase their typing speed, and solve learning problems. It can also help the teachers write lesson plans and communicate with their students through chat rooms and Web forums. ICTs have changed the learning model from a teacher-centered model to a learner-centered model, but financial problems and poor internet connection have become the largest obstacle for the usage of the computers and internet in rural schools. It will take some time to use the ICTs in the education system of Ghana widely.

Ghana was one of the first countries in sub-Saharan Africa to establish internet connections, but penetration rates for internet and broadband remain at very low levels. By talking with internet service providers (ISPs), Fosu (2011) determined the cause of high broadband costs in Ghana. In Ghana, ISPs use the SAT-3 submarine cable system to connect to the internet, which is extremely expensive. In addition, high license and regulatory fees, and high cost of end user equipment are also the cause of high broadband costs. Fosu (2011) discussed how ISPs could successfully solve the cost issues, and provided a proposal to reduce the cost of broadband. Firstly, more submarine fiber optic networks should be established, and the infrastructure should be shared in order to reduce the costs. Secondly, another way to help lower costs is demanding the government reduce the taxes, licensing and regulatory fees. Furthermore, the complete elimination of import duties is an efficient method to reduce end user equipment cost.

Cellular Telephone Usage in Ghana

Tagoe and Abakah (2014) investigated distance education students' views on mobile learning in Ghana, and explained how student beliefs influenced intention to use mobile learning through the theory of Planned Behavior. The theory of Planned Behavior helps us to understand how people change their behavior models. The behavior of people is a thoughtful and planned outcome. It means the more the students know about mobile learning, the more positive attitude they will have toward mobile learning. With the questionnaire method, Tagoe and Abakah (2014) investigated 400 students of the University of Ghana distance education program. The results showed that student intentions to adopt mobile learning were affected by attitude, subjective norm and behavioral control. The mobile phone is a device easily carried, which makes study more convenient. At the same time, mobile technologies give the students flexibility. They appreciate the opportunity to take courses whenever they want.

Closed user groups (CUG) are groups of cellular telephone users who can make and receive calls from members within the group for free. Kaonga and Others (2013) collected bill statements and call data from a health team who is using a CUG in the Millennium Villages Project site of Bonsaaso, Ghana, in order to analyze the utilization and acceptability of the CUG and encourage the use of cellular telephone CUG. Kaonga and Others (2013) adopted social network analysis methods to explore the flow of communication in this group. They gathered and analyzed extensive data from March 2011 to September 2011. This revealed that most group members frequently used their cellular telephone for work. CUG helped improve their communication and work efficiency, while bringing benefit to the communities they serve. However, several obstacles of using CUG still remain, such as blocked subscriber identity modules (SIMs) and poor connections.

Research Questions

Based on development of communication technology and an increasing demand for the cellular telephone and internet usage in Ghana, the following demographic, descriptive and social impact questions about cellular telephone and internet use in Ghana have emerged:

- RQ1: What age, gender and education demographics use electronic devices in Ghana?
- RQ2: What types of electronic communication are used in Ghana?
- RQ3: Do males use chatroom more often than females?
- RQ4: How long have people used cellular telephones and internet in Ghana?
- RQ5: What are the main reasons for using the internet in Ghana?

METHOD

This survey is based on the quantitative data of respondents' personal opinions. Questionnaire method was used to collect and analyze the data, and to study (1) internet and cellphones usage, age and media choices, (2) internet and cellphones usage and education level, (3) chatroom usage and positive internet comments, (4) media usage types, and (5) internet and cellphones usage time.

Instrument

The instrument was a 40-item questionnaire in English for Ghana citizens. It allowed 9 open-ended responses, 8 demographic queries, and 23 Likert scale items. It was hoped to collect at least 30 completed surveys. Age, gender and ethnicity were not screening factors. All the respondents were invited from email, Twitter, or Facebook to fill out an online survey. The survey was converted into electronic webpages and placed on the SurveyMonkey website. A link from the homepage of the SurveyMonkey was sent to respondents by email to complete the questionnaire, and direct them to a brief explanation of the research. All the respondents could open these links and complete the questionnaire online. Compliance wholly or in part determined consent.

Participants

Participants were adults 18-55 years of age who lived in Ghana, Ghanaians who permanently live in the US, and those who had studied abroad. They were invited to answer and think about their usage of cellular telephones and internet communication.

Procedures

Upon agreement to participate a choice of language was given to each respondent, and explanation was offered if items were not understood. No illiteracy was encountered. There were no foreseeable risks to participants who could stop at any time, skip questions, or disregard the survey request. No questions were offensive, confidential, invasive or embarrassing. No name or distinguishing data were assigned to the surveys.

RESULTS

RQ1: What age, gender and education demographics use electronic devices in Ghana?

The greatest age cohort for electronic use in Ghana by Ghanaians was 25-35 years with 42.4% response. The next highest age cohort was 18-25 years with 39.4% response, making age 18-35 the largest population of users at 81.8%. Ghana gender demographics (n=33) were 54.5% female and 45.5% male. Age demographics can be seen in figure 2.1.

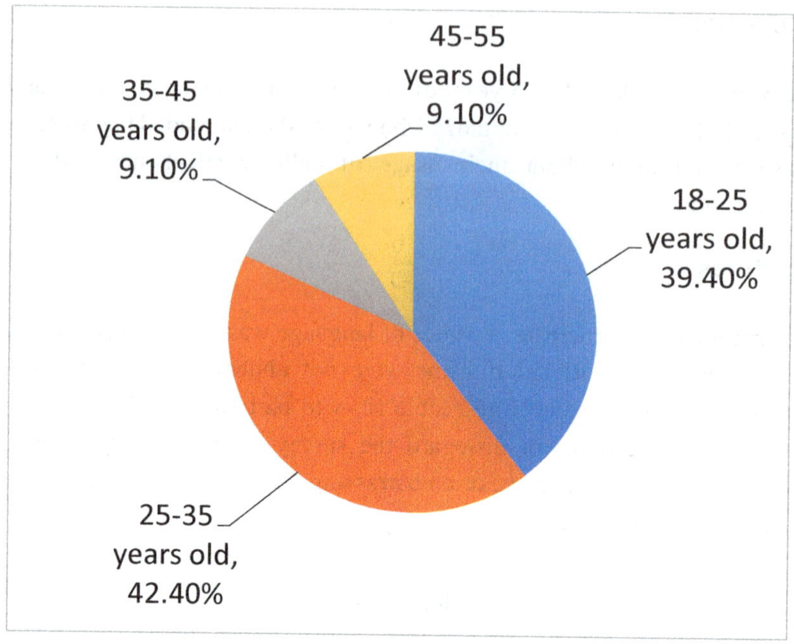

Figure 2.1: Ghanaian Age Demographics (n=33)

Education demographics of the use of electronic devices in Ghana cover all possible means of formal education. Completion of an institute, college or university degree represented the greatest percentage of respondents (54.5%). About thirty percent of the respondents had completed a master's degree or a doctor's degree. See figure 2.2.

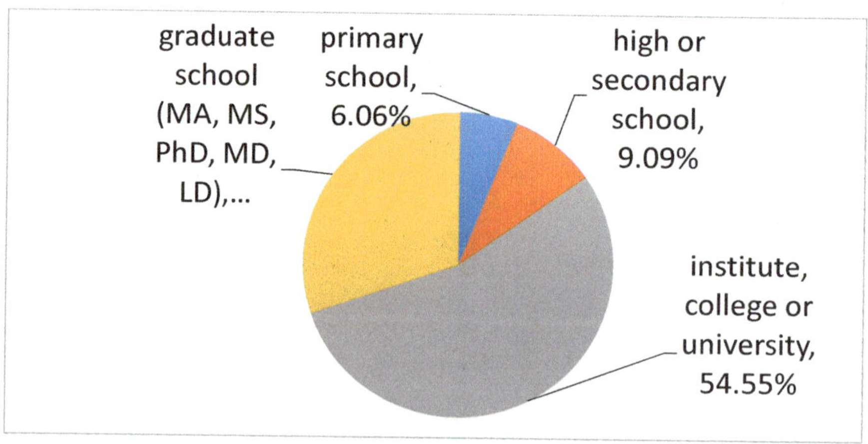

Figure 2.2: Ghanaian Education Level Demographics (n=33)

RQ2: What types of electronic communication are used in Ghana?

Fourteen electronic devices and means of communication were listed for participants to check if they had been used at least once. The devices included cellular (mobile or moto) telephone, chatroom, radio, TV, VCR, satellite dish, digital camera, electronic commerce website (shopping, business), banking kiosk or money card, internet listserv, internet bulletin board, email account, voicemail, and presentation devices like Proxima. One and four choices of media garnered the highest percentages of media choices in Ghana. See figure 2.3.

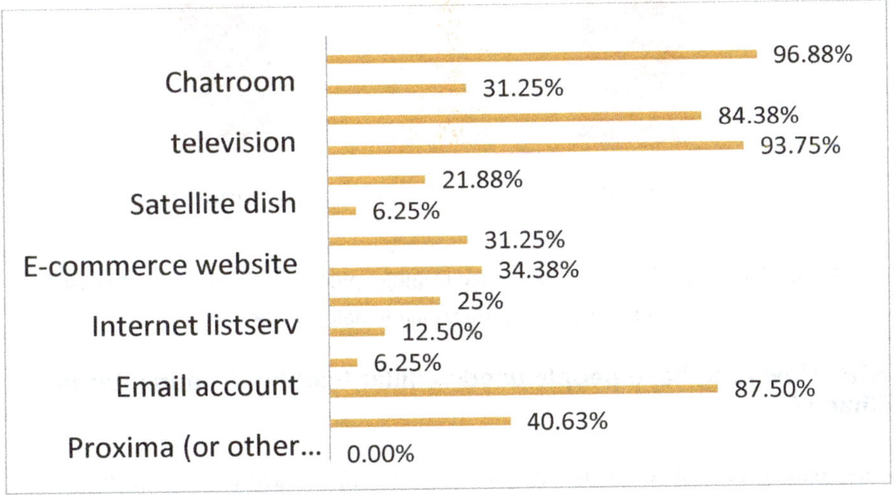

Figure 2.3: Ghanaian Media Usage Demographics (n=33)

Cellular telephone was listed as used by all respondents of the subset except one. No one has ever used presentation devices like Proxima before. These were the dominant means of electronic communication for Ghana nationals and sojourners operating within the culture.

RQ3: Do males use chatroom more often than females?

Yes, Ghanaian males used chatroom 7.93% more often than females. See figure 2.4.

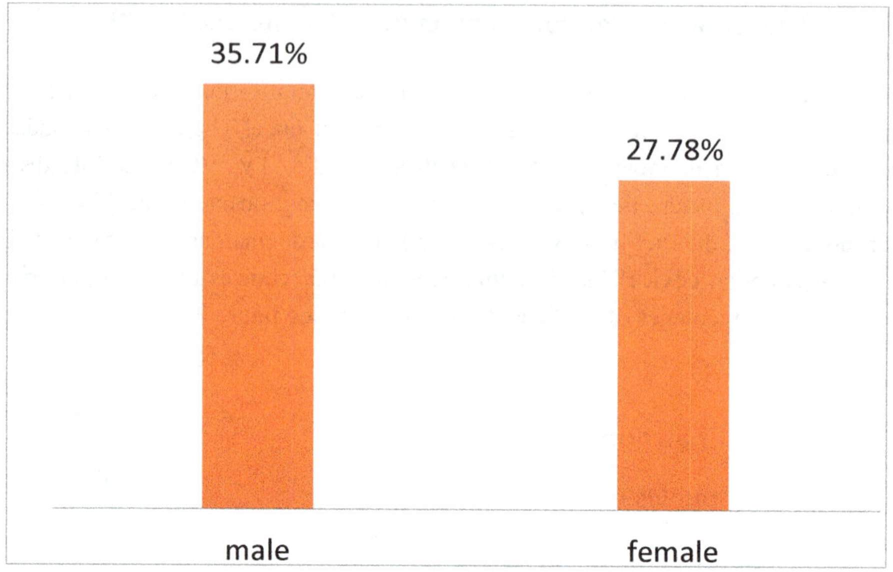

Figure 2.4: Ghanaian Chatroom Usage Demographics (n=33) Males (n=15) used chatroom more often than females (n=18).

RQ4: How long have people used cellular telephones and internet in Ghana?

Ghanaian users of cellular telephone were more acclimated with 42.42% reporting more than five years of use. A scant 6.06% reported less than one year of use. See figure 2.5.

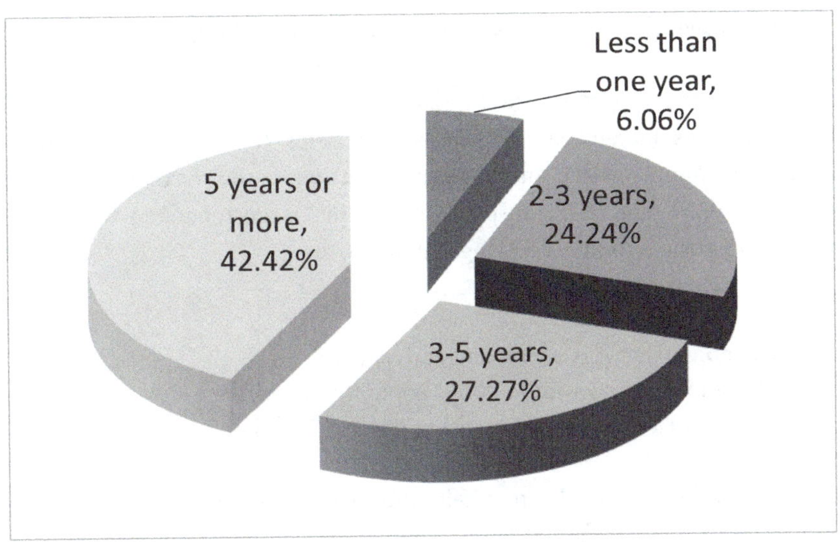

Figure 2.5: Ghanaian Cellular Telephone Usage Demographics (n=33)

Ghanaian users of internet were more acclimated with 39.39% reporting three to five years of use. A scant 6.06% reported less than one year of use. See figure 2.6.

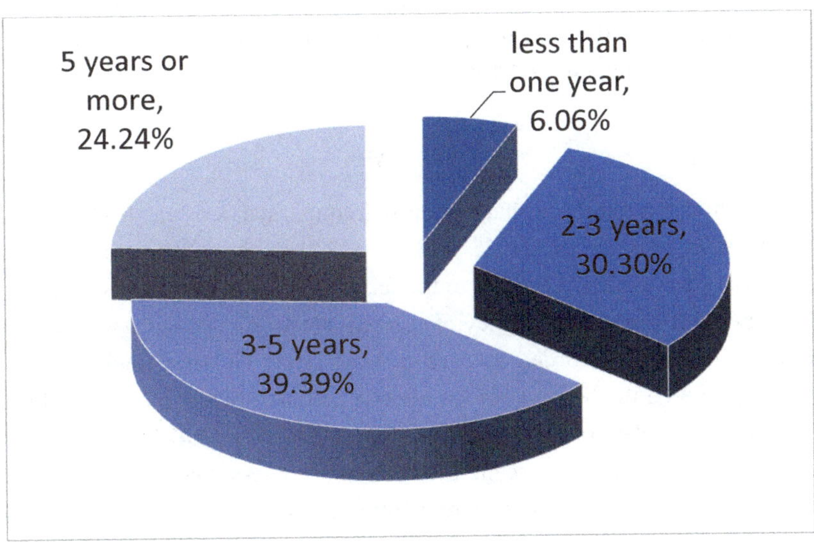

Figure 2.6: Ghanaian Internet Usage Demographics (n=33)

RQ 5: What are the main reasons for using the internet in Ghana?

Ghana qualitative responses (n=26) indicate two main reasons for using the internet: It is useful and convenient, and it helps to obtain information and resources. Communication and entertainment were the next most recorded reasons at 19.2% and 7.7% respectively. This is corroborated by voluntary qualitative responses to the following statement.

I like to use the Internet because…

38.5% replied that it is useful and convenient.
    "It helps me know more things about the world."
    "I can do a lot of things by using internet in a short time."

34.6% replied that it helps to obtain information and resources.
    "It has all the information."
    "There are lots of resources can help me improve myself.

19.2% replied that they use internet to communicate.
    "I can communicate with my family wherever I am."

7.7% replied that they use internet for entertainment.
    "I like playing computer games."

SUMMARY

Results of this study showed that the most frequent users of electronic means of communication were between 25 and 35 (42.4%). According to the survey, younger users were more receptive to the technology than older users, and use the technology with greater frequency. Fourteen electronic devices and means of communication were recorded as used at least once. Cellular (mobile or moto) telephone garnered the highest percentages of media choices in Ghana. Ghanaian users of cellular telephone were more acclimated with 42.42% reporting more than five years of use. A scant 6.06% reported less than one year of use. Ghanaian users of internet were more acclimated with 39.39% reporting three to five years of use. A scant 6.06% reported less than one year of use.

In the usage of cellular telephone, the penetration rate is very high in Ghana, especially among the young. The results indicate that people in Ghana using the cellular telephone with internet capabilities very weak. The top two cellular brands are Samsung and Blackberry. Ghanaians use iPhone less frequently.

People prefer to buy a cellular telephone less than $100. A possible reason would be the high purchase price and high taxes of smartphones.

In the usage of internet, the internet penetration of Ghana is above average for Africa. Ghana qualitative responses (n=26) indicate two main reasons for using the internet: It is useful and convenient, and it helps to obtain information and resources. Communication and entertainment were the next most recorded reasons at 19.2% and 7.7% respectively. The survey finds that people in Ghana use internet cafes and seldom use WIFI because of the high broadband costs but with the development of telecommunications infrastructure in recent years, it has become much cheaper than before. As the network coverage continues to improve, it will increase internet penetration in Ghana.

REFERENCES

Amenyedzi, Frank W. K. , Mary N. Lartey, and Beloved M. Dzomeku. "The Use of Computers and Internet as Supplementary Source of Educational Material: A Case Study of the Senior High Schools in the Tema Metropolis in Ghana." *Contemporary Educational Technology* 2, no. 2 (2011): 151-162.

Donner, Jonathan. "Research Approaches to Mobile Use in the Developing World: A Review of the Literature." *Information Society* 24, no. 3 (2008): 140-159.

Fosu, Ignatius. "Broadband Costs and Impact on Universal Internet Access: The Case of Ghana." *Southwestern Mass Communication Journal* 26, no. 2 (2011): 29-40.

Hamilton, Craig A, and Ralf Schneider. "From Iser to Turner and Beyond: Reception Theory Meets Cognitive Criticism." *Style* 36, no. 4 (2002): 640-658.

Kaonga, Nadi Nina, Alain Labrique, Patricia Mechael, Eric Akosah, Seth Ohemeng-Dapaah, Joseph Sakyi Baah, Richmond Kodie, Andrew S. Kanter, and Orin Levine. "Mobile Phones and Social Structures: An Exploration of a Closed User Group in Rural Ghana." *BMC Medical Informatics and Decision Making* 13, no. 1 (2013): 1-9.

Kumar, Anuj, Anuj Tewari, Geeta Shroff, Deepti Chittamuru, Matthew Kam, and John Canny. "An Exploratory Study of Unsupervised Mobile Learning in Rural India." *Conference Papers -- International Communication Association* 2011 Annual Meeting (2011): 1-30.

Libreroa, Felix, Angelo Juan Ramos, Adelina I. Ranga, Jerome Triñona, and David Lambert. "Uses of the Cell Phone for Education in the Philippines and Mongolia." *Distance Education* 28, no. 2 (2007): 231-244.

Lv, Liangqiu, and Puyu Ning. "EST Translation Guided by Reception Theory." *Open Journal of Modern Linguistics* 3, no. 2 (2013): 114-118.

Shi, Yanling. "Review of Wolfgang Iser and his Reception Theory." *Theory and Practice in Language Studies* 3, no. 6 (2013): 982-986.

Tagoe, Michael, and Ellen Abakah. "Determining Distance Education Students' Readiness for Mobile Learning at University of Ghana Using the Theory of Planned Behavior." *International Journal of Education and Development Using Information & Communication Technology* 10, no. 1 (2014): 91-106.

Thompson, Martyn P. "Reception Theory and the Interpretation of Historical Meaning." *History and Theory* 32, no. 3 (1993): 248-250.

Yeow, Paul H. P. , Yee Yen Yuen, and Regina Connolly. "Mobile Phone Use in a Developing Country: A Malaysian Empirical Study." *Journal of Urban Technology* 15, no. 1 (2008): 85-116.

CHAPTER 3

Diffusion of Cellphone and Internet Technologies in Guinea

Brandon Steinert

As a rapidly growing but socially and politically unstable nation with recent access to technologies like cellphones and internet, Guinea is a country of significant interest. The research focuses on identifying the prevalence of antecedent conditions to adoption of cellphones and internet like perceived benefit, obstacles to adoption and gender differences in adoption tendencies.

LITERATURE REVIEW

About Guinea

The population of Guinea is estimated to be 10.5 million, which is triple the population from only 40 years ago. Of the developing countries, Guinea seems to have one of the largest gaps between its current situation and more desirable conditions. It is rife with violence and civil unrest, as evidenced by recent protests resulting in multiple deaths and hundreds of injuries. The United Nations Development Programme ranks Guinea 179th out of 186 using the Human Development Index (Khalid 2014).

Guinea also harbors severe gender inequality. For example, female genital mutilation (FGM) is prevalent (Rossem & Gage 2009). Though it might have little impact on the adoption and use of internet and mobile phones, the presence of FGM inspires the question of whether an increase in the availability of communication technologies will influence a population's tolerance toward and understanding of such inhumane acts.

Availability of Internet and Mobile Phones in Guinea

Despite its political and societal instabilities, the availability and use of communication technologies such as mobile phones and internet are on the rise in Guinea, and its iron-mining potential has generated renewed interest in the country's natural resources. These elements are a driving force behind the country's development:

> With five competing mobile networks, its telecommunications sector has shown triple-digit growth rates for three years in a row… Broadband services are still very limited and expensive ... The country slipped into a mild recession in 2009, but stable GDP growth of around 5% per year is expected from 2014, possibly reaching up to 20% in 2015 and 2016 (BuddeComm2014).

Weslowski and Others (2014) found these trends to be accurate. A map constructed to reveal the connection between ebola outbreaks and mobile connectivity shows West Africa and Sub-Saharan Africa categorized as having "high connectivity."

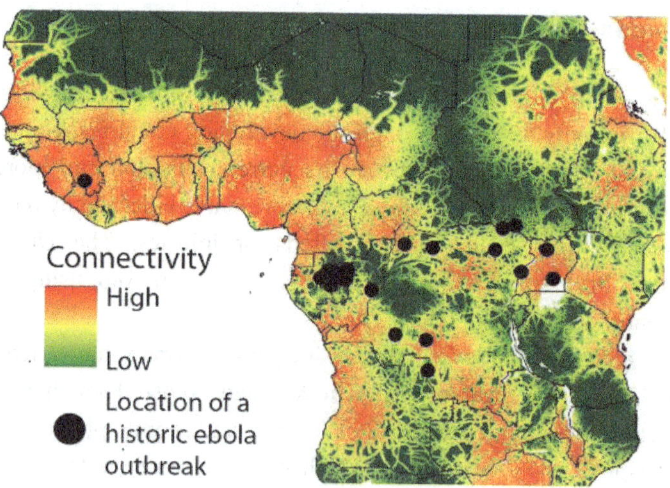

Figure 2.1: Mobility patterns and connectivity in West Africa
Weslowski and Others (2014)

Kende (2014) states internet use in Sub-Saharan Africa increased at a rate of 32 percent from 2007 to 2012, compared to developed portions of the world like

North America, which are on the far right of the Diffusion of Innovations "S" curve.

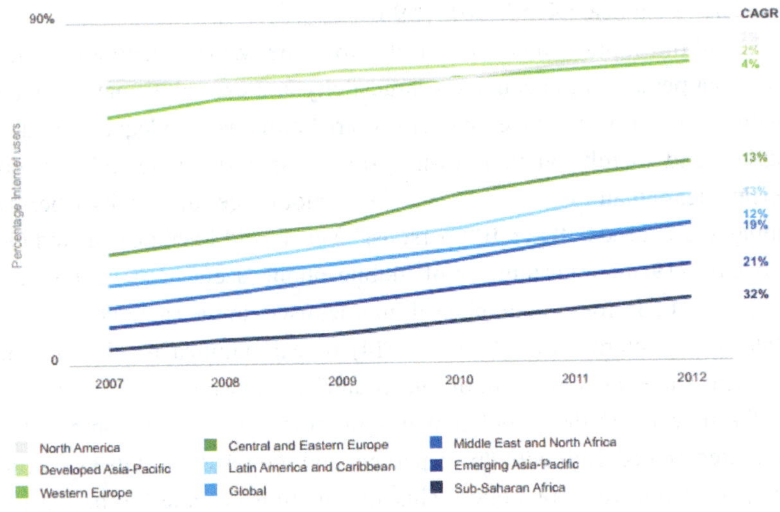

Figure 2.2: Proportion of Population Using the Internet
Kende (2014, 22)

Though the rate of internet use is increasing substantially, the penetration of mobile broadband use is only just beginning for Sub-Saharan Africa, as illustrated in Figure 3.

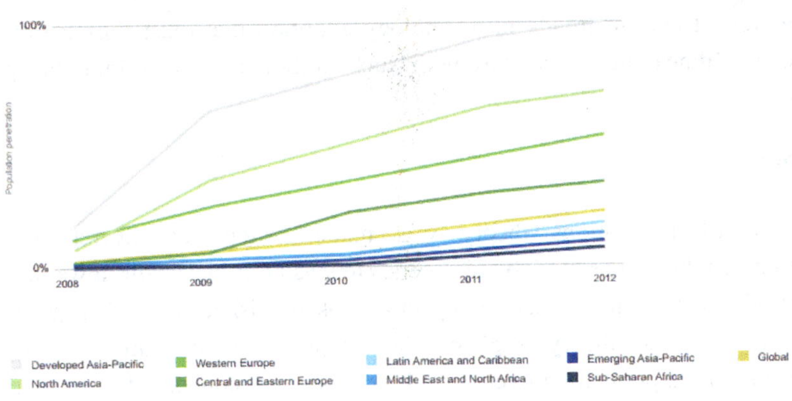

Figure 2.3: Mobile Broadband Population Penetration, Kende (2014, 22)

The same report identifies the African Internet Exchange System (AXIS) as a force to expand access to reliable internet service providers. The primary tools used are Best Practices Workshops and Technical Assistance Workshops. Both are available in Guinea. (Kende 2014, 59-61)

Most of the rapid growth, according to "The World Factbook - Guinea" (2012), is happening in the country's largest city and capital, Conakry. "Conakry reasonably well-served; coverage elsewhere remains inadequate and large companies tend to rely on their own systems for nationwide links; fixed-line teledensity less than 1 per 100 persons; mobile-cellular subscribership is expanding and exceeds 40 per 100 persons." The World Factbook ranked Guinea 115th out of 219 for total number of mobile phone users in 2012. Guinea was ranked 223rd of 235 for total number of internet users in the same year.

Data from "Internet Live Stats" (2014) reveals Guinea is still early in the development stage, ranked 190th of 198 countries for penetration of internet use. About 1.7 percent of the population has internet access. For comparison, the United States ranked 25th with 86.75 percent and the top 10 countries reached 93 percent penetration rates or better. This information provides some clarity and context for what would otherwise appear to be rapid and widespread growth across the country; the availability of mobile phones and internet in Guinea is quite limited.

The combination of Guinea's relatively low human development index ranking and rapidly growing mobile market could reveal communication technology's impact, or lack of impact, on a developing nation, or if the reverse is true: whether political and social instability and lack of development stifles or fuels diffusion of innovations. This paper does not seek to find a correlation, but it does reveal current trends in technology adoption that exist during times of unrest in Guinea. Future studies with such ambitions should find this paper useful.

Theory

Diffusion of Innovations theory was made famous by Everett Rogers, affectionately known as "Ev" by his peers. He was born in Carroll, Iowa in 1931. He introduced his widely used theory in a book published in 1962, *Diffusion of Innovations*. The theory explains the adoption of innovations, primarily technology, by a population over time. It covers many contexts, from geography to culture, and is the origin of many commonly used terms, including "early adopter." The theory claims change agents create and introduce an innovation, and its perceived benefits will be the driving factor behind whether or not the innovation will be successfully disseminated, and those who adopt the innovation will influence others to do the same. The theory argues the spread of a new innovation is represented by an S curve. It begins slowly as early adopters take in an innovation, then the adoption hits a period of swift growth followed by a tapering effect as late adopters jump on the bandwagon (Rogers 2003). This research project examines the perceived benefits of internet and mobile phones by Guineans that might affect diffusion.

McEachern and Others (2008) spent a great deal of time and effort studying the diffusion process by observing the rate of adoption of a government-subsidized solar-energy system for rural communities unlikely to receive electricity in Sri Lanka, a small island country off the southeast coast of India. The authors were quite thorough in their research, taking into consideration numerous factors, including social status of early adopters, a village's relationship with policy-makers, physical infrastructure and distance.

The authors found villages where diffusion was more pervasive "were tolerant of non-conformist behavior, frequently advocate for village development, have a primary school, are accessible by paved road and are situated in flat, rice and other lowland crop areas" (McEachern and Others 2008, 2582). The authors also found villages closer to population-dense areas are more likely to see rapid diffusion. This project will focus primarily on Guinea's largest population center, a coastal city called Conakry, to determine the level of awareness of internet access where diffusion typically starts. It will be up to future studies to determine the rate of diffusion to surrounding communities.

The *World Factbook* (2012) reveals most of Guinea's internet and cellphone users reside in the country's largest city and capital, Conakry. Based on Diffusion of Innovations theory, the technology should spread to less population-dense areas from Guinea as the epicenter of adoption. It will be up to future studies to reveal if this has occurred.

Islam (2013) conducted a study of the likelihood that Bangladesh residents will use advanced mobile phone services, like internet browsing, m-commerce and other technologies:

> ...social influence can be considered to make a significant difference to the consumer adoption decision ([29] Lopez-Nicolas *et al.*, 2008). Rogers' Diffusion of Innovation paradigm helps to address this gap because the concept of diffusion can be described as "the process by which an innovation is communicated through certain channels over time among the members (consumers) of a social system. (828)

Bricolo and Others (2013) look at the use of internet and computers among Italian families, with a rather large sample size of about 5,000. The average adult male in Italy spends more than 100 hours per month on the internet, while the average adult female spends about 60 hours per month. Given Guinea's gender inequality issues, a question of interest is the difference between male and female habits regarding cell phone and internet use.

Annafari and Others (2013), Pearcey and Draper (1995) and Wang and Liao (2008) apply the Diffusion of Innovation theory in a variety of contexts and locations. Annafari and Others took a reverse look at Diffusion of Innovations and found many individuals might not find a need for mobile devices, or do not value their functions due to their life situation. For example, an elderly person who does not have a large network of friends might not find value in using a mobile phone to stay in touch. Pearcey and Draper (1995) used a case study to show conditions in which an innovation is more readily accepted using a team of nurses. Wang and Liao (2008) briefly covered the influence of multiple perceived factors can have on one's behavioral intention, including perceived financial resources, usefulness, ease of use and compatibility.

Eigenmann (2014) is a compilation of several scholarly articles covering topics highly relevant to the one pursued by this project, namely the prevalence of mobile phones and internet use in developing countries. She identifies that South Africans use the internet primarily for online shopping or entertainment, and it is not a mainstream tool for education, banking, business and other uses common in the United States today (106). This observation inspires a question of whether Guineans have fully integrated mobile phones and internet into their lives, or if similar issues like lack of education have stifled its implementation.

Unfortunately, little research pertains to Guinea specifically. Many addressed issues within Guinea-Bissau, a neighboring country. Though Papua New Guinea is not in Africa, several studies have been done in PNG specifically related to mobile phone and internet usage that appeared relevant to Guinea's situation, as mobile phone infrastructure was introduced at about the same time, circa 2007. Watson (2013) found about half of respondents owned mobile phones with slightly more male respondents than female. The study focuses specifically on mobile phones and media in Papua New Guinea. "The current research project examines contemporary trends in communication practices in PNG. Specifically, it looks at the recent introduction of mobile telephone services into rural areas of the country". (159-160) One factor influencing the diffusion of innovations is perceived benefit. Watson found the most commonly identified benefits of mobile phones is communication with distant family members, and ease and speed of communication. These findings inspire the qualitative question of whether Guineans perceive similar benefits, or if the nation's unstable nature will reveal more pragmatic or emergency-oriented benefits as priority. Several obstacles to adoption of mobile phones in Papua New Guinea included access by poor people and availability of power to charge batteries (171). This project intends to consider similar obstacles via qualitative analysis.

Research Questions

- RQ1: What is the current level of awareness and use of mobile phones and internet in Guinea?
- RQ2: What are the perceived benefits of mobile phones and internet among Guineans?
- RQ3: What are the most common uses of mobile phones and internet by Guineans?
- RQ4: What are the differences in adoption and use among genders in Guinea?
- RQ5: What are the obstacles to adoption in Guinea?

METHOD

Instrument

A survey containing 40 questions was designed to gather both quantitative and qualitative data. It was sent to numerous contacts in Guinea. The survey contained multiple choice questions and eight short answer questions. It was purposive in gathering information about cellphone and internet usage among both male and female residents in Guinea. It was presented in English and in French, which is the country's official national language and is taught in Guinean schools.

Participants

Participants were reached via Lutheran Church Missouri Synod missionaries and contacts at universities in Guinea. A technology focused retailer called Hi-Tech Store Guinea shared a link to the survey on its Facebook page, which boasts more than 20,000 likes. The survey was also advertised via Facebook to people throughout the entire country, though the lack of internet availability in rural Guinea limited the target geographic area to the capital city of Conakry.

Participants must have lived in Guinea for at least three months or be Guinea nationals. They must also be at least 18 years old, but not older than 65. No incentive to respond was presented to potential participants. Participation in the survey was completely voluntary and anonymous, and required about 15 minutes to complete. Submitting a completed survey implied consent to use the aggregated data in a published study. Locating potential respondents required systematic effort and repeated communication effort.

RESULTS

RQ 1: What is the current level of awareness and use of cellphones and internet in Guinea?

One hundred percent of survey respondents had used a cell phone before. Answers to other questions on the survey revealed a more detailed picture. One individual selected "I do not use a cellphone" when asked how many calls per day are made via cellphones. About 27 percent of respondents chose not to answer the question. The remaining respondents are represented in figure 2.4. Most respondents make two to ten calls per day on a cellphone.

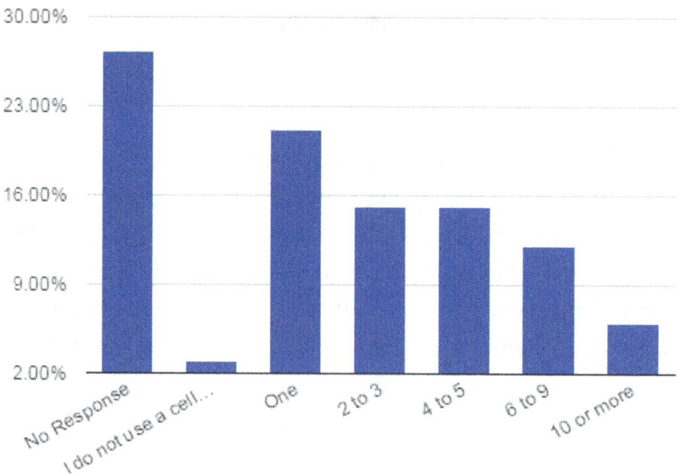

Figure 2.4: Calls made daily via cellphone by Guinea residents.

The amount of time spent using cellphones is also relevant, and is represented by figure 2.5.

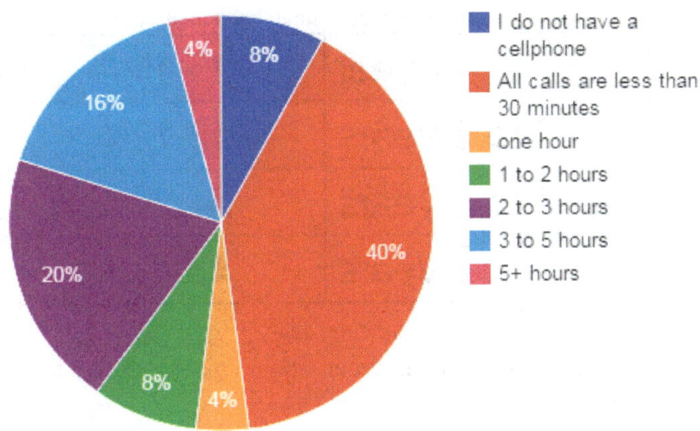

Figure 2.5: Longest duration respondents have spent talking on a cellphone.

Figure 2.5 reveals 40 percent of respondents keep their calls to less than 30 minutes. More than 50 percent reported having spent anywhere from one to five hours on a single call. Qualitative responses reveal many Guineans utilize electronic communication primarily for connecting with distant family members because travel is prohibitively expensive. This could be one of the reasons for the long phone calls.

Results show most respondents have been using a cellphone for more than five years. See table 2.1.

Table 2.1 Duration Guineans have been using cellphones

< 1 year	**3.03%**
2-3 years	**6.06%**
3-5 years	**15.15%**
5+ years	**72.73%**

Results show most Guineans have been aware of the internet for more than three years. Almost 50 percent have used the internet for more than five years. See table 2.2.

Table 2.2 Duration Guineans have been using the internet.

< 1 year	15.15%
2 to 3 years	12.12%
3 to 5 years	24.24%
5+ years	48.48%

Among respondents who use the internet, about half spend five to 20 hours online each week. See figure 2.6.

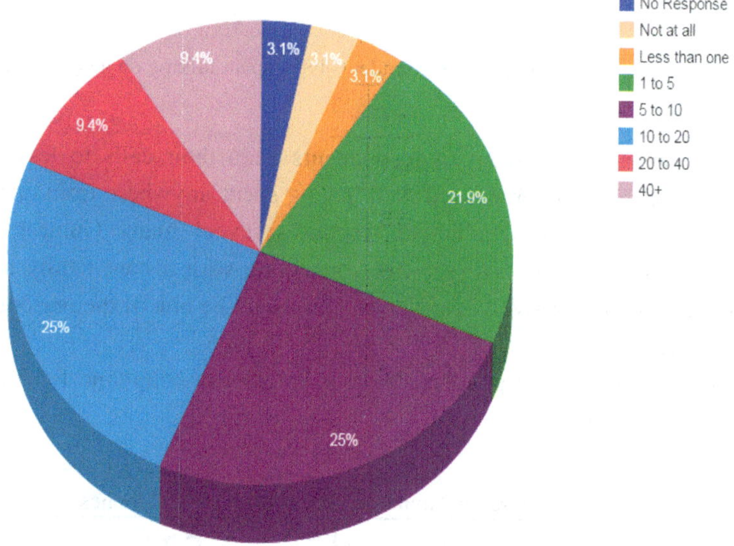

Figure 2.6: Hours Guineans spend online each week.

One of the most basic functions and uses of the internet is email. Nearly 50 percent of respondents reported checking their email once per day or more, with six percent checking email four times or more daily, as illustrated in table 2.3.

Table 2.3 Frequency with which Guineans check email.

I check email ______.						
No response	Never	Hardly Ever	Sometimes	Daily	2-3 times daily	4 or more times daily
21.21%	12.12%	3.03%	15.15%	33.33%	9.09%	6.06%

Based on the data, awareness of internet and cellphone technologies is widespread. However, adoption appears to still be in the rapid growth portion of the S curve that defines the process of diffusion of innovations, as a significant number of laggards have yet to adopt.

RQ2: What are the perceived benefits of mobile phones and internet among Guineans?

Ease of communication was one of the primary reasons Guineans provided for their adoption of electronic communication technologies. Table 2.4 shows whether Guineans prefer to communicate electronically versus face-to-face.

Table 2.4 Guineans' preference to communicate electronically versus face-to-face.

never	9.38%
hardly ever	3.13%
sometimes	59.38%
no difference	6.25%
quite often	18.75%
always	3.13%

As a follow-up to the question represented in table 2.10, respondents were asked to provide a reason for their preference. Qualitative responses included:

"Sometimes it is easier to deal with something awkward on the internet."
"Saves time to communicate electronically."
"Mainly 'cause of distance. I prefer to talk face-to-face, but so many of those I need to contact are not near me geographically, I need the phone or internet."
"The internet allows me to communicate with many people at the same time."
"Well, I'm not in the same country as my family and friends. So it is through the internet that I communicate."
"Because it lets you communicate better in terms of cost and time."

Table 2.5 shows Guineans' perception of how much electronic communication has improved their lives.

Table 2.5 Electronic communication has _____ improved my life.

not	7.41%
hardly ever	11.11%
sometimes	62.96%
tremendously	18.52%

The follow-up question was also posed: "Why do you communicate electronically?"

Qualitative responses include:

"Fast and cheap."
"Family."
"Geographic distance."
"This is a very fast way."
"Reduce displacement."
"It is simple."
"It is the simple and easy way."
"I like it."
"Fast communication."
"To spend time with my friends."
"It's better for me."
"It is fast."
"To exchange news."

Speed, convenience and cost savings versus traveling were the most common responses.

RQ 3: What are the most common uses of mobile phones and internet by Guineans?

Qualitative responses regarding use of cellphones include:

"Staying in touch with distant friends and family. Easier access to help in an emergency, like when I threw out my back and couldn't move but was able to call a friend who is a nurse."
"Talk to friends and family far away."
"I use it for everything without disturbing my relatives."

"Easily reaching remote people."
"I can call my family at any time."

Qualitative responses regarding use of internet include:

"I can communicate with the states and get information."
"Information available. Connect with those I love who don't live near me.
"Creative outlet."
"It gives me ways to learn things, and to contact others."
"Staying in touch with distant friends and family. Easy access to a wealth of information."
"Access to information. Access to people I would not otherwise have access to. Access to businesses that would be much more complicated to contact via phone. Share things going on in the world with my kids in a way they can connect with in reality. Communication with friend and family."
"It helps me do research on my studies. Introduces me into forums and established relationships with people I do not know!"
"I want to stay in this global village."
"It helps me to do research."
"Communicate with distant people."
"To be with the world."
"Search easy and fast."

RQ 4: What are the differences in adoption and use among genders in Guinea?

Awareness and adoption among both male and female respondents was near 100 percent for both cellphones and internet. One measure of use of cellphones includes the frequency of calls made. The data shows there is little difference among males and females regarding number of calls made per day via cellphone. See figure 2.7.

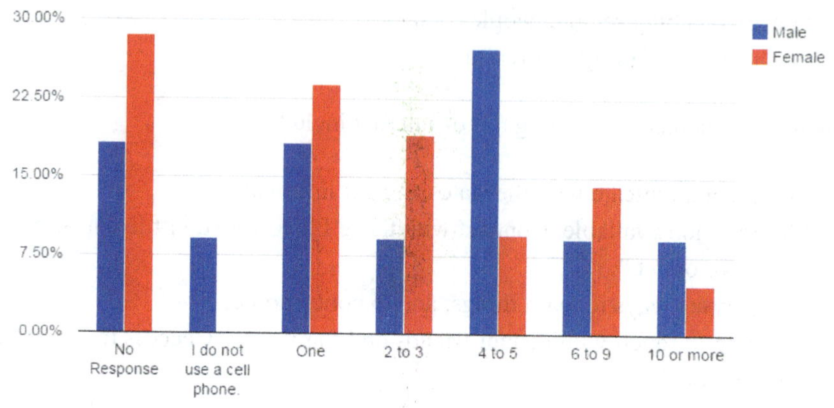

Figure 2.7:   Number of calls made via cell phone by gender.

Call duration showed differences between male and females. Females lingered far longer in phone conversations than their male counterparts. See Figure 2.8.

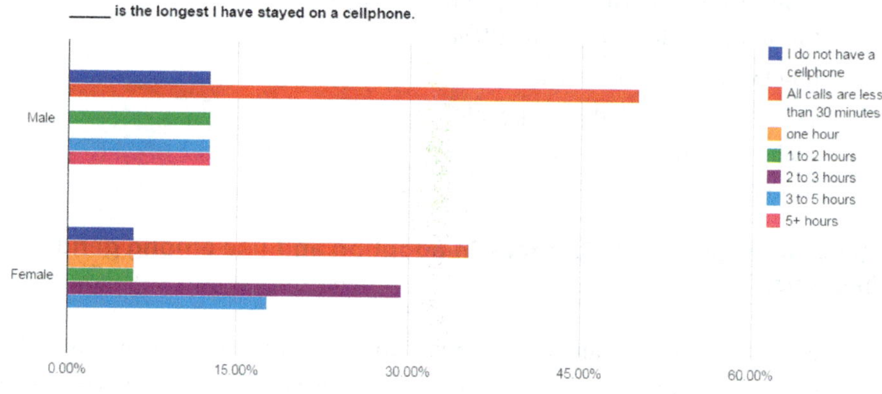

Figure 2.8:   Longest call duration by gender.

The amount of hours spent online per week did not vary greatly between male and female respondents. The male population did, however, have a greater percentage of individuals spending more than 40 hours per week online. See figure 2.9.

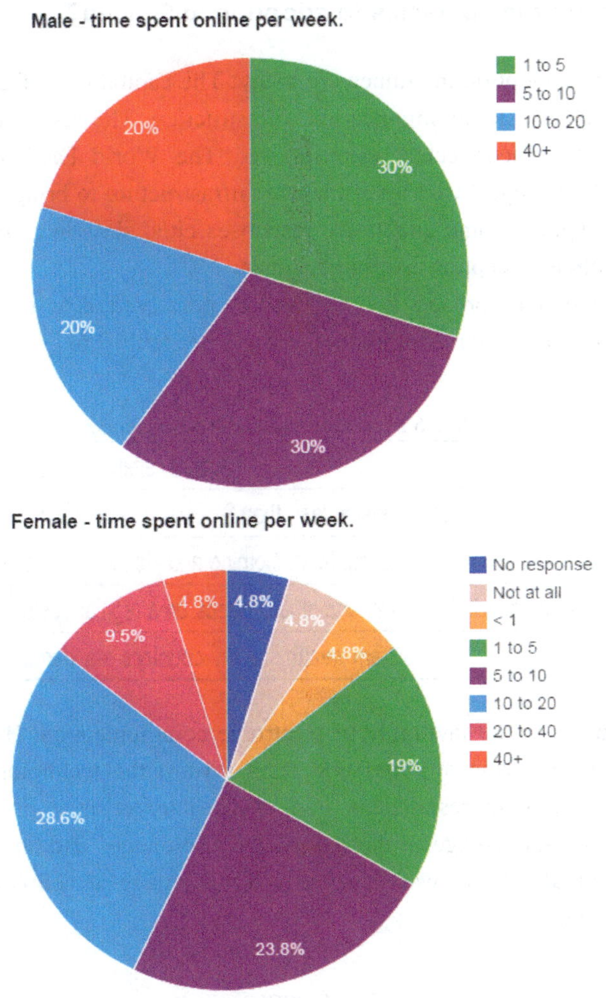

Figure 2.9:   Hours per week spent online by gender.

Qualitative responses revealed very little difference among common cell phone and internet uses by gender, but the types of responses varied slightly. Notably, responses from females mention communication with family on multiple occasions in addition to pragmatic uses. Male responses focus primarily on pragmatic uses such as research and saving money on travel.

RQ 5: What are the obstacles to adoption in Guinea?

The obstacles to adoption in Guinea are many. The capital city of Conakry is the hub for most electronic communication infrastructure. The rest of the country is rural and largely unconnected according to "The World Factbook - Guinea" (2012). It will take significant investment in infrastructure to bring the rest of the country up to speed. Some qualitative responses cited slow internet connections and poor cellphone reception as disadvantages.

Lack of economic prosperity could also be perceived as an obstacle. Half of respondents reported having no reliable income. See table 2.6.

Table 2.6 Economic status of respondents.

I have no reliable income.	50.00%
I have a fixed income/pension less than $1,000 USD per month.	7.69%
I have a job or business providing housing, food, clothing and life necessities.	15.38%
I have enough to provide for myself and others' needs on a regular basis.	23.08%
I am wealthy by my country's standards.	3.85%

Most respondents spoke favorably of electronic communication. However, there could be some lingering attitudes and fears toward the technology that could hamper adoption. One respondent said, "The internet is a necessary evil." Another said a disadvantage of the internet is "espionage" and another, "lack of privacy." "Cranial cancer" was listed as a disadvantage of cellphones. Another cited "adverse health."

DISCUSSION

Observations and Implications

The French translator who worked on the survey, John Wilkos, mentioned many of the questions would not apply to Guinea residents, as internet and cell phone use are not common in the country. The diffusion of cellphones and internet use in Guinea should theoretically flow inland from the coastal city Conakry, which is the hub for electronic communication. The current adoption and use of electronic communication technology is still very much in its infancy, but is spreading rapidly and has been widely adopted by residents of Conakry.

Facebook.com's advertising dashboard showed a potential reach of 460,000 people age 18-65 in Guinea.

Limitation

The intended number of recipients for this study was at least fifty. Due to limitations including lack of internet access in Guinea, reaching potential survey respondents proved challenging despite numerous points of contact in the United States with connections in Guinea. Contacts within the country that would have been key to disseminating the survey returned to the United States due to limited funding before the survey was ready, which reduced the number of contacts within the country.

It took several days for a trained translator to get to the task of converting the survey to French. This delay limited the amount of time available for collecting data. In addition, most of Guinea's 11.75 million people live in rural homes. Conakry is the country's capital and has been noted as the hub for communication technology and was the primary source of survey respondents. With a population of 1.7 million, Conakry residents are not representative of the rest of the country, which is far less developed. Less than one third of respondents reside in rural areas.

Fort Hays State University issued a notice instructing researchers not to travel to Guinea during the survey dissemination period. In addition, the institution required individuals who came into contact with someone from Guinea to report to the campus health center.

In the interest of transparency, it should be noted that a disproportionate number of females responded to the survey. Initially, most of the responses coming in were from males. To counter this trend, the Facebook ads targeted primarily females. The ads were more effective than predicted; therefore the survey was answered by about twice as many females as males.

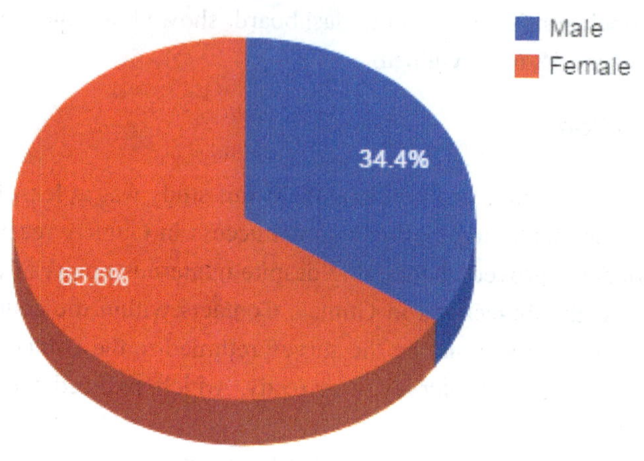

Figure 2.10: Gender of survey respondents.

Suggestions for Future Research

A research team with funding and the ability to travel to Guinea would allow for a more thorough and accurate study of the country's use of cellphones and internet, particularly in rural areas where only printed surveys would be possible to collect.

Only four respondents reported using the internet for e-commerce, and zero listed online shopping as a use for the internet. Such uses are highly common in more developed countries. These uses might be valuable indicators of a country's progress in the diffusion of innovations.

REFERENCES

Annafari, Mohammad T., Ann-Sofie Axelsson, and Erik Bohlin. "A Socio-Economic Exploration of Mobile Phone Service Have-Nots in Sweden." *New Media & Society* 16 no. 3 (2013): 415-433. doi: 10.1177/1461444813487954.

Bricolo, Francesco, Douglas A. Gentile, Rachel L. Smelser, and Giovanni Serpelloni. "Use of the Computer and Internet Among Italian Families: First National Study." *CyberPsychology & Behavior* 10 no. 6 (2007): 789-798. doi:10.1089/cpb.2007.9952.

BuddeComm. 2014. "Guinea - Telecoms, Mobile and Broadband - Market Insights and Statistics." *The Largest Telecommunications Research Site*

on the Internet. Paul Budde Communication Pty Ltd.
www.budde.com.au.

Eigenmann, Connie. *Electronic Communication in Developing Countries*.
Champaign, IL: Common Ground Publishing LLC, 2014.

"Guinea." 2014. *U.S. Department of State*. U.S. Department of State.
http://www.state.gov/p/af/ci/gv/.

"Guinea Massacre Toll Put at 157." 2009. BBC News. BBC. September 29.
http://news.bbc.co.uk/2/hi/8280603.stm.

Islam, Md. Zahidul, Patrick Kim Cheng Low, and Ikramul Hasan. "Intention to
Use Advanced Mobile Phone Services (AMPS)." *Management Decision*
51 no. 4 (2013): 824-838. doi:10.1108/00251741311326590.

"Internet Users by Country 2014." 2014. *Internet Live Stats*.
http://www.internetlivestats.com/.

Khalid, Malik. 2014. *Human Development Report 2014*. New York: United
Nations Development Programme.
http://hdr.undp.org/en/content/human-development-report-2014.

Kende, Michael. 2014. "Global Internet Report 2014." *Internet Society*. Internet
Society. January 9. www.internetsociety.org.

Mceachern, Menzie, and Susan Hanson. "Socio-geographic Perception in the
Diffusion of Innovation: Solar Energy Technology in Sri Lanka."
Energy Policy 36 no. 7 (2008): 2578-2590.
doi:10.1016/j.enpol.2008.03.020.

Pearcey, Pat, and Peter Draper. "Using the Diffusion of Innovation Model to
Influence Practice: a Case Study." *Journal of Advanced Nursing* 23 no. 4
(1996): 714-721. doi:10.1111/j.1365-2648.1996.tb00042.x.

Robinson, Les. 2009. *A Summary of Diffusion of Innovations*.
http://www.enablingchange.com.au/summary_diffusion_theory.pdf.

Rogers, Everett M. *Diffusion of Innovations*. 5th. New York, NY: Free Press,
2003.

"UN Rights Office Calls on Guinea to Protect Civilians Following Violent
Clashes." 2013. *UN News Center*. UN. May 3.
http://www.un.org/apps/news/story.asp?newsid=44283&cr=conakry&cr
1=#.vev3gvl4roz.

Rossem, Ronan Van, and Anastasia J. Gage. "The Effects of Female Genital
Mutilation on the Onset of Sexual Activity and Marriage in Guinea."

Archives of Sexual Behavior 38 no. 2 (2009): 178-185. doi: 10.1007/s10508-007-9237-5.

"The World Factbook - Guinea." 2012. *Central Intelligence Agency*. Central Intelligence Agency. https://www.cia.gov/library/publications/the-world-factbook/geos/gv.html.

Wang, Yi-Shun, and Yi-Wen Liao. "Understanding Individual Adoption of Mobile Booking Service: An Empirical Investigation." *CyberPsychology & Behavior* 11 no. 5 (2008): 603-605. doi:10.1089/cpb.2007.0203.

Watson, Amanda. "Mobile Phones and Media Use in Madang Province of Papua New Guinea." *Pacific Journalism Review* 19 no. 2 (2013): 156-175.

Wesolowski, Amy, Caroline O. Buckee, Linus Bengtsson, Erik Wetter, Xin Lu, and Andrew J. Tatem. 2014. "Commentary: Containing the Ebola Outbreak – the Potential and Challenge of Mobile Network Data – PLOS Currents Outbreaks." *PLOS Currents Outbreaks*. Public Library of Science. September 29. http://currents.plos.org/outbreaks/article/containing-the-ebola-outbreak-the-potential-and-challenge-of-mobile-network-data/.

CHAPTER 4

The Usage of Electronic Means of Communication in Indonesia

Xiaotian Marcus Gao

Indonesia is a developing country located in Southeast Asia. The industry sector plays the most important role in the country's GDP. Throughout the history of Indonesia, it has been famous for international trade in its ports so the connections between Indonesia and other countries have been very tight over the years. There are 300 distinct ethnic groups in Indonesia; however, the Javanese ethnic group is culturally and politically dominant. A modern communication infrastructure has been established in Indonesia. Since there are so many islands in this country, the media companies focus on the largest, most populous island. Those living on the smaller islands rely on satellite in order to communicate electronically. Indonesia has remote areas to improve as its economy allows. By June, 2011, all the places in Indonesia were connected to the internet. There used to be internet and media censorship in Indonesia, but the current government encourages media freedom.

LITERATURE REVIEW

Convergence Theories

Indonesian industry has an international focus. Figueroa (2002) asserts that the Convergence theory plays an important role in shaping the social change process. To describe the process of social change, a model of communication is required to depict how Indonesian goods and information are sent, how people receive culture from one another, and how people mutually understand each other; eventually, they take steps to make real changes. The groups of people that are involved in social change are distinct from each other.

Kincaid (2002) argues that different groups have various interests and political goals. This leads us to assume that it is never easy to apply the

Convergence theory to this complicated process. Kincaid (2012) provides the reader with two basic models of Convergence theory in the social context. The first one has the basic components of the Convergence Model of Communication. This model explains how simple information at the beginning leads to the collective efforts of all the community members to make real changes. The second model is the integrated model of communication for social changes. This model is about combining the complicated social change process with the communication model. Scholars cannot fully understand the social change process unless they can fully understand the function of these two models.

Cross-cultural communication could be achieved by understanding the Convergence theory. Barbero (2009) points out that, by the rapid development of technology, especially digital technology, the boundary between different cultures becomes weaker. Digital convergence, which is referred to as communication transparency, provides people with a new way to interact cross-culturally and exchange new ideas. Barbero (2009) indicates that an increasing number of communication methods are used in a digital form, and this changes the old pattern of cultural communication. The pace and speed in exchanging new cultural issues has never been seen before in the history of human beings.

The creator of the Convergence theory, Lawrence Kincaid, (1983) applied his brainchild to the cultural convergence of Korean immigrants in Hawaii. By analyzing the process of how the Korean immigrants finally fit in the local community of Hawaii, the Convergence Model of Communication was connected with the second law of thermodynamics. The Convergence Model of Communication has been deeply explained and analyzed. When the Korean immigrants finally settled in Hawaii, they had to exchange information and knowledge with the local residents. Because of the cultural difference between the Korean immigrants and other ethnic groups, sometimes this process was smooth whereas sometimes it did not go very well. Kincaid (1983) implies that when the information was being conveyed, a network was established, and the communication converged. And finally, different ethnic groups reached a mutual understanding to be able to co-exist well.

Our old way of looking at culture and the knowledge behind it has been changed by the emergence of the profit-oriented market. Canclini (2009) asserts that the convergence of communication which is reflected by digital convergence accelerates this trend in the cultural theories. Digital means of communication dominate the way we acquire new cultural information. Our old means of communication such as printed books, theaters and films are being challenged and altered. The consequences of this phenomenon on communication are being discussed and manifested in academic publication.

If we limit our views of cultural communication to the category of drama, the Convergence Model of Communication can also be explored by Drama theory. The process and story-telling function of a drama can be achieved and observed by the Convergence Theory. The Convergence theory can be applied in any situation that includes information sharing and exchanging. Kincaid (2002) asserts that the content of drama which includes the goal that drama needs to convey information to the audience members is about information sharing. While the storylines of the drama move forward, audiences are expected to converge to the points of view of the characters. Therefore, Convergence theory helps explains why drama is able to change the emotion and reaction of the audience.

In addition to its impact on social change and cross-cultural communication, Convergence theory can also be applied in interpersonal communication. In this perspective, the submissive partner in an interpersonal relationship looks at the world in the view of their dominant partners. In other words, the person who is in charge of the daily communication determines the other partner's perspective. Miller-Day and Jackson (2012) indicate that convergence communication is discussed in three dimensions, which are disequilibrium, interpersonal deference, and motivation. When the submissive partner thinks that the dominant partner has the power to manipulate the emotional resources, the coercive power and reward power of the dominant partner will serve as the motivation for the submissive partner to follow this will.

Electronic Communication

The electronic communication method has been employed in the education sector in Indonesia. Burns (2010) has stated that since there are over 17,000 islands in the country of Indonesia, it is extremely difficult for educators to allocate resources to educate the students. Therefore, the educators of Indonesia decided to utilize online resources to teach and communicate with each other, making the best of the students' learning experience. However, Burns (2010) has also pointed out that because of the geographic characteristics of Indonesia, the country is one of the few countries that have the lowest internet penetration rate. However; participants in online learning have overcome the limitation of technology, and adopted this learning and teaching method. In addition, distance education has also been introduced to the classroom because of the scarce resources of education in Indonesia. Early on, Rye (2008) pointed out that many institutions in Indonesia have used distance education to some degree, but because internet connection could not be guaranteed, only half of the students received proper education online. Since the quality of the internet depends on the development level of the islands, the more affluent islands can receive better internet connections; and therefore, better education.

Other than utilizing the online resources for teaching, Lelong and Fearnley-Sander (1999) have reported a true project that has happened between the students in Indonesia and the students in Australia. In this event, the students in Indonesia were paired up with the students from Australia and the communication method for them was email. All the pairs in the project had to complete research and conduct surveys together; and based on their difference in geographic locations, they each do what they are good at in the project. The project is a good demonstration of how the electronic means of communication has helped the students in Indonesia open their eyes to the bigger world, and how they can enlarge their personal connections with people from other countries.

Electronic communication has also been witnessed in regular people's daily life. This phenomenon can be divided into parts: the electronic communication usage of people below the age of 18 and adults.

Child Usage

Hollander (2011) has conducted surveys about the media use of children under the age of 15 who mainly live on the Jakarta Island. These children are in possession of rich media applications such as internet, TV, and other electronic communication methods. There is extensive electronic media coverage for the children in this particular area. However, it is also worth noticing that the reason why these children have easy access to electronic media is because the island Jakarta is far more affluent in comparison to the other islands of Indonesia, and the families that these children are from possess high social status. Beentjes (2011) has also pointed out that based on their particular target of the group on Jakarta Island gender has also played a role in the ways that electronic media and communication channels are being used. Boys are more in favor of playing online computer games whereas girls are more into communicating with their circle of friends electronically.

Adult Usage

Besides the electronic usage of communication for children, previous research also focused on adults in Indonesia. When we think about the daily adult life in Indonesia, we inevitably think about the influence of religion. There are two major religions in Indonesia, which are Islam and Christian. Hui (2010) has argued that the internet usage cannot be separated from religion. Religious websites develop as the technology of the internet develops.

The impact of radical websites in Indonesia has been discussed thoroughly. Hui (2010) has demonstrated that many radical Islamic websites have been utilizing the internet as a communication method to let more people be aware of their proposition and belief. The radical Islamic groups wish to expand their influence in Indonesia by electronic communication methods.

In order to protect the integrity of people's religions in Indonesia, the government has used the internet censorship to prevent netizens (average internet users who are nationals) to browse certain types of content. For example, Lim (2013) has demonstrated an event in Indonesia that in the Islamic holy month of Ramadhan some websites from both in and outside of the country were blocked in order to protect the holiness of the spirit of Islam. As a result, pornographic and anti-Islamic websites were censored for the entire holy month. This measure of internet censorship has raised speculations of the fact that the advent of electronic communication has led to moral panic of the religious conservatives.

Health care workers have used electronic communication methods to help facilitate medical resources. The geographic feature of Indonesia is incredibly complicated because of its 17,000 islands. Moreover, the population spreads to nearly all islands. Therefore, it is fairly complex to make sure all the people can have equal access to medical care. Lee, Chib, and Kim (2011) have stated that cell phones have been used among the health care workers in Indonesia. The midwives in the rural areas of Indonesia have used cell phones as a way to acquire health knowledge from health care workers at hospital centers. Because of cell usage, the lack of health knowledge for people who live in rural islands of Indonesia is improving. Furthermore, Lee, Chib and Kim (2011) have also said that ambulances are not available in some rural islands in Indonesia, so despite emergent situations, people can use cell phones to make reservations with the health care workers in order to obtain relevant professional help. Therefore, with the help of cell phones, people who live in rural islands of Indonesia have shown significant improvement in physical health.

Electronic Communication in Political Movement

Electronic communication has also been used in the political movement of Indonesia. Since the introduction of the internet to Indonesia in the year of 1983, several changes in the political stage have taken place that have changed the development of Indonesia. Gazali (2014) has discussed the usage of social media and social networks used by political activists to expand their influence in the field of politics. People believe that the sense of democracy should be reflected in many aspects of the society including social media. Social media should be a platform for democracy. Hill and Sen (2006) have shown this as an example of the impact of electronic communication in the political dynamics of Indonesia. In the year of 1998, many people had received a message that calls for people to protest against President Soeharto. This message was written in the poetic form and was sent through email. At that point, this message embodied the spirit for students to oppose the leadership of the president. As we can see from this incident, social media had already become an important factor in the political movement of Indonesia. Hill and Sen (2006) have argued that after the fall of President Soeharto, the internet became a public space for citizens to discuss politics and express their political opinions. This brief background lays the foundation for the following eight research questions:

- RQ1: What age, gender and education demographics use electronic means of communication in Indonesia?
- RQ2: What categories of electronic means of communication are more popular in Indonesia?
- RQ3: How long have Indonesians used cellular telephones?
- RQ4: How long have Indonesians used the internet?
- RQ5: What are the main reasons for using the internet for people in Indonesia?
- RQ6: Has internet usage ever been a problem for Indonesians?
- RQ7: Are electronic means of communication being utilized in education in Indonesia?
- RQ8: Do people form new electronic relationships in Indonesia?

METHOD

Data will be collected through the means of electronic survey online. As part of the research, questionnaires will be disbursed in different forms such as Survey Monkey and webpage surveys on Facebook. The primary ways to approach the Indonesian subjects are through email and Facebook. After I gain contact with the Indonesian subjects, it is up to them to decide if they are going to answer the survey.

Survey subjects willing to take the initiative to answer the survey questions volunteer to do it for free. The Indonesian respondents are notified in advance that this survey is intended to be harm-free; if they think the survey content is offensive or may cause any kind of distress, it is absolutely okay for them to withdraw at any time.

For the respondents' information, they will also be notified about the purpose of the survey, which is the electronic use of communication of their developing country, Indonesia. In addition, I will also inform the respondents about how their collective responses are going to be used in academic presentation and publication, and the survey being used in the research has been approved by Fort Hays State University Institutional Review Board. Most of the Indonesians, who are able to finish this survey, are familiar with the electronic means of communication, and some of them have been educated in the United States.

The instrument is a survey with questions with the topic of electronic means of communication usage in Indonesia. The survey is conducted in English because most of the survey subjects understand English and are able to respond in English. There are different types of questions in the survey including several multiple choices questions and subjective questions. Participants will be told to feel free to skip several questions if it is in their best interest. In order to recruit more respondents in Indonesia, the snowball method asks our test subjects to look for more people from Indonesia to be respondents in the research.

The participants that are being targeted are mostly from educational backgrounds. Some of the Indonesians have gained their Bachelor's degree and Master's degree in the United States. These individuals mostly have good jobs and are well-paid, so they possess a fairly high social status. The participants are Indonesians, who are currently living in Indonesia and in other foreign countries.

For the purpose of recruiting participants, I contacted several Indonesian students, who are in China. Meanwhile, I also keep in touch with American college professors in China who have previous experience of teaching Indonesian students in the United States. The complexity and variety of the research respondents' backgrounds are also being taken into account. With that being considered, Indonesian students who currently are studying at mid-western universities were recruited. These Indonesian students are in touch with Indonesians from different economic backgrounds in Indonesia. With the semi-purposive survey results from people of different backgrounds, it can be more effective to compare data and gain a deeper insight into the usage of the electronic means of communication in Indonesia. The survey distribution will add perspectives to balance gender and geographic response.

RESULTS

RQ1: What age, gender and education demographics use electronic means of communication in Indonesia?

Indonesian respondents between the ages of 18-25 years old made up 54.29% of the total respondents, and the ages of 25-35 make up 45.71%. No one under 18 or over 65 was contacted for response. This result indicates that the respondents who participated were college age or adults less than 35 years old. Other age cohorts were unobtainable by electronic survey means. See figure 3.1.

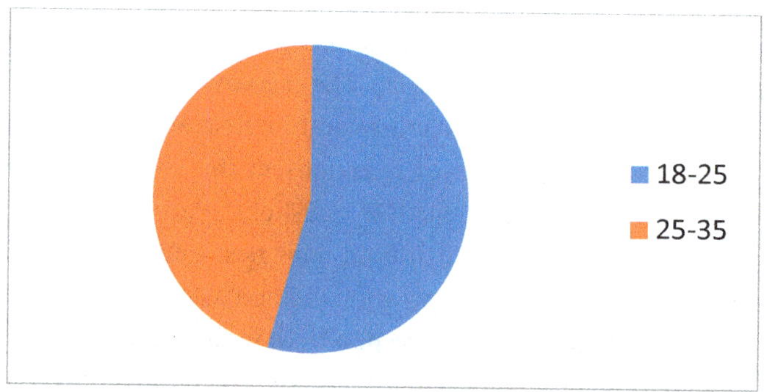

Figure 3.1: Indonesian Respondent Age Demographics (n=35)

Legend for Pie Chart Figure 3.1

Color	Age Cohort	Percentage	Respondents
Blue	18-25 years	54.29%	19
Red	25-35 years	45.71%	16

According to the research data collected by the survey, the gender demographics are that male respondents make up 60% of the entire respondents whereas female respondents make up 40%. No attempt at equalizing gender respondents was made. The random result indicates that somewhat more males use electronic means of communication than females.

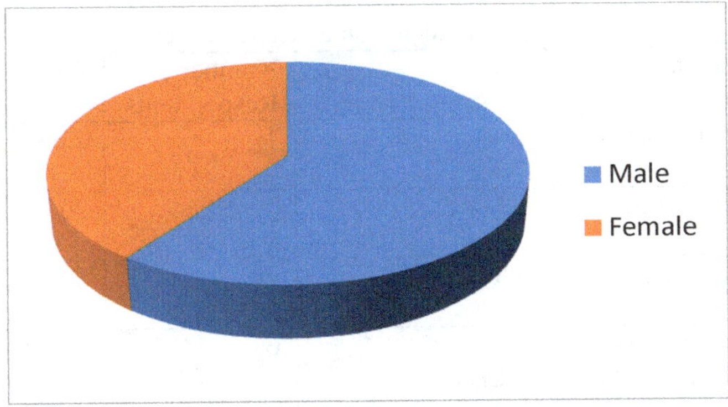

Figure 3.2: Indonesian Respondents Gender Demographics.

Legend for Pie Chart Figure 3.2

Color	Gender Cohort	Percentage	Respondents
Blue	Male	60%	21
Red	Female	40%	14

In accordance with the age demographic, there is only one respondent who is primary school level which makes up 2.68% of the entire respondents. There are 7 respondents who are high school or secondary school level which make up 20.00% of the entire respondent population. However, the number of respondents who have been in an institute, college or university is 21 which make up 60.00% of the whole respondents. The number of people who have finished graduate school (MA, MS, PhD, MD, and LD) is 6 and they make up 17.14% of the entire population. As we can see from the results, the vast majority of the respondents who have participated in the survey have been through higher education. See table 3.1 and figure 3.3.

Table 3.1 Indonesian Education Level

Education Level	Number of Respondents	Percentage
Primary School	1	2.68%
High or Secondary School	7	20.00%
Institute, College or University	21	60.00%
Graduate School (MA, MS, PhD, MD, LD)	6	17.14%

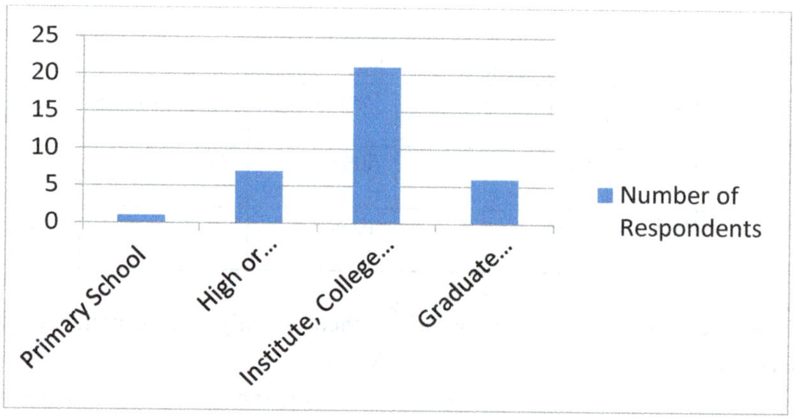

Figure 3.3: Indonesian Respondents Education Demographics

RQ2: What categories of electronic means of communication are more popular in Indonesia? Cellular Phone, Moto, Mobile Phone.

Most of the people (85.71%) prefer cellular phone, moto, or mobile phone as their favorite means of electronic communication. Other than cellular phone, moto, and mobile phone, people also use chat rooms (37.14%), internet list serve (20.00%) and internet bulletin board (17.14%). Respondents argue that cellular phone, moto, and mobile phone offer people the most flexibility and they can use them whenever they can. Qualitative responses include the following:

"Smartphone allows me to do basically whatever I want to do under several situations."

"I love using cellphone; they make me feel it is easier to get connected with the outside world."

"I use mobile phone to call my boss every day."

"I have been using cell phone for five years."

See figure 3.4 for quantitative data.

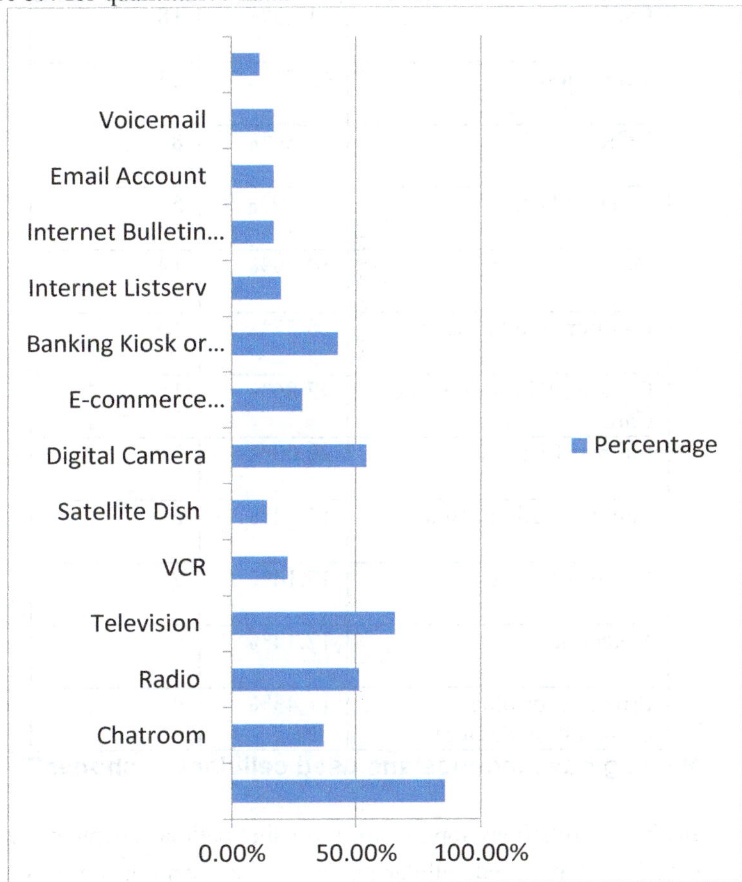

Figure 3.4: Popularity of Electronic Means of Communication

Legend for Bar Chart Figure 3.4

Means of Communication	Percentage	Respondents
Cellular Phone, Moto, Mobile Phone	85.71%	30
Chatroom	37.14%	13
Radio	51.43%	18
Television	65.71%	23
VCR	22.86%	8
Satellite Dish	14.29%	5
Digital Camera	54.29%	19
E-commerce Website	28.57%	10
Banking Kiosk or Money Card	42.86%	15
Internet Listserv	20.00%	7
Internet Bulletin Board	17.14%	6
Email Account	17.14%	27
Voicemail	17.14%	6
Proxima (or other presentation device)	11.43%	4

RQ3: How long have Indonesians used cellular telephones?

Indonesians have a relatively long history of using cellular telephones as 69.44% of the respondents have used cellular phone for five years or more. Nearly 20%, or 19.44% of the respondents have used cellular phones for 3-5 years, and 8.33% of the Indonesian respondents have used cellular phones for 2-3 years. Only 2.78% of the survey participants have used cellular phones for less than one year. As we can see from the data collected, most of the Indonesians have been using cellular phones for a significant amount of time. It is reported as common knowledge that cellular phones became extremely popular around 2008 or six years before the study; therefore, most of these Indonesian participants have been using cellular telephony since then. See figure 3.5.

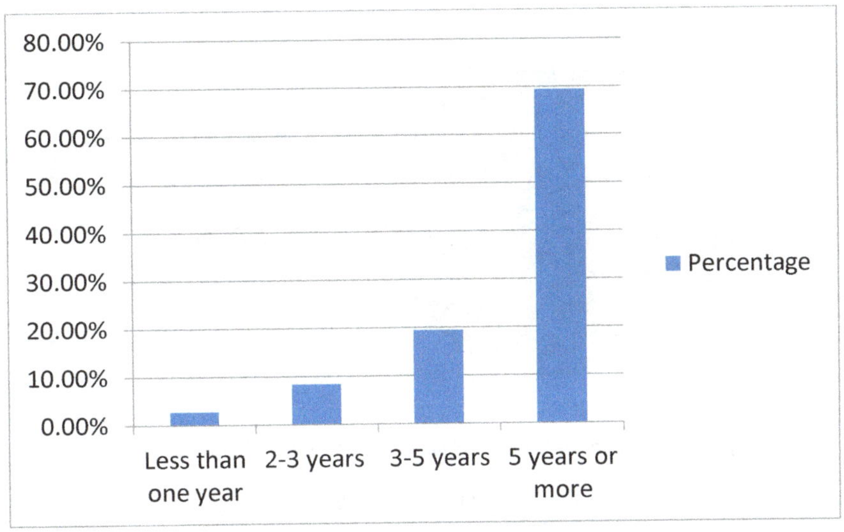

Figure 3.5: Years of Cellular Telephone Usage among Indonesian Respondents

Legend for Bar Chart Figure 3.5

Years of Cell Usage	Percentage	Respondents
Less than one year	2.78%	1
2-3 years	8.33%	3
3-5 years	19.44%	7
5 years or more	69.44%	25

RQ4: How long have Indonesians used the internet?

Indonesians have a fairly long history of using the internet, and 63.89% of the respondents reported they have used the internet for five years or more. About a third, 22.22% of the respondents have used internet for 3-5 years. Slightly over a tenth, 11.11% of the Indonesian respondents have used internet for 2-3 years. Only 2.78% of the survey participants have used internet for less than one year. Based on the data collected, most of the people in Indonesia have used the internet for a long time (three years or more). Internet usage in Indonesia has been pervasive since 2010 in over half the population.

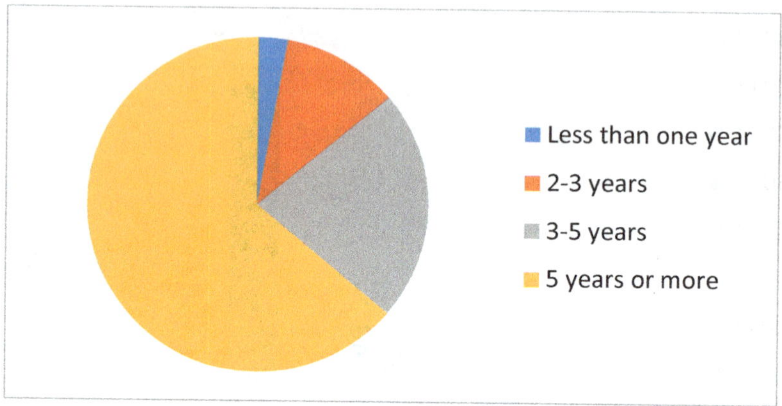

Figure 3.6: Years of Internet Usage among Indonesian Respondents

Legend for Pie Chart Figure 3.6

Years	Percentage	Respondents
Less than one year	2.78%	1
2-3 years	11.11%	4
3-5 years	22.22%	8
5 years or more	63.89%	23

RQ5: What are the main reasons for using the internet for people in Indonesia?

The main reasons for using the internet for people in Indonesia are that the internet is convenient and reachable. According to the answers of the open-ended questions on the survey about the internet, respondents think that internet makes everything more effective and easy to do. They are adapting the fast-paced life of their countries; therefore, they have to make sure that they can keep pace with their country's development. Some think that internet technology can help them with their school assignments, and there are also some participants who think that they can search for more information from the internet.

Responses include:

> "I could finish my PPT or assignments with it. I can research the information or source which is necessary for my work. Furthermore, I would like to watch TV series on the internet in the dorm. It saves money for me, instead of watching movies in the cinema."

Other respondents replied:

> "I can finish a variety of assignments and research information through the internet. It is a window for me to look over the news as it happens every day anytime."

> "Can get exposed to lots of info and make friends on the internet. Makes life colorful and enjoyable."

> "It helps me to find anything about news, world, trends, etc."

RQ6: Has internet usage ever been a problem for Indonesians?

Yes, the internet usage has been a problem for Indonesians. There were several negative effects for many Indonesian participants. One of its negative effects is that if people use it for too long, it will hurt people's eyes and body. People will develop poor eyesight if they watch the screen for too long. Moreover, some people also asserted that if they do too many things on the internet, they will finally become addicted to it. In addition, some reported focusing too much on the internet instead of the real world so they gradually became isolated from reality, and cannot regularly communicate with people under normal social circumstances. In addition, the internet can also become the target of cybercrime.

Individual responses include:

> "It hurts my eyes."

> "It is harmful to our eyes and cervical vertebra when we seat in the front of computer for a long time."

> "Sometimes it crashes and has viruses."

> "It is vulnerable to cybercrime."

RQ7: Are electronic means of communication being utilized in education in Indonesia?

Yes, they are. Electronic means of communication has been utilized in education in many ways. According to the responses of the participants, some people use the internet for doing their assignments from school and conducting research. Some schools in Indonesia have formed international partnerships with schools in the United States, so they have online classes which are taught through the internet. Moreover, there are a lot of online materials being distributed and shared by Indonesian instructors through the internet.

Examples of voluntary qualitative response:

"I use the internet to do my homework."

"American professors grade my papers through the internet."

"In my university, announcements are made on the internet."

RQ8: Do people form new relationships through electronic communication methods in Indonesia?

Yes, they do. But this is not very common among the participants. The advent of the electronic means of communication provides people with new platforms of communication that they could not even imagine before. It is very easy and fast to meet someone on the internet. People utilize electronic means of communication to share ideas and express their opinions. However, it is amazing that about a third do and a third do not like to form new relationships via electronic means of communication. The survey question was, "Do you meet new electronic friends face-to-face?" Response indicated 29.17% of the participants never form any new relationships on the internet, and 12.50% of the participants hardly ever form any new relationships on the internet. People who sometimes form new relationships make up 20.83% of the population and 33.33% of the participants quite often form new relationships on the internet. However, only 4.17% of the participants always form new relationships. People in Indonesia do form new relationships on the internet, but it is not prevalent behavior despite its ease and low cost. See figure 3.7.

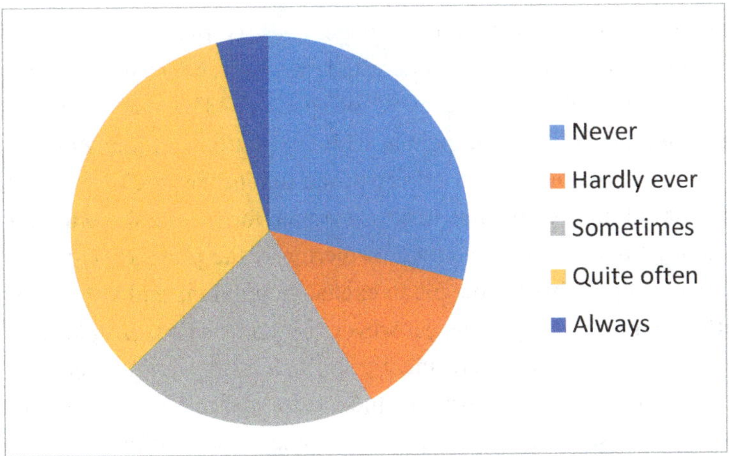

Figure 3.7: Frequency of Forming New Relationships through Electronic Means of Communication among Indonesian Respondents

Legend for Pie Chart Figure 3.7

Frequency	Percentage	Respondents
Never	29.17%	7
Hardly ever	12.50%	3
Sometimes	20.83%	5
Quite often	33.33%	8
Always	4.17%	1

DISCUSSION

The Convergence Theory shows that people, in order to properly coexist in one country, have to adjust to other behaviors to adapt to the norms of the society as a whole. Indonesians from different demographic groups need to reduce misunderstanding, find mutual interests and ultimately strive for common goals. This is especially important for Indonesia because it consists of a great number of islands and a variety of ethnic groups. The usage of electronic means of communication plays an important role in determining the ramification of the convergence of people.

The population which is of college ages is the most active group in communication. They are more interested in new and trendy communication means than any other age groups of people. The electronic means of communication provides young people with the platform to know each other and understand the development and culture of the Indonesian society as a whole. The popularity of cellular telephones makes it easier for people to communicate and interact at any time and any location. Furthermore, the prevalence of the usage of internet makes it possible for people to exchange opinions and acquire knowledge from different groups of people. However, in some cases, people do not quite trust the relationships which are merely built through the internet. Overall, the convergence process that exists in Indonesia is mostly on the right track; nevertheless, it still takes time for people from different demographic backgrounds to utilize all kinds of communication to converge towards the general behavior of the society.

For most individuals, they would like to converge toward the behaviors and thinking patterns of other identity groups of people when they understand the meanings and messages behind what others do. According to these research results, people who are over 35 years old still have a long way to go to embrace the lifestyle of electronic means of communication. Moreover, due to the lack of an all-encompassing infrastructure for internet and signal towers for cellular telephones, people still find it difficult to utilize the resources of electronic means of communication in Indonesia. Therefore, relevant laws and policies should be made to encourage the development of the technology of electronic communication. Technology engineers should devise yet more ways to lower the cost of electronic communication.

According to the Convergence Theory, the common interests developed by different demographic groups are the key to the cohesion and unity of the country. Electronic communication should be used for finding the shared goal of the people, and be used as a way for people to collaborate with each other for business, education, religious pursuits, and pleasure. Meanwhile, people can utilize the evolving electronic communication platform to express their opinions and standpoints.

When the surveys that were used for research questions were disbursed, a variety of electronic means of communication were reported as widely used in Indonesia. Many people were able to answer the surveys which were sent through the internet. This indicates that many Indonesians were able to use electronic means of communication to finish certain tasks and missions. According to the results, age, gender and education demographics played an important role in the subject matter.

RQ1. What age, gender and education demographics use electronic means of communication in Indonesia?

People who are between the age of 18 and the age of 25 are the largest portion of research subjects who use electronic means of communication in Indonesia. This demonstrates that electronic communication is particularly popular among young people. They are more open to new trends and are able to use them. Sixty percent of the research subjects were male. Currently, more males use electronic means of communication than females in Indonesia. The phenomenon among respondents shows that there is a gender difference in terms of electronic communication in Indonesia. In Indonesia, gender difference is still reflective of a male-dominant society. Sixty percent of the participants had been educated in a higher education level institution. People who have been in a college, university or institute have easier access to electronic means of communication. The higher the education the Indonesians have, the easier it is for them to access electronic communication.

RQ2. What categories of electronic means of communication are more popular in Indonesia?

There is a great number of electronic means of communication available in Indonesia. People have many options to choose. According to the survey results, cellular phone, moto, and mobile phone are the most popular communication method for Indonesians. The flexible, convenient features of a mobile phone meet the needs of people who have to adapt to the rapid speed of development in this developing, international country. Other than a mobile phone, the chatroom which is made possible by the internet is also very popular among Indonesians. Chatting through the internet is free of charge; and most of the time, very fast and efficient. Since Indonesia is a developing country, people need to work hard and become more competitive in the workplace, so they have to choose whatever option saves time and does not cost much. This is the reason why internet and cellular phones are more popular among Indonesians.

RQ3. How long have Indonesians used cellular telephones?

Based on the data that was collected through the survey, 69.44% of the Indonesians have used the cellular telephone for more than five years. Cellular telephones prevail among the society even though they have a high cost. Cellular telephones require the construction of an infrastructure to provide the cellular phone users with signals in order to receive data, text messages and phone calls. A third of Indonesians have used the cellular phone for five years or more. This is because Indonesia is a country which is made of many islands, and it is almost impossible for residential islands to be equipped with the required infrastructure in order to support their cellular phone usage.

RQ4. How long have Indonesians used the internet?

A large majority, 89% of the Indonesians, have used the internet for more than five years. The popularity of the internet among Indonesians is intense. The basic infrastructure of internet technology provides Indonesians with firm support of internet service. The application of internet ensures that Indonesians can stay connected within their own country and the outside world.

RQ5. What are the main reasons for using the internet for people in Indonesia?

Internet has become an irreplaceable part of people's lives in Indonesia. There are many reasons for Indonesians to use the internet due to the colorful functions of its nature. There are three main reasons for people to use the internet which are education, work and the flexibility that is offered by internet. In some schools, education is partially achieved through the internet. For example, assignments need to be finished through the internet and papers need to be graded by professors outside of Indonesia. Workers have to use internet to connect with the company when they are doing field work. Moreover, people use the internet whenever they need to because internet is fast and convenient.

RQ6. Has internet usage ever been a problem for Indonesians?

Based on the data that was collected, Indonesians commonly complain about the internet due to its health reasons and the addiction of the internet. People argue that if they keep staring at the computer for a long time, it is going to hurt their eyes and body. In addition radiation is caused by both the internet and computers. Other than that, internet addiction may take away people's social skills. People may behave abnormally if they rely too much on the interpersonal relationships which are achieved through the internet.

RQ7. Are electronic means of communication being utilized in education in Indonesia?

Education usage has already become an important part of internet usage in Indonesia. Indonesia has neighbors such as Australia which have built partnerships with Indonesians in terms of education. Some schools in Indonesia have formed international partnerships with schools in the United States, so they have online classes which are taught through the internet. Moreover, there are a lot of online materials being distributed and shared by the instructors through the internet.

RQ8. Do people form new relationships through electronic communication methods in Indonesia?

Only 4.17% of Indonesians always form new relationships which are formed on the internet. Most of the people have a suspicious and skeptical attitude toward this issue. Most of these Indonesians do not trust new relationships which are merely formed on the internet. It lies in the Indonesian culture that they cannot make sure if anyone is reliable unless they can meet him or her in person. When it comes to interpersonal relationships, Indonesians still believe that face-to-face interaction is the most valuable form of personal communication.

REFERENCES

Burns, Mary. "17,000 Islands, One Goal: Indonesia Turns to Online Resources to Create a Network of School-based Coaches." *Journal of Staff Development* 31, no. 1 (2010): 18-23.

Gazali, Effendi. "Learning By Clicking: An Experiment with Social Media Democracy in Indonesia." *International Communication Gazette* (2014): 1748048514524119.

Hill, David T., and Krishna Sen. *The Internet in Indonesia's New Democracy.* New York: Routledge, 2005.

Hill, David T., and Krishna Sen. "Netizens in Combat: Conflict on the Internet in Indonesia." *Asian Studies Review* 26, no. 2 (2002): 165-188.

Hendriyani, Ed Hollander, Leen d'Haenens, and Johannes WJ Beentjes. "Children's Media Use in Indonesia." *Asian Journal of Communication* 22, no. 3 (2012): 304-319.

Huus, Kari. "Two Years of Live Electronically: Covering Breaking Foreign News for the Internet." *Nieman Reports* 52, no. 4 (1998): 63-64.

Lee, Seungyoon, Arul Chib, and Jeong-Nam Kim. "Midwives' Cell Phone Use and Health Knowledge in Rural Communities." *Journal of Health Communication* 16, no. 9 (2011): 1006-1023.

Lelong, Peter, and Mary Fearnley-Sander. "E-mail Communities—A Story of Collaboration between Students in Australia and Indonesia." *Social Studies* 90, no. 3 (1999): 114-120.

Rye, Ståle Angen. "Digital Communication, Transportation and Urban Structuation in Students' Daily Life." *Geografiska Annaler: Series B, Human Geography* 92, no. 1 (2010): 81-96.

Rye, Ståle Angen. "Exploring the Gap of the Digital Divide." *GeoJournal* 71, no. 2-3 (2008): 171-184.

Yogaswara, A., Abuh Muhammad Waskito, Akbar Kaelola, Dikdik M. Arief Mansur, Elisastris Gultom, and Ellis Lestari Pambayun. "The Internet and Everyday Life in Indonesia: A New Moral Panic?" *Bijdragen tot de Taal-, Landen Volkenkunde* 169 (2013): 133-147.

Yang, Jennifer Hui. *The Internet in Indonesia: Development and Impact of Radical Websites.* SEAsia and ASEAN Working Papers, WP16T, Last Updated 01/07/2014. Nanyong, Singapore: Center of Excellence for National Security, 2018.

CHAPTER 5

Internet, Cellular Telephone, and Social Networking Usage in Jamaica

Chen Chen

Because of its need of economics, science and technology, Jamaica is a developing country in our modern world. But now, Jamaica has already stepped into the digital information age. The purpose of this research using qualitative and quantitative methods is to examine the development of e-communication in Jamaica, including the usage of cellular telephones, the internet, and social networking.

Jamaica, part of the Greater Antilles, is multicultural island country that lies in the northwest of the Caribbean Sea. In recent years, the application of e-communication has shown a significant increase in Jamaica. As a developing country, Jamaica is a huge potential market for culture, trade, and travel, so cross-cultural communication becomes more and more important in developing Jamaica. This brief study examines the distribution and usage of cellular telephones, the internet, and social networking to ascertain the development of e-communication in Jamaica.

The impact of technology towards human affairs is undeniable. The byproducts of technology such as computer and internet usage are essential, and these computers and the internet become tools in human progress. The emergence of science and technology creates an impact on people and their human concerns, with internet, cellular telephone, and social networking that are connecting people, and yet disconnecting familiar face-to-face communications among them. The technology in the forms of internet and cellular telephones aid any human affairs work and grow toward personal, business, academic, and social accomplishments. Its existence becomes the answer in many industrial and technological developments that facilitate global communication. The internet, which tools global communication, also has the capability to redesign educational systems, business transactions, and even political status. According to Hearst (1999), she pronounced that "the interconnectivity and global accessibility of the

Web has also given the rise of some unexpected ways in which people can take advantage of the technology at the expense of other people" (10). This idea suggests that the birth of internet can alter the culture of life and the very fabric of human affairs. The main objective of this study aimed at determining the distribution of the usage of cellular telephones, the internet, and social networking in Jamaica, especially comparable distribution among the different genders.

This research relies on the Feminism theory and Face Negotiation theory to explain several questions, such as why the female internet users are growing rapidly in some developing countries; what types of social networking are most used in Jamaica; and if internet usage has become a problem for Jamaican users. Nearly a century later, the same feminist issues remain as hot topics for the whole world, particularly in developing countries. Therefore, gender issues may also exist in a booming internet technology. In order to research perineal gender issues, the distribution and individual usage of cellular telephones, the internet, and social networking in Jamaica are examined by descriptive and experimental design.   Literature review and a stratified random sampling approach sought answers to report current developmental progress in Jamaican citizens and sojourners.

OVERVIEW OF JAMAICA

Jamaica, part of the Greater Antilles, is multicultural island country that lies northwest in the Caribbean Sea, and is the largest English-speaking Caribbean island. It is about 700 miles south of Miami, Florida, USA, and 90 miles south of Cuba. Jamaica has a warm tropical maritime climate. It is extremely mountainous, with a central chain of mountains running east to west. The highest point is Blue Mountain Peak where abounds the most expensive and sought-after coffee in the world----Jamaican Blue Mountain coffee. The largest river in Jamaica is Black River, located in the parish of St. Elizabeth. Jamaica is divided into three counties----Cornwall, Middlesex and Surrey. The capital and commercial center of Jamaica is Kingston, situated on the southeast coast of the island. The island also has a rich history in the field of sports, and has distinguished itself especially in the area of cricket. Jamaica can be described as a multi-ethnic island, yet its population primarily comprises persons of African descent. People of European, East Indian and Chinese origin also make up a portion of the population.

The E-communication of Jamaica

In recent years, the application of e-communication has had a significant increase in Jamaica. According to a survey conducted by the Mona School of Business (MSB) at the University of the West Indies (UWI), Professor Hopeton Dunn (2010) reported that internet users in Jamaica numbered 715,414. A total of 24 percent of the population have access to a computer at home. The most popular uses of the internet were for sending and receiving email, which 76.9 percent reported doing in the past 12 months of 2010. Social networking was the second most popular activity with 71.7 percent of respondents reporting using websites such as Facebook and Twitter. In addition, educational use was third with 65.5 percent.

	Internet users (per 100 people)			
	2008	2009	2010	2011
Jamaica	23.6	24.3	27.7	37.9

Dunn (2010)

Internet users are increasing and becoming widespread in Jamaica as people tend to get connected in groups. Dunn (2010) emphasized the majority of Jamaican internet users are aged between 16 and 35 years old. More than 80 percent of Jamaican internet users are young people, including high school students, college students, young workers, and business men. Among all internet users in Jamaica, 45 percent were shown by Dunn (2010) to have access to fixed broadband, while 65 percent have access to mobile devices, such as cellular telephones, auto PCs, and iPads. As the cellular telephone is becoming an indispensable tool in Jamaican lives, 45 percent of cellular telephone users said they access the internet via their cellular telephones. The trend is for more Jamaicans to utilize the internet via cellular telephones.

Over the years, people have struggled against and experienced a slow-moving process of any work tasks that were done at school, in business, and in private and public offices. Many assigned tasks were accomplished within a day or so. Establishments such as hotels, schools, hospitals, public agencies, bureaus, and other private offices now demand efficient digital systems to speed up transactions and services to minutes instead of days. When the wonders of technology such as the internet, cellular telephone, and other electronic devices arrive, average people and their ways of life change. This includes all individuals, and the social, political, economic, academic, business, and healthcare services; along with religious affairs. Such developments on these various fields have been recognized and have been given enough attention in developed countries and the world on a global scale. However, the progress of developing countries and their relationship to the world might expand understanding of technology to its horizons. Every country has its own adaptations and uses of technology that may benefit the world in totally new and helpful ways.

The depth of product utilization of these technological inventions has grown in prominence in the world-wide society. Lives of the people most isolated and in greatest need are dependent on these technological innovations. People have been learning to use computers, and these devices are used in schools, hospitals, and government establishments. No country in the world today exists that has never been touched by technological innovation. In other words, all countries are aware of meeting and keeping their pace towards globalization. This technological overture of progress and human cooperation offers more benefit to people than risk. Many imagine a world of peace, sharing of information, and prosperity through technology. Questioning and searching for possible answers and expansions of technological innovations could be a proof to help people, especially young children today and in the future.

Nowadays, the expansion of technology and its use reshapes individual lives by educating children. For instance, children who go to schools and who are inclined to technology such as the internet, laptops, and iPods become adept to the use of the internet before adulthood, not knowing a world without e-communication. They can listen to music, watch television, and read information from e-book files and other electronic devices. Laptop and tablet computers have become commonplace for children or students in academia for practical use. In this scenario, teachers and educators are astonished at the impact on young children and students in the classrooms. In many educational facilities, teachers and other educators have adopted the use of laptop and tablet computers in the classroom to replace books and chalkboards. Each student is no longer forced to learn lock-step with the entire group.

The use of these technological devices has been studied and analyzed for years in developed countries, and the effects remain acceptable (Jackson et al., 2003) for transference to developing countries. The use of the internet and computers on a daily basis does not harm children, nor does it disorient nor displace individual teaching methods in academia. Yes, it is true that the technology does not harm, and the children learn in multiple ways because the technological inventions display a vast array of possibilities for academic use and for other purposes. The discovery of these technological innovations helps human beings in work tasks and learning.

The Social Media of Jamaica

As the network movement is in full swing, the social network companies such as Facebook, Twitter and MSN try to open new markets in Jamaica. Facebook is the most popular online social networking service in North America, which launched in 2004 by US college student Mark Zuckerberg. While Facebook's popularity has fallen in North America, Richardson (2010) announced that it has almost doubled in Jamaica in less than a year. With the fast increase of network information, the network application is expanding daily; the number of internet users in Jamaica is growing rapidly. Facebook users in the US dropped by six million to 149.4 million in May 2011, but Jamaican local internet marketing consultant Sandor Panton (2011) noted that more than a fifth of the population of Jamaica was using Facebook, including 497,660 persons age 18 or older, within the same timeframe.

Macroscopically, internet penetration is still relatively low in Jamaica, so Jamaica has become a market for social network companies. According to Panton (2011), local companies of all sizes are now using social networks to reach out to new businesses and build brand awareness. For a developing country like Jamaica, social networking is a relatively new frontier. However, with trends like these, Jamaica will soon catch more developed countries in terms of internet penetration – particularly in social networking.

As internet penetration grows, huge numbers of people may be less likely to pay attention to health; such as getting ample sleep, eating right, and exercise. Excessive users may only enjoy online offerings. Thus, overweight and obesity are becoming common problems around the world, especially in developing countries. Therefore, the rate of excessive use across the spectrum of high to low-income countries is examined by comparing the rates of change in weight of sampled populations from low (Nigeria), middle (Jamaica), to high (US) income countries. Multi-level regression models and pattern mixture models were used to approximate weight change and the potential effect of missing data respectively. Results indicated that the obese across the three samples experienced weight loss, and also the missing data patterns affected the rates at which weights changed. Evidently, more rapid weight gains were observed among individuals from the middle-income country (Jamaica), although the prevalence of obesity remained highest in the US and lowest in Nigeria. Weight related health problems are expected to increase in the future since the report indicates that the rate at which lean individuals gain weight is comparatively 9 times higher than that of which obese people lose weight.

However, the report's accuracy is questionable since inferring the sampled data to the entire population heavily depends on the assumption that the cohorts are representative. Moreover, corroborating information regarding causal factors that relate to the risk posed by weight gain is not available in the report. In conclusion, changes in weight of sample cohorts from Jamaica were relatively greater than that of other cohorts from Nigeria and US. This is attributed to the increasing effects of behavioral and cultural shifts which are mostly experienced in transitional societies or middle income countries like Jamaica. This report is; therefore, consistent with its view that obesity is a global phenomenon since weight gain is observable in all countries regardless of economic stability or even social networking activities.

Feminism Theory

Nearly a century after feminist issues were made public in developed countries, they continue for the whole world, particularly in Jamaica and other developing countries. Much theory was created by feminists to balance the rights between males and females. Feminist Standpoint theory is only one theory in the group of standpoint theories. It is an epistemological theory that focuses on how knowledge is shaped by social location. Wood (2005) claims there are five concise elements in Feminist Standpoint theory:

1. Society is structured by power relations.
2. Subordinate social locations are more likely than privileged social locations to generate knowledge.
3. The outsider-within is a privileged epistemological position.
4. Standpoint infers a critical understanding of location and experience.
5. Any individual can have multiple standpoints.

According to this analysis Wood (2005) summarized that Feminist Standpoint theory offers a critique of existing power relations and the inequality they produce in the lives of women and men. Women's lives are much different from men's lives, as influenced by physiological characteristics and psychological patterns. No one can conclude which gender is better than the other. In modern times, knowledge as a vital part of society can help people to gain different social status. It seems a fair way to be successful, but no one can assure that everyone has the same opportunity to receive equal education. In my opinion, Feminist Standpoint theory offers a critique of existing power relations and the inequality they produce. Even so, women still struggle for equality in every country of our world, including the right of voting and religious participation. According to Seedat (2013), he argues that because work toward sexual equality is easily assimilated into a feminist analytic paradigm, Islamic feminism may appear to be the inevitable result of this convergence. However, Islamic feminism as an analytic construct is also inadequate to voice concerns for sexual equality in Islam. Feminism is a tool for gender struggles and equality work in Muslim societies; however, feminism has not fully combined with the varieties of Muslim culture. The new form of feminism emerging is labeled Islamic Feminism.

Seedat (2013) emphasizes the convergence of Islam and feminism can be separated into four parts. First, the convergence of Islam and feminism is inevitable; Second, Islamic Feminism is most evident in culture and politics. Third is the reason that the convergence of Islam and feminism was presented. Finally, the application of Islamic Feminism indeed influenced women's state and living in Muslim societies. Meanwhile, there also are several challenges to Islam. Seedat (2013) hopes Islamic Feminism is only one product of a negotiation between feminism and Islam as two intellectual traditions. In summary, Islamic Feminism as an analyzed construct precludes new understandings of sexual difference originating in non-Western and anticolonial cultural paradigms.

Female jurisprudence can be seen through an approach of the feminist interpretation to metaphors of mediation. Female heroes and other women mediators offer a female jurisprudential clarification as seen in the "Chocolat" and "Fried Green Tomatoes" movies. These movies empower women through portraying female characters that are independent, strong, powerful, and successful, to portray their jurisprudential abilities. This article summarizes the ideas of feminist mediation. Schultz (2009) claims that the feminist informative methods to arbitration metaphors are portrayed in the two movies. The difference with a non-feminist movie is that; in feminist movies, the protagonist seeks help from other people while in non-feminist movies the ideas are usually an internal revelation.

Much as people think of romantic movies as chick-flicks, there is vital information relayed through them. The jurisprudence portrayed in the "Chocolat" and "Fried Green Tomatoes" movies show that women turn to mediation rather than seeking out the law; which in most cases, is in the leadership of a bad patriarchal representation. The protagonists of "Fried Green Tomatoes," Ninny and Idgie, are observed to be in battle against a wicked mayor. With the village against the half-French woman, she seeks help from another woman rather than the law. The message portrayed is not defiance of the power of the law, rather trust and confidence in the feminist mediation approach; which in turn propagates the jurisprudence. These feminist films are successful in their quest of resolving a conflict. The films are meant to bring to light the female mediation abilities and educate on their viewpoints. The two movies are perfect for illuminating the female mediation powers and the empowering of the women portrayed.

Face-negotiation

In addition, Face-negotiation theory will also be a support theory for my research, which is a concept developed by Stella Ting-Toomey (2004). It deals with how different people and cultures place importance and value on identity, and how conflict is handled by individuals within those cultures. According to the old idea in some Asian culture, "Face" in this theory refers to a metaphor for the public image people should display within certain cultures. Thus, this theory also deals with how people gain "positive" or "negative" face, based on how others perceive them. Nowadays, Face-negotiation theory is largely used in conflict negotiation and understanding how different cultures handle conflict. The basic concept behind Face-negotiation theory is the idea that each person's identity is represented by a "face" that he or she shows to others. This is then expanded upon so that the society a person exists within also has a group "face" of which he or she is a part. Individualist cultures are those that place greater importance on individual face than on the group face, while collectivist cultures are those that place the importance of group face above the individual faces. By using the theory of Face-negotiation to understand the nature of a particular culture, it is often easier to understand how conflict can best be resolved within that culture.

The following research questions have emerged from these two theories, electronic communication, and current social media usage in Jamaica.

Research Questions

- RQ1: What types of online activity are most used in Jamaica?
- RQ2: What age and gender demographics use internet in Jamaica?
- RQ3: Do males use cellular telephones more often than females in Jamaica?
- RQ4: Has internet usage become a problem for Jamaican users?

METHOD

Examination of the research questions was conducted using a stratified random sampling approach. Surveys were sent to various public, government and educational locations electronically. Volunteers were requested in English to participate in a survey related to cellular and internet use. Each region of the country was canvassed by purposive sampling to include all geographical areas. Government and private workplaces; government, international, and private schools; and urban and rural locales were offered surveys. One hundred surveys were distributed to respondents in Jamaica to be used for comparison to other developing countries. Systematic effort to locate certain age categories; female, illiterate, or unemployed respondents may be necessary. Ages 18-55 were contacted for possible participation.

The 100 e-questionnaires were provided to local people living in Jamaica via Facebook, e-mail, and an online chat channel. There were 60 effective responses, with 45 providing comprehensive data that could be used for analysis. The following results were obtained by and about e-communication as an emergent property which is growing rapidly in Jamaica.

RESULTS

Jamaicans provided the data pool of valid responses. Total response was 60 fully completed questionnaires collected in two months. Research questions were answered and corroborated by multiple inquiry methods.

RQ 1: What types of online activity are most used in Jamaica?

According the valid statistics, 50 percent of the internet users in Jamaica consider social networking as the most popular activity in their daily lives. See figure 1.

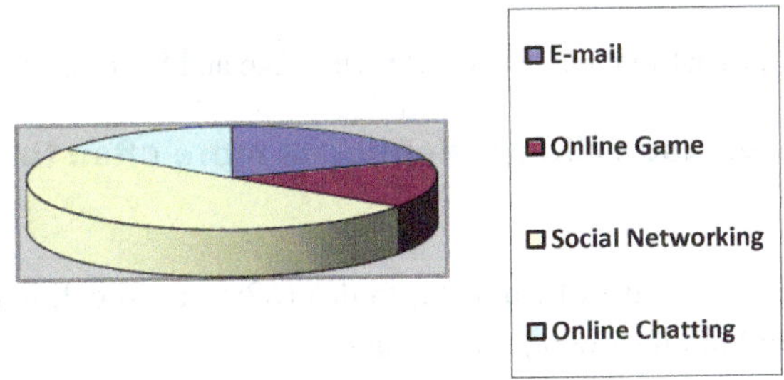

Figure 1:Reported Activities of Internet Users

E-mail was reported at 20% use, and Online Gaming and Chatting were reported at 15% each.

RQ 2: What age, gender, and socio-economic demographics use internet in Jamaica?

Jamaican females (52%) used the internet 4% more often than males (48%), which means the women in Jamaica are slightly more likely to use the internet. See figure 2. The ages 18-28 are the major group of internet users in Jamaica. Those of the Jamaican upper middle class generally tended to have online lives. See figure 3.

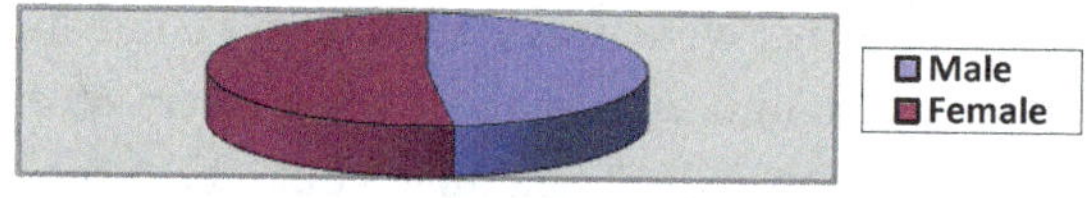

Figure 2: Distribution of Internet Users' Gender

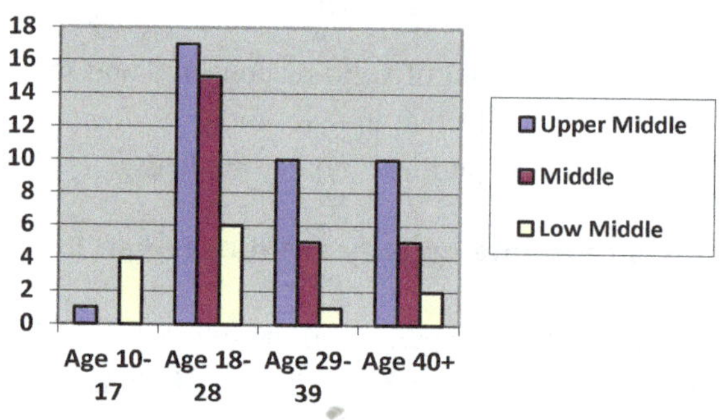

Figure 3: Distribution of Internet Users' Age and Socio-economic Class

RQ 3: Do males use cellular telephones more often than females in Jamaica?

In the developing country of Jamaica, males (62%) used cellular telephones 24% more often than females (38%). See figure 4.

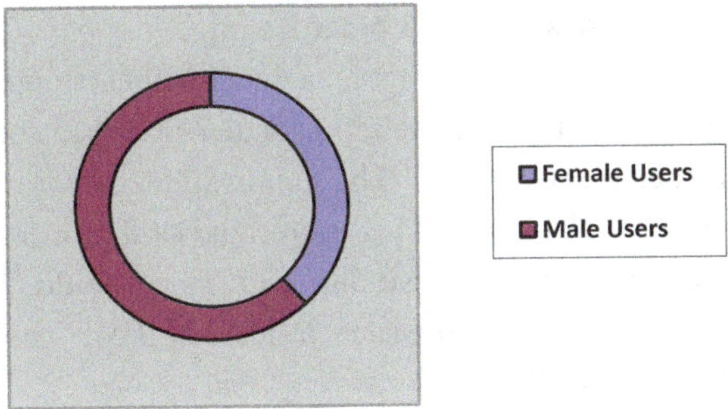

Figure 4: Distribution of Cellular Telephone Users' Gender

RQ 4: Has internet usage become a problem for Jamaican users?

No, while Facebook's popularity has fallen in North America, Julian (2010) as a surveyor announced that it has almost doubled in Jamaica in less than a year. According to valid data I collected, all the respondents were surveyed via internet, which means internet has already become indispensable in Jamaican's lives. No reportage for excessive use by the survey participants was noted.

CONCLUSION

In a nutshell, Jamaica as a developing country is a potential market for internet and cellular telephones. Use of the internet penetration is growing rapidly in recent years in Jamaica. In addition, more and more Jamaican internet users tend to enjoy online lives available via their cellular telephones. By stratified random sampling approach, the surprising result is that Jamaican females used the internet 4% more often than males, which means the lives of women in Jamaica are often in connectivity. Comparing with other developing countries, the application of Feminism theory explains a balance e-usage has made in Jamaica.

Results of this study showed that the most frequent users of internet in Jamaica are aged between 18 and 28. Only 10% of the response (removed from the study) were under 17. It means the youngest, upcoming generation has grown up with the internet, and seems to more readily adapt to, and use the internet with greater frequency. A staggering 50 percent of the internet users in Jamaica consider that social networking is the most popular and necessary activity in daily life. Most obviously, social networking, such as Facebook and Twitter, were already very popular in Jamaica before this study began in 2014. Panton (2011) noted that more than a fifth of the population of Jamaica was using Facebook, including 497,660 persons age 18 or older.

In summary, with trends like these, Jamaica will catch up to more developed countries in terms of internet penetration and cellular telephone penetration – particularly in social networking. The implications are that those developed countries may need social networking more; and utilize it less, than developing countries such as Jamaica. Accessibility and speed could flip communication dominance among the world's countries, at least in those whose cultural patterns necessitate constant surveillance.

REFERENCES

Carpenter, Karen, and Gavin Walters. 2011. "A So Di Ting Set: Conceptions of Male and Female in Jamaica and Barbados." Sexuality & Culture 15(4): 345-360.

Crispin, Jessa. "Feminism and Folklore, Braided Into Modern Life." February 16, 2010. Npr.org
http://www.npr.org/templates/story/story.php?storyId=123771512

Dolenc, Karl. "Twitter, Feminism and Race: Who Gets A Seat At The Table?" September 05, 2013.
http://www.npr.org/blogs/codeswitch/2013/09/05/219278156/twitter-feminism-and-race-who-gets-a-seat-at-the-table

Dunn, Hopeton. 2010. "38% of Jamaicans Now Daily Internet Users, but Home Access Still Low." *Jamaica Observer.*
http://www.jamaicaobserver.com/news/38--Jamaicans-now-Internet-users_(accessed April 14, 2011)

Durazo-Arvizu, Ramón A., Amy Luke, Richard S. Cooper, Guichan Cao, Lara Dugas, Adebowale Adeyemo, Michael Boyne, and Terrence Forrester. 2008. "Rapid Increases in Obesity in Jamaica Compared to Nigeria and the United States." *BMC Public Health* (8): 133-143.

Hearst, Marti A. (1999). "The Changing Relationship between IT and the Society." Retrieved April 11, 2014, from <http://people.ischool.berkeley.edu/~hearst/papers/ieee-social.pdf>.

Hindus, Debby. (1999). "The Importance of Homes in Technology Research." Springer Berlin Heidelberg, (pp. 199-207).

Lott, Bernice, and Heather Bullock. 2010. "Social Class and Women's Lives." *Psychology of Women Quarterly* 34(3): 421-424.

Martinez, Jacqueline M. 2014. "Culture, Communication, and Latina Feminist Philosophy: Toward a Critical Phenomenology of Culture." *Hypatia* 29(1): 221-236.

Mindy, Finn. "Feminist Hero of 2008." September 04, 2008. Npr.org http://www.npr.org/blogs/sundaysoapbox/2008/09/feminist_hero_of_20 08.html

Richardson, Julian. 2010. "Facebook Soars Past 600,000 Users in Jamaica, Social network Sites Represent 'Big' Investment for Local Firms." *Jamaica Observer.*http://www.jamaicaobserver.com/business/Facebook-soars-past-600-000-users-in-Jamaica_9050536 (accessed June 26, 2011)

Schultz, Jennifer L. 2009. "The Cook, the Mediator, the Feminist, and the Hero." *Canadian Journal of Women & the Law* (21): 177-195.

Schussler, Aura. 2012. "The Relation between Feminism and Pornography." Scientific Journal of Humanistic Studies (46): 66-71.

Seedat, Fatima. 2013. "Islam, Feminism, and Islamic Feminism: Between Inadequacy and Inevitability." *Journal of Feminist Studies in Religion* 29(2): 25-45.

Stephenson, J'Nelle. 2012. "Violence Prevention Targeted at Women: What Can Be Done to Improve The Life of Women in Jamaica?" *American Journal of Community Psychology*: 39-41.

Ting-Toomey, Stella. (2004). "Translating Conflict Face-negotiation Theory into Practice." In Landis, D. R., Bennett, J. M., and Bennett, M. J. (Eds*.). Handbook of Iintercultural Training*. Thousand Oaks, CA: Sage.

Wood, Julia T. 2005. "Feminist Standpoint Theory and Muted Group Theory: Commonalities and Divergences." *Women and Language* (28): 61-4.

CHAPTER 6

M-Pesa, the Mobile Payment System in Kenya: Leapfrogging or Diffusion of Innovations?

Jane-Yajuan Zhou

Kenya, located in Eastern Africa, is typically considered as a developing country; in reality, it is significantly under-developed. Even so, it is more developed than most countries in Africa, especially in the area of electronic technologies. Kenya's recent advances in electronic technology have opened the world's eyes. The country's innovative use of cell phone technology, such as its applications to personal banking, has helped turn mobile devices into mobile banks, weather stations, doctor's offices, atlases, compasses, textbooks, radio and TV stations, and more. As a result, Kenyans are less reliant on infrastructure such as highways, clinics, and schools. Focusing on cell phone usage in Kenya; and based on gender, age, residence, educational background and socio-economic status, this research addresses the factors behind Kenya's direct involvement in the cellular age, the development of M-Pesa (in Swahili M stands for mobile and Pesa means money), and their Safaricom monopoly during the technology's leapfrogging stage. Based on both Joseph Schumpeter's economic development theory of Leapfrogging and Everett Rogers' communication theory of Diffusion of Innovations, this chapter examines Kenya's electronic technology leapfrogging and the diffusion of innovations phenomena. The research employs both qualitative and quantitative methods of analysis to determine the demographics, impact of cell phone usage, and applications during the development of the advanced electronic technologies.

LITERATURE REVIEW

The mobile phone has undergone an adoption rate unprecedented to that of any consumer technology in history (William and Tavneet 2011). Africa is the fastest growing market of this technology in this last decade – with Kenya at the forefront with a rapidly growing population of 46 million. At the end of the

1990's, 3% of the households in Kenya owned a telephone and fewer than 0.1 % percent of Kenyans used mobile phone services. In contrast, by 2011 93% of Kenyan households used mobile phones, and by the end of 2012, Kenya had nearly 31 million mobile phone users (CCK 2012, 13).

The sector of information and communication technology has seen explosive growth in the last decade, resulting in a significant transformation of the operation and processes of Kenyan businesses. What is most significant is how the mobile phone has been put to use in innovative ways to transform the lives of Kenyans and their business operations. For example, mobile phones have enabled millions of poor Kenyans to have access to Med-Africa, M-Farm, I-Cow, My Social Media, Africa Travel Guide, M-Shop, and M-Pesa mobile banking, for whom such services would have previously been out of reach.

M-Pesa launched in 2007, to offer a significantly more cost-effective, reliable, and inexpensive consumer platform. It enabled its users to complete electronic banking and person-to-person transfers. In a few months after the debut of M-Pesa, it was already widely used by a majority of residents to remit money to the rural homeland (Morawczynski 2008). It was well received and has been rapidly adopted for these applications. Many Kenyans use M-Pesa to pay their bills, to shop, and in place of day-to-day banking functions such as using credit cards, paying by cash, and as a replacement to time wasters such as waiting in line at the bank or using ATM's. With just a simple phone number and a PIN code, Kenyans can effectively accomplish much of their daily communication and banking tasks. The mobile-phone-based money service has helped turn Kenyan's mobile devices into mobile banks. By the end of 2010, M-Pesa was already attracting 10,000 new customers per day. M-Pesa had quickly grown to 13 million registered customers (Safari 2011) – or about 60% of Kenya's population, who are at least 15 years of age (CIA 2011).

As a result, Kenya has less of a need to install cables for conventional landline telephones. The country's innovative uses of cell phone technology with its applications to personal banking have enabled Kenyans to be less reliant on traditional infrastructure such as transportation, institutions and businesses. Mobile phones have become a surrogate for many, becoming a powerful and cost-effective replacement for many functions.

Theoretical Background

Both the Leapfrogging theory and the Diffusion of Innovation theory are utilized to guide this research concerning the mobile phone usage in Kenya. Some studies on the adoption of mobile banking tend to consider M-Pesa as a product of an economic developmental stage, leapfrogging; other studies believe that M-Pesa is a technological innovation. The Leap Frogging theory is commonly used to explain why Kenya, a still underdeveloped country in many aspects, is leading the world with M-Pesa mobile banking. The Diffusion of Innovations theory is one of the popular theories explaining M-Pesa consumers' behavior on adoption and acceptance patterns in the process. A snapshot of the theoretical perspectives follows.

Leapfrogging

Based on Joseph Schumpeter's (1962) notion in *Gales of Creative Destruction*, originally leapfrogging was used in economic growth theories that highlighted competition among companies. The hypothesis proposes that firms exercising monopolies based on incumbent technologies are less innovative than their competitors. As a result, they eventually lose their leading role due to technological innovations that are adopted by new firms, which are willing to take on higher risks. Schumpeter's creative destruction is a cyclical leapfrogging process characterized by those falling behind and triumphing ahead in a predictable pattern (Burlamaqui 2011). Gerschenkron (1962) postulated that the ability to leapfrog by a less developed nation created an advantage for rapid economic growth. Utilizing advanced technologies through less expensive labor causes this effect (Hanna, Guy, and Arnold 1995).

Recently, this theory has been built upon by modern economic growth models, which posit that developing nations could skip a generation by leveraging technological spillovers and knowledge (Grossman and Helpman 1991; Temple 1999). Adequate market size, demand references, and the initial quality gap can boost less developed nations ahead of more advanced ones (Sharif 1989; Huang and Tilton 1990; Motta, et al 1997; Klundert and Smulders 2001). Strategic moves (Thoening and Verdier 2003), capital and labor costs (Krugman 1981), new technologies yielding higher productivity (Brezis, Krugman, and Tsiddon 1993), historical examples (Soete 1985), and those generating less pollution (Warhurst and Bridge 1997) have been cited as examples.

More recently, the idea of leapfrogging has been used to describe sustainable development which may enable organizations and nations to accelerate development by skipping less efficient, obsolete or higher polluting industries and technologies by moving directly to more advanced ones. As a recent example in India, Pakistan, and many African countries, the widespread adoption of community solar power is much easier than in comparable western countries, which are burdened with the problems associated with the integration of the existing power infrastructure. An illustration of this is China, which shifted to fuel cell vehicles, is compared to countries in the West. China's much smaller network of gas stations did not need to be converted to hydrogen. Another example is Kenya, which is not burdened by the now obsolete network of telephone lines as are most highly industrialized countries. Instead, Kenya has skipped directly into the infrastructure model of the cellular age.

Diffusion of Innovation

The Diffusion of Innovation theory explains the process by which an idea, practice, or object that is perceived as new and is adopted quickly into a social system from its discovery by early adopters, to formation of an early majority, to finally attracting a late majority and laggards. The theory has its roots in Tarde's 1903 book, *The Laws of Imitation.* Forty years later, a more concerted development of this approach occurred when a study by Bryce Ryan and Neal Gross (1943), two rural sociologists, published research on the diffusion of innovations in planting of hybrid corn in Iowa (Rogers 2003). Everett M. Rogers was the most prominent developer of Diffusion of Innovations. His book, *Diffusion of Innovations* was first published in 1961. The theory has spread to many different fields, and numerous studies support its principles. The theory has been applied to different academic disciplines such as communication (Rogers and Shoemaker 1971), business (Greenhalgh, Robert, Macfarlane, Bate, and Kyriakidou 2004), anthropology (Linton, 1936), education (Mort 1964), geography (Spolaore and Wacziarg 2011), sociology (Katz, Levin, and Hamilton 1963), marketing (Mahajan, Muller, and Bass 1990; Sarvary, Parker, and Dekimpe 2000), political science (Walker 1969), economics (Dopfer et al 2004), public health (Moseley 2004; Rogers 2004; Sanson-Fisher 2004), and more. The five primary characteristics of innovation that influence the rates of diffusion are 1.relative advantage, 2.complexity, 3.compatibility or simplicity, 4.ability for trial, and 5.observability. Relative advantage is the most important factor because it influences the rate of adoption, including the innovation's perceived relative advantage over previous ideas or thoughts. Complexity, or ease of comprehension, reveals its compatibility within the potential adopter's values and needs. Trial ability is the ability to be tested by the potential adopter, while observability is the visibility of its result (Budman, Portnoy, and Villapiano 2003; Rogers 2003).

Often, innovations that are adopted more rapidly are the ones with greater relative advantage, less complexity, compatibility, trial ability, and observability. As an example, M-Pesa was first introduced to low income individuals in Kenyan areas that did not have money transfer services. Compared to opening a bank account, the registration process of becoming an M-Pesa customer was easy. Customers needed only to have a phone and ID to subscribe to Safaricom mobile phone service. Kenyans were in need of affordable money transfer services that were conveniently located. The M-Pesa application was easy to use, and compatible with mobile phones that were already owned by Safaricom customers. Consequently, the innovators and an early majority were testimony of successfully sending and receiving money, and influencers of the skeptical ones. As a result, the late majority and laggards were able to observe the results, and eventually adopt this innovation. Furthermore, Rogers recognized that the innovators /early adopters tend to be better educated, and of higher social-economic status which affords them the ability to take greater risks.

Research Questions and Hypotheses

The success of the M-Pesa service in Kenya has benefited, in large part, from the unique dynamics of an emerging economy. Based on trends of technology usage, access of information, and patterns of mobile phone communication in Kenya, the mobile phone's function has evolved to serve as a platform of online banking services. Consequently, the following demographic, descriptive, and social impact research questions and hypotheses about cellular telephone use in Kenya are essential:

- RQ 1: What types of electronic communication are commonly used in Kenya?
- RQ 2: What gender, age, race, and education demographics use the mobile phone in Kenya?
- RQ 3: What are the main reasons for mobile phone usage?
- RQ 4: How long have Kenyans been using cell phones?
- RQ 5: Do urban/rural residences use mobile phones for financial transaction purposes?
- RQ 6: Are economic status and financial transaction uses of mobile phones related to each other?

- H1: Socio-economic status and M-Pesa usage are not correalated in Kenya.
- H2: Educational level and M-Pesa transaction usage are negitively correlated in Kenya.

- H3: Social economic status and M-Pesa transaction usage are negatively correlated in Kenya.

METHOD

This research aims to predict the consumer behavior to adopt M-Pesa service in Kenya by extending the leapfrogging and diffusion of innovation models. Both qualitative and quantitative methods were employed in this research. Coefficient of correlation and coefficient regression analyses are used to correlate significance among (1) educational level and M-Pesa transaction use, (2) social economic status and M-Pesa transaction use.

Quantitative data are documented by a survey developed for the data collection. The survey was conducted in Kenyan context by Kenyan research assistants. The survey instrument contains a 41 question questionnaire in English for Kenyans and sojourners of at least three months. It allows a combination of open-ended responses, demographic queries, and Likert scale items. A combination of paper and electronic surveys were distributed to participants via e-mail, as well as being uploaded to the Survey Monkey website. The website invited volunteers in conjunction with a brief explanation of the research. A total 100 questionnaires were distributed without compensation for respondents. All participants were adults, between 18-65 years old who live in Kenya, or the US. Survey responses were anonymous and used for research in aggregate form.

RESULTS

RQ 1: What types of electronic communication are commonly used in Kenya?

All 30 respondents are regular users of mobile phones and television. The majority of respondents have email accounts and have utilized bank koisks, cash cards, and voice mail. Internet list service, internet bulletin board, e-commer website were reported as least commonly used in Kenya.

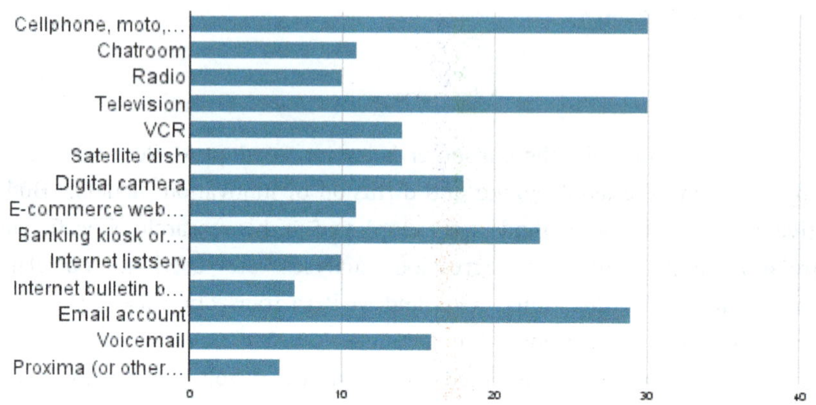

Figure 5.1: Types of Electronic Communication Used by Kenya Respondents

RQ 2: What gender, age, race, and education demographics use the mobile phone in Kenya?

See figures 5.2-5.5 for the 30 respondents profiled between 18-65 years old. Gender demographics were similar. The greatest age cohort was 25-35 at nearly 37% of all respondents. Racial composition was predominantly black at nearly 97%. Education level was reported by 13 respondents who completed high or secondary school; 9 completed institute, college or university degrees, and 8 had completed graduate school degrees.

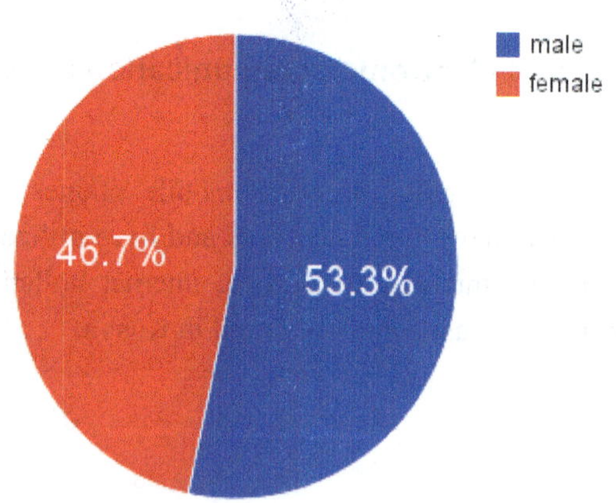

Figure 5.2: Respondent Gender Demographics (n=30)

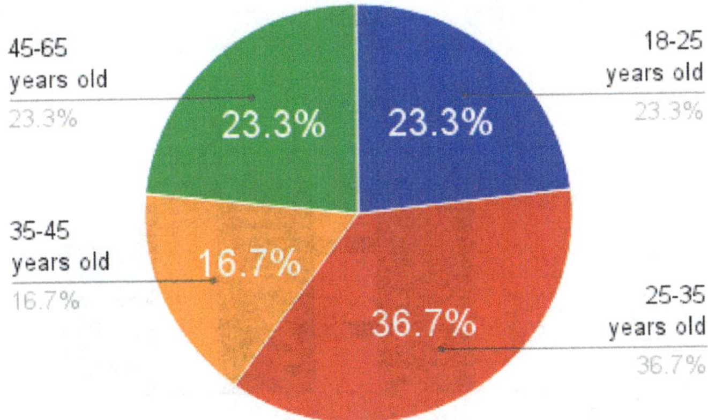

Figure 5.3: Respondent Age Demographics (n=30)

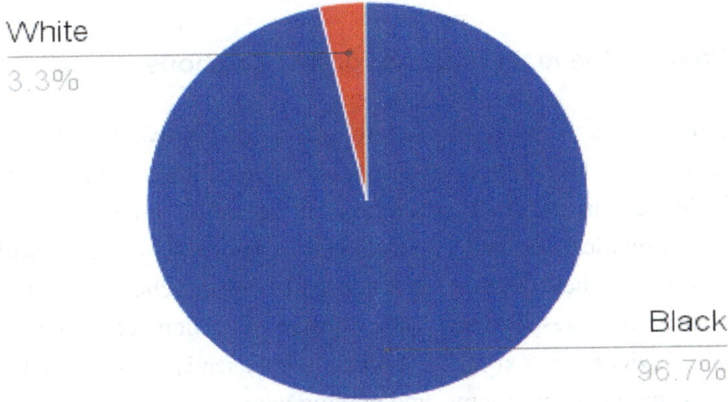

Figure 5.4: Respondent Race Demographics (n=30)

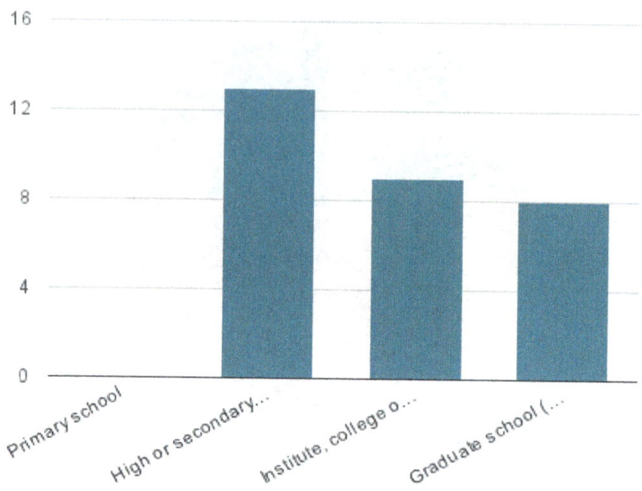

Figure 5.5: Respondent Education Levels (n=30)

RQ 3: What are the main reasons for mobile phone usage?

In-person interviews were conducted to gather qualitative responses for this research question. The respondents' main reasons for utilizing mobile phone service included making and receiving phone calls, financial transactions, education, communication, and entertainment. Those respondents with higher education also used their devices for reading (e-books), scheduling and listening to music while those respondents with a lower education were inclined to use their mobile devices for texting, e chatting with friends on Facebook, meeting virtual friends on local chat rooms, and playing games.

RQ 4: How long have Kenyans been using cell phones?

Of 30 respondents, 29 indicated cell phone usage for greater than 5 years while 1 respondent in age group 18-25 answered in the 3-5 years category.

RQ 5: Did the urban residents use mobile phones for financial transactions?

No, among 30 respondents, 13 have lived more than 6 months of the year in an urban area while 17 have lived in rural areas. Five of the 13 urban dwellers answered that transacting financially was one of the reasons they used mobile phones, and 15 of the 17 rural Kenyans responded that one of the reasons they liked to use a mobile phone was transacting financially.

RQ6: Are education level and M-Pesa transaction usage positively related to each other?

No. Among 30 respondents, 20 of 30 use cellphone for M-Pesa transactions, including 10 of 13 respondents with high or secondary school education, 6 of 9 with college education, and 4 of 8 with graduate degrees. The negative correlation of 0.9907 strongly indicates that higher educated people in Kenya use M-Pesa transactions less. See tables 5.1-5.3. The raw data is summarized in table 5.1. Statistics and comparisons follow in tables 5.2-5.4.

Table 5.1 M-Pesa Usage by Education

	High/Secondary School	College/University	Graduate/Post Graduate	Subtotals
M-Pesa Used	10	6	4	20
Not Used	3	3	4	10
Subtotals	13	9	8	30

Table 5.2 Coefficient of Correlation

	X (Education Code)	Y (Ratio)
High/Secondary School	1	0.7692
College/University	2	0.6667
Graduate/Post Graduate	3	0.5000
	Correlation(p)=	-0.9907

Tables 5.3 Regression Statistics Summary and ANOVA Tables

Regression Statistics	
Multiple R	0.9907
R Square	0.9814
Adjusted R Square	0.9628
Standard Error	0.0262
Observations	3

ANOVA	df	SS	MS	F	Significance F
Regression	1	0.0362	0.0362	52.7474	0.0871
Residual	1	0.0007	0.0007		
Total	2	0.0369			

	Coefficient	Standard Error	t Stat	P-value	Lower 95%	Upper 95%	Lower 95.0%	Upper 95.0%
Intercept	0.9145	0.0400	22.8421	0.0279	0.4058	1.4232	0.4058	1.4232
X Variable	-0.1346	0.0185	-7.2627	0.0184	-0.3701	-0.1009	0.3701	-0.1009

Table 5.4 Residual Plot

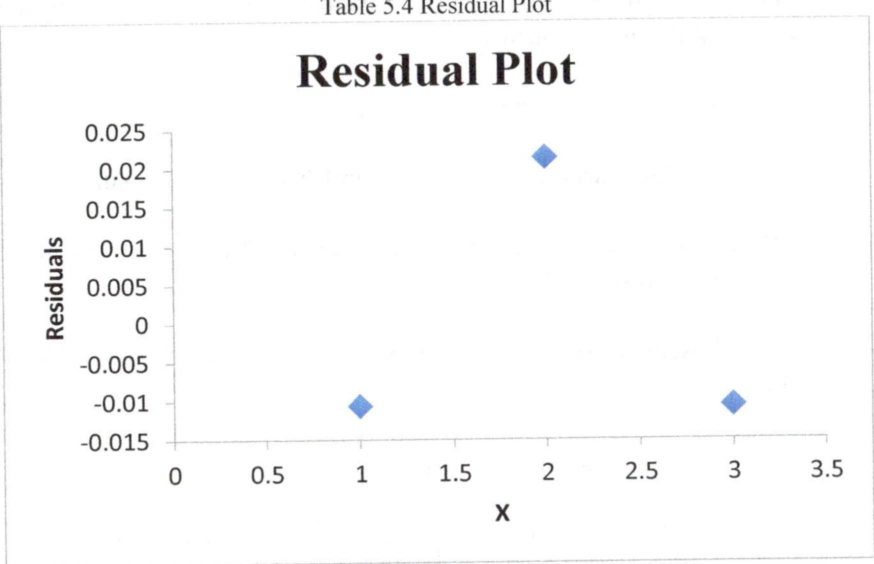

The coefficient of correlation in Table 5.2 indicates a strong negative correlation between education level and M-Pesa usage. A Multiple R of 0.9907 in the Regression Summary of table 5.3 confirms a strong negative correlation. Both R Square of 0.9814 and Adjusted R Square of 0.9628 in table 5.3 indicate a high confidence level of predictability. The low Standard Error (standard deviation) of 0.0262 in table 5.3 produces a sufficiently narrow prediction interval of 95%. The P-value in table 5.3 of 0.0184 indicates a rejection of the null hypothesis; therefore, residuals in the Residual Plot table 5.4 show no patterns. Both models explain those with lower education levels have higher M-Pesa rates of usage.

RQ 7: Are economic status and usage of the M-Pesa application for mobile phones positively related to each other?

No. Of 30 respondents, 10 of 14 who reported no reliable income used their mobile devices for M-Pesa transactions. All six respondents who reported a fixed income/pension less than 1,000 USD per month used their mobile devices for M-Pesa transactions. Fifty percent who have a job or business providing housing, food, clothing, and life necessities used their devices for M-Pesa transactions. One third who have enough to provide for their own and others' needs on a regular basis stated that conducting financial transactions was one of the main reasons for usage. No respondents who reported being wealthy by Kenya's standards made financial transactions on their mobile phones.

Twenty M-Pesa transaction respondents indicated a lower social-economic status, with the following comments:

"I have no reliable income."

"I have a fixed income/pension less than 1,000 USD per month."

"I use M-Pesa to send and receive money from family members, relatives, and friends."

The raw data is summarized in table 5.5. Statistics and comparisons follow in tables 5.6-5.8.

Table 5.5 M-Pesa Usage by Income

	No income	Low Income	Moderate income	Enough Income	Subtotals
M-Pesa Used	10	6	2	2	20
Not Used	4	0	2	4	10
Subtotals	14	6	4	6	30

Table 5.6 Coefficient of Correlation

	X (Income code)	Y (Ratio)
No Income	1	0.7143
Low Income	2	1.0000
Moderate Income	3	0.5000
Enough Income	4	0.3333
	Correlation(p)=	-0.7366

Tables 5.7 Regression Statistics Summary and ANOVA Tables

Regression Statistics	
Multiple R	0.7366
R Square	0.5426
Adjusted R Square	0.3138
Standard Error	0.2385
Observations	4

ANOVA

	df	*SS*	*MS*	*F*	*Significance F*
Regression	1	0.1350	0.1349	2.3726	0.2634
Residual	2	0.1138	0.0569		
Total	3	0.2487			

	Coefficients	*Standard Error*	*t Stat*	*P-value*	*Lower 95%*	*Upper 95%*	*Lower 95.0%*	*Upper 95.0%*
Intercept	1.0477	0.2921	3.5864	0.0697	-0.2092	2.3045	-0.2092	2.3045
X Variable 1	-0.1643	0.1067	-1.5402	0.2634	-0.6232	0.2946	-0.6232	0.2946

Table 5.8 Residual Plot

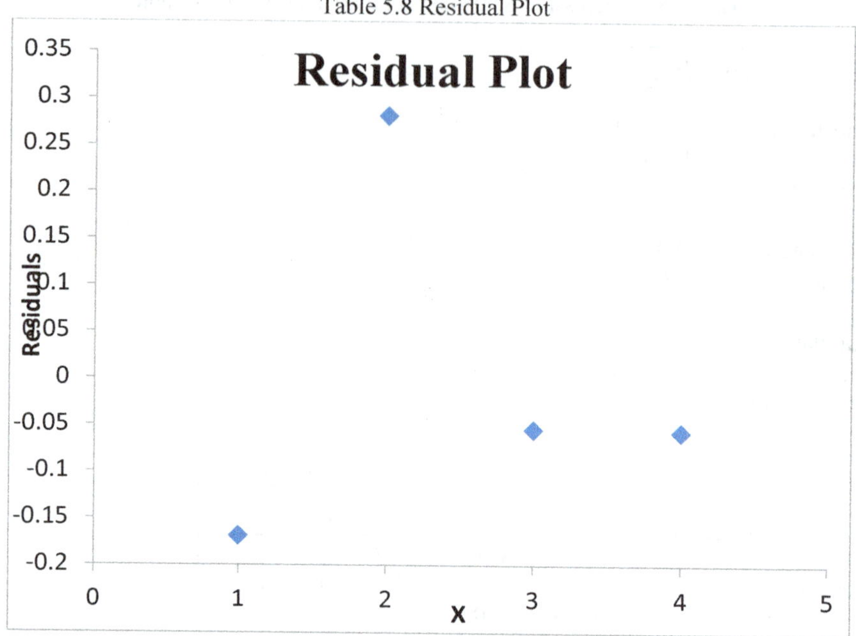

The negative correlation, -0.7366 is a relatively strong indication that people with both lower income and no income have a higher usage rate of M-Pesa. See table 5.6. The R Square value of 0.5426 in table 5.7 indicates a moderate level of predictability, which is acceptable in social science studies. The F-value of 2.3726 and corresponding P-value of 0.0697 indicate the validation of results. See table 5.7. Residuals in table 5.8 show no patterns. Regression analysis presented in tables 5.7 and 5.8 confirms that economic status and M-Pesa rate of usage are negatively correlated to each other.

DISCUSSION

Mobile phone users in Kenya demonstrate innovative use of mobile phones such as M-Pesa which can affect other developing countries, and developed countries. Seven related research questions were answered, and three hypotheses were tested in this study.

H1: Socio-economic status and M-Pesa usage are not correlated.

This hypothesis has been rejected. Both the coefficient of correlation and regression analyses show that education level has very strong negative correlation to M-Pesa usage in Kenya, and that social-economic status is strongly negatively correlated as well.

H2:   Educational level and M-Pesa usage are negatively correlated in Kenya.

The hypothesis was supported. Both analyses, coefficiient of correlation and regression, indicate strong negative correlation.

H3:   Socio-economic status and M-Pesa transaction usage are negatively correlated to each other in Kenya.

This hypothesis is supported. According to analyses, social economic status and M-Pesa transaction usage are moderately negatively correlated.

Implications

The cell phone is merely a wireless communication device which can transmit audio, visual and electronic data; and the M-Pesa system, in reality is an application. In this study, we can see the finger-prints of Diffusion of Innovation as the technology were quickly embraced by the rural population, not only as a phone, but also as a means of banking. In a developing country such as Kenya the transfer of funds and other banking services as access to traditional banking outlets were not previously available in rural areas.

Kenya's innovative uses of cell phone technology with its applications to personal banking have enabled Kenyans to be less reliant on traditional infrastructure such as highways, clinics and schools. Mobile phones have become a surrogate for many, becoming a powerful and cost-effective replacement for necessary life functions. Regarding the development and adoption of M-Pesa in Kenya, we can see both Leapfrogging and the theory of Diffusion of Innovations. Upon in-depth study of M-Pesa, the paradox is resolved by understanding the co-development of two technologies. The first one was the infrastructure the Kenyan government adopted. The institutional adoption of a cell phone technology opposed or replaced the older, more expensive installments of telephone poles, an early 20th century solution dominant in the west. The second was the local, or grassroots, adoption of a banking software application enabled by the newly adopted cell phone technology.

When studying the adoption of M-Pesa in Kenya, we need understand there are really two technologies and different theories of adoption with an asymmetrical co-dependence. The M-Pesa banking protocol is a portal into banking and money transfer by use of a cellular application. The first technology, the widespread adoption of wireless phone technology, is a classic example of Leapfrogging theory. The rapid adoption of M-Pesa software, or app for banking and money transfer, is in itself an excellent example of Diffusion of Innovations theory.

The theory of Leapfrogging posits that monopolistic technologies seldom advance as barriers of entry. This usually lessens the edge of technology as their promulgators become satisfied with usable technology. Unlike the West, the Kenya of 30 years ago did not have the infrastructure of hard line transmission to support a Western style telephone system carried on telephone poles. While the West has been converting to wireless systems, the conversion has been implemented at a slow pace. Strung wire had little competition to compel it to either quickly update or be replaced by fiber optics or wireless systems. Kenya on the other hand, had little infrastructure and developed; in its stead, wireless technologies at a fraction of the cost of stringing wires, especially in Asian and African remote areas of dense population.

The theory of Diffusion of Innovations defines the adoption of technologies on a bottoms-up approach, first gaining traction in the street and eventually spreading to displace competing technologies. The M-Pesa solution has clearly followed the path of Diffusion of Innovations to replace a centuries-old system of bricks and mortar banks that served as the spinal cord of the country's banking system. As physical banking locations are not necessarily convenient in Kenya, especially in the countryside, many people were not even in the system. In short order, M-Pesa; as easy to use and understand software, or an app, allowed retail banking to send money with need for neither physical locations nor third parties engaged in what is basically peer-to-peer transactions. M-Pesa is dependent on cellular technologies as the infrastructure to deliver its services, but cellular technologies are not necessarily dependent upon banking systems.

In comparing the two vis-à-vis in theory, we can clearly see that theories and instances have a broad applicability, or scope. Each fulfills a major need, not only in Kenya, but also globally. Regarding the appropriateness, the simplification of limited-step technologies are grasped and received well because of their simplicity. What is so difficult to explain how a familiar technology, the cellular telephone, is used to perform another familiar technical act, the ways and means of depositing and withdrawing funds? The simplicity of the technology and its widespread adoption allow for the better collection and analysis of statistical models for current and future researchers.

One merely need look to Kenya, China, Vietnam, Burma and others which all leapfrogged from virtually no hard wire communication infrastructure to totally embracing cellular technologies. To see diffusion, one need only look to the US and Western Europe, both areas with a heavy infrastructure footprint of hard wire communications, who have been phasing in cellular infrastructure and service. Both real-time living experiment theories have been consistent in their predictable results. Both theories' simple explanations are easily drawn by predictive, intuitive results. If a country has no hard wire infrastructure, it is only natural to pursue the next generation, i.e. cellular, technology which is both advanced and cheaper to implement. Regarding, the M-Pesa app/software, its diffusion in Kenyan society was catalyzed by the countryside users who had little or no banking opportunities as a matter of geography; and potentially, a dearth of services for the non-metropolitan consumers needing banks and banking services.

As to openness, or whether a theory should not be dogma, the underlying dynamics of the governing, and smaller micro economics building blocks for more comprehensive theories tend to be more dogmatic. However, the uncertainty of any system can be a victim of unforeseen consequences from the resulting complexity of layered building blocks.

According to Diffusion theory and studies of innovators /early adopters, innovators/early adopters tend to be better educated and come from the higher social-economic status. They can afford to take the risks; however, this study does not support Rogers' claim that innovators/early adopters of cellular banking are the ones with both higher educational background and social-economic status. As such, innovators/early adopters of M-Pesa in Kenya were the ones who tended to be less educated, and of lower social-economic status. Many in the US and other developed countries have become laggards in cellular banking.

Limitation

The research was conducted in the US with most respondents residing in Kenya. This made it difficult to establish continuing, fruitful relationships with the respondents. A systematic, theoretical framework to address research subjects in depth was not possible. In addition, 200 respondents were expected. Due to the limited time to collect data, there were only 30 Kenyans who participated in this survey. The sample size was relatively small and it was less representative of the population. This could also result in being less generalized, with accuracy and reliability. In addition, a few questions were too broad, providing limited answers. The Likert scale question, "I have used a cell phone for more than..." allowed responses indicating the time frame the respondent had been using a cell phone; however, it could not show for what function they had used mobile phones. Furthermore, not all the research questions are well described by regression models. This may cause inaccuracy and inappropriateness because regression focuses on an outcome variable.

Suggestions for Future Research

Mobile phone usage has improved the lives of Kenyans, especially those living in the rural areas. In Kenya, most people use mobile phones, with a majority using M-Pesa. The financial component of Kenyans' lives has improved since Safaricom first introduced M-Pesa in 2007. M-Pesa was adopted so rapidly to the point of finding utility in mobile banking, again to also eliminate infrastructure scarcity by on-site locations and ATM machines. Obviously, this is a significant milestone in the financial and mobile industry as at least 50% of the globe's population has limited-to-no access to a convenient and usable banking system. Further study might describe which of the five primary characteristics of innovation influence the diffusion, relative advantage, complexity, compatibility or simplicity, trial ability, and/or observability of rural banking. Further study might also include what other services and functions can be facilitated by, or delivered through the cell phone, and to determine if the cell phone will be a major factor in both developed and developing countries in the near future.

REFERENCES

Budman, Simon. H., David Portnoy, and Albert. J. Villapiano. "How to Get Technological Innovation Used in Behavioral Healthcare: Build It and They Still Might not Come." *Psychotherapy: Theory, Research, Practice, Training* 40, no. 1-2 (2003): 45-54.

Burlamaqui, Leonard. "Development Theory: Convergence, Catch-up, or Leapfrogging? A Schumperterian Approach. " Paper presented at New Developmentalism and Structuralist Development Macroeconomics Workshop, Sao Paulo, Brazil, August 15-16, 2011.

Brezis, Elise S., and Paul R. Krugman. *Leapfrogging: A Theory of Cycles in National Technological Leadership.* Cambridge, MA: National Bureau of Economic Research, 1991.

CIA. Kenya Statistics. *CIA World Factbook,* 2011.

Communications Commission of Kenya (CCK). "Sector Statistics Report 2012/13."

Dopfer, Kurt, John Foster and Jason Potts. "Micro-Meso-Macro." *Journal of Evolutionary Economics.* 14, (2004): 263-279.

Gerschenkron, Alexander. *Economic Backwardness in Historical Perspective: A Book of Essays.* Cambridge: Belknap Press of Harvard University Press, 1962.

Greenhalgh, Trisha, Robert, Glenn, MacFarlane, Fraser, and Kyriakidou, Olympia. "Diffusion of Innovations in Service Organizations: Systematic Review and Recommendations." *The Millbank Quarterly.* 82, (2004): 581-629.

Grossman, Gene M., and Elhanan Helpman. "Trade, Knowledge Spillovers, and Growth." *European Economic Review* 35, no. 2-3 (1991): 517-526.

Hwang, Kee Hyung, and John E. Tilton. *Leapfrogging, Consumer Preferences, International Trade and the Intensity of Metal Use in the Less Developed Countries: The Case of Steel in Korea.* Golden, Colorado: Colorado School of Mines, Dept. of Mineral Economics, 1990.

Jack, William and Tavneet Suri. "Mobile Money: The Economics of M-Pesa." NBER Working Paper, 16721, January 2011.

Katz, Elihu, Martin I. Levin, and Herbert Hamilton. Traditions of Research on the Diffusion of Innovations. *American Sociological Review,* 28, no. 4 (1963): 237-252.

Klundert, Theo Van De, and Sjak Smulders. "Loss of Technological Leadership of Rentier Economies: A Two-Country Endogenous Growth Model." *Journal of International Economics* 54, no. 1 (2001): 211-231.

Krugman, Paul. "Trade, Accumulation, and Uneven Development." *Journal of Development Economics* 8, no. 2 (1981): 149-161.

Linton, Ralph. *The Study of Man.* New York: Appleton-Century-Crofts, 1936.

Mahajan, Vijay, Eitan Muller, and Frank M. Bass. "New Product Diffusion Models in Marketing: A Review and Directions for Research." *Journal of Marketing* 54, No.1 (1990): 1-26.

Moseley, Stephen. F. "Everett Rogers' Diffusion of Innovations Theory: Its Utility and Value in Public Health. *Journal of Health Communication* 9, (2004): 59-69.

Morawczynski, Olga. "Surviving in the Dual System: How M-Pesa is Fostering Urban-to-Rural Remittances in a Kenyan Slum 2". University of Edinburgh, Social Studies Unit, Working Paper, 2008.

Motta, Massimo, Jacques-Francois Thisse, and Antonio Cabrales. "On the Persistence of Leadership or Leapfrogging in International Trade." *International Economic Review* 38, no. 4 (1997): 809.

Mort, R. Paul, *Educational Adaptability*. New York: Metropolitan School Study Council, 1953.

Nagy, Hanna, Ken Guy, and Erik Arnold. "The Diffusion of Information Technology: Experience of Industrial Countries and Lessons for Developing Countries, A World Bank Discussion Paper, No. 281." *Prometheus* 15, no 3 (1997): 416-419.

Rogers, Everett, M. "A Prospective and Retrospective Look at the Diffusion Model." *Journal of Health Communication* 9, (2004): 13-19.

Rogers, Everett, M. and F, Floyd Shoemaker. *Communication of Innovations: A Cross Cultural Approach*. New York: The Free Press. 1971.

Rogers, Everett, M. *Diffusion of Innovations*. 5th ed. New York: The Free Press, 2003.

Sanson-Fisher, R. W. "Diffusion of Innovations Theory for Clinical Change." *The Medical Journal of Australia*. 180, (2004): S55-S66.

Sarvary, Miklos, Philip M. Parker, and Marnik G. Dekimpe. "Global Diffusion of Techno Logical Innovations: A Coupled-hazard Approach." *Journal of Marketing Research* 37, no. 2 (2000): 47-59.

Spolaore, Enrico, and Romain Wacziarg. *Long-Term Barriers to International Diffusion of Innovations*. Cambridge, Massachusetts: National Bureau of Economic Research, 2011.

Schumpeter, Joseph Alois. *Capitalism, Socialism, and Democracy*. 2d ed. New York: Harper & Brothers, 1947.

Sharif, M. Nawaz. "Technological Leapfrogging: Implications for Developing Countries." *Technological Forecasting and Social Change* 36, no. 1-2 (1989): 201-208.

Soete, Luc. "International Diffusion of Technology, Industrial Development and Technological Leapfrogging." *World Development* 13, no. 3 (1985): 409-422.

Tarde, Gabriel. *The Laws of Imitation*. New York: Holt, 1903.

Temple, Jonathan. "The New Growth Evidence." *Journal of Economic Literature* 37, no 1 (1999):112-156.

Thoenig, Mathias, and Thierry Verdier. "A Theory of Defensive Skill-biased Innovation and Globalization." *American Economic Review* 93, no. 3 (2003): 709-728.

Walker, L. Jack Jr. "The Diffusion of Innovations among the American States." *American Political Science Review*. 63, no.9 (1969): 880-899.

Warhurst, Alyson, and Gavin Bridge. "Economic Liberalisation, Innovation, and Technology Transfer: Opportunities for Cleaner Production in the Minerals Industry." *Natural Resources Forum* 21, no. 1 (1997): 1-12.

I thank and praise my Almighty God for His unfailing love, shower of blessings, and support.

CHAPTER 7

The Application of Electronic Communication, Cellular Telephone and Internet Usage in Laos

Bo Yang

Electronic communication is changing people's life gradually, and the cellular telephone and internet usage has become more and more important to measure the development and economy in the advancement of society. This research employs intercultural communication theory, and qualitative and quantitative methods of inquiry to gather current information on the application of electronic communication, internet and cellular telephone use in Laos.

In these days, the various applications of electronic communication have become increasingly important in Laos. As a developing country in Eastern Asia, people from Laos have more and more opportunities to access the internet and use cellphones, and their lives have been improved by information communication technology. Many intercultural communication theories have been applied to understand communication and electronic communication. This study examines the impacts of electronic communication within a framework of Acculturation theory. Demographics and qualitative descriptors are from questionnaires (n=43) responded to by people from Laos and near regions. The data will be answering five basic research questions.

Communication is the basis of collective human activity. It is important to note that people cannot survive without effective communication. In recent years, electronic communication has become increasingly important in intercultural communication. An increasing number of individuals use electronic communication tools (ICT) instead of traditional tools, such as mailing letters and talking. Although the flourishing development and advancement of ICTs can make up the shortage of communication among different people, the distinction of culture has always been a barrier for individuals from different areas to communicate. The information communication technology brought many advantages to intercultural communication; for instance, people can communicate with others at home. It is important to note that the electronic communication brings convenience; however, the differences of cultures and customs among country regions are adverse factors affecting communication. This chapter introduces the Acculturation theory, which is an intercultural communication theory that addresses many communication problems. The Acculturation theory helps people accept and understand other cultures and customs, which will ease the intercultural communication problems. Various applications of the Acculturation theory can help people to acculturate with others who are from different regions. The problems different cultures engender can be addressed effectively with the Acculturation theory. In addition, this research illustrates current applications of electronic communication in Laos, including the applications in business, organizational communication, and migration. This research and analysis can add to the growing body of literature on electronic communication in a key Asian population. The situation of using the internet and cellphones in developing countries can present an important facet of the development of information communication technology world-wide.

ACCULTURATION THEORY

As electronic communication becomes prevalent in intercultural communication, the need to understand and adapt to this timely, yet limited form becomes necessary. Although the development of the internet can make up the time, distance and cost of communication among different people, the differences of culture and language have always been a barrier for people from different regions to communicate. The information technology brings convenience for people to communicate, but there are so many adverse factors affecting communication itself. However, Acculturation theory is essentially important and beneficial for easing this issue. According to Redfield, Linton, and Herskovits (1936),

> Acculturation comprehends those phenomena which result when groups of individuals having different cultures come into continuous first-hand contact with subsequent changes in the original culture patterns of either or both groups. (149)

It means different ethnic groups may be affected by different cultures superimposed upon them, such as the customs, language, and religion of others. Although different regions have different cultures, Acculturation theory can help people to acculturate with others which will be beneficial for intercultural communication in education, business, travel, family life, and government.

Acculturation affects people in not only behavioral adjustment, but also psychological adjustment, which means acculturation has strong penetrability.

According to Berry (1997), "psychological adaptations to acculturation are considered to be a matter of learning a new behavioral repertoire that is appropriate for the new cultural context". (13) Berry thinks that the process of immigration adaption has become an important topic in communication research of social problems. He introduced acculturation and adaptation, and indicated issues to be analyzed for a framework of acculturation from the perspective of immigrants, sojourners, indigenous peoples, and ethno-cultural groups.

For instance, if some people immigrate to Laos, they will be affected by local culture and custom. As time passes, they will acculturate with Laotian life style from a perspective of psychology. If these people accept local culture from their heart, the communication between immigrants and local people will become easier and more convenient.

Moreover, Acculturation theory has a positive influence on some health behaviors. Landrine and Klonoff (2004) used operational research to seek a composite, reasonable argument from existing arguments. An increasing number of studies show that acculturation affects health behavior among different groups negatively. Landrine and Klonoff (2004) tested the accuracy of this information. Landrine and Klonoff (2004) mentioned Behavior theory and Learning theory, then tested five hypothesis to prove that acculturation does not affect health behavior among different groups negatively. The result of their hypothesis indicates the acculturation can have either positive or negative influence on health behavior, depending on different issues, like cigarette smoking, and fruit or vegetable consumption.

Acculturation can be seen as a positive aspect in people's life, but sometimes it brings negative influence to others. For instance, a bad living habit like smoking will be harmful for other individuals. Assuming people living in China like to smoke, and if this bad living habit is brought into a non-smoking area by acculturating immigrants, it could become a negative living habit that spreads to the host country.

In addition, acculturation also has many other functions and a range of application, like illegal immigration. According to Mudambi (2011), "It relied on Berry's (1990) and Kim's (2005) theories of acculturation, which have not been previously applied to undocumented immigrants" (1). Mudambi goes on to analyze the factors affecting the acculturation of illegal immigrants in America. First, Mudambi (2011) indicated two frameworks of Acculturation theory. Second, Mudambi analyzed Berry's and Kim's Acculturation theory. Last,

Mudambi (2011) introduced a method and sample results to decide acculturation can be applied to illegal immigration.

Acculturation theory is very appropriate when applied to intercultural communication. For instance, Greenland and Brown (2005) researched how 35 Japanese students studying in the United Kingdom acculturated with British culture. Their study indicates the advantages of this theory. However, the negative side of Acculturation theory is lack of integration in groups. Self-criticism of Acculturation theory and the experiment are presented in the outcome and discussion step by step.

Communication is an indispensable part of people's lives. Appropriate communication can promote the relationships of people, improving people's style of necessary interaction, and other displays such as entertainment. Electronic communication has become so popular in the world even developing countries with limited resources see it as necessary. With development and information technology, an increasing number of individuals will enjoy satisfying communication, and adapt reciprocally to the world as Acculturation theory describes.

THE APPLICATION OF ELECTRONIC COMMUNICATION TO THE CONTEMPORARY ERA

Nowadays, many larger enterprises have changed to an organizational form which is closely related with electronic communication. Fulk and Gerardine (1995) introduced the interaction between communication technology and new forms of organization, and the special issues this interaction raises. The researchers believed inter-organization and intra-organization are associated with electronic communication. The purpose of their article is to test the relationship, and encourage future research. Fulk and Gerardine (1995) introduced changes in organizational form, and indicated what a new organizational form looks like in the beginning. They also indicated special, on-going issues making an impact in the final stages of organizational change. The use of electronic messaging systems has been widely applied as a new form of organization itself.

However, Ku (1996) made a survey for a small telecommunicated company in Maryland in 1992. The outcome showed electronic messaging systems are not widely used in certain societies, but they are widely used in some particular groups. This article focuses on the determining factors and impacts of electronic messaging systems in social and nonsocial usage. Ku (1996) found that the young workers were more likely to use electronic messages to communicate with other workers in this small company. However, the managers preferred to use electronic messages to manage employees because of the distances they bridged, and the record of specific content.

In addition, the wide application of electronic mail has become the most popular communication form among many enterprises. Ngwenyama and Lee (1997) indicated why and how communication becomes rich when using electronic mail. They examined how electronic mail is exchanged between managers in a company. To sample this effect through a general and empirical way, Ngwenyama and Lee (1997) also indicated that whether communication is rich or not does not depend on external factors, but email's inherent properties. The key in their research article is the interaction between electronic mail and organizational context. Ngwenyama and Lee (1997) demonstrated the concept of Information Richness theory, and distinguished the value of an interpretive perspective in general and particular forms.

With the significant development of electronic communication, new communication theories have been used to explain, predict or control electronic communication; like Critical Social theory and Information Richness theory. Ngwenyama and Lee (1997) indicate that Information Richness theory (IRT) has seen a dramatic shift. Due to this shift, IRT is no longer accepted by all information systems; therefore, they present a new theory to replace the old IRT. Ngwenyama and Lee think IRT has been limited to the perspective of positivism. Therefore, Critical Social theory (CST) has been introduced.

An increasing number of corporations favor electronic communication as their organization form. Moore (2001) found that more and more companies prefer to use electronic communication to link with their commercial partners. They want to build stronger relationships within their commercial interests through electronic data interchange (EDI). However, using EDI is not an easy and inexpensive way for these companies to communicate within their patterns. Therefore, Moore (2001) proposed an application that is based on technology to add flexibility and capability into this system in order to break down the restrictions of EDI, which will be beneficial for helping companies deal with massive data in their systems.

Moreover, electronic communication could be applied in various fields. For instance, in some developed countries, electronic communication has been used in sign language. Schneider, Kozak, Santiago and Stephen (2012) explore how new methods of communication, like electronic communication, affect sign language of America. They believe that new technological innovations are closely associated with language, especially sign language. These researchers emphasize current tools of electronic communication, such as e-mail, smart cellphone, and laptop. In addition, Schneider and Others (2012) focus on demographic social factors affecting American Sign Language, such as gender, race, and age.

As the emerging form of communication, the function of electronic communication is the same as traditional communication. Sarbaugh-Thompson and Feldman (1998) test the importance of greetings in electronic communication. Many researchers indicate that using electronic mail has a positive effect on organizational communication, but Sarbaugh-Thompson and Feldman (1998) think it has a negative influence on organizational communication. In their article, they constructed a communication matrix which is based on absence versus presence to research how important greetings are in electronic mail. Sarbaugh-Thompson and Feldman (1998) distributed three questionnaires during a four-year period to research the difference between communication that is face-to-face, and communication by electronic mail.

Electronic communication has been widely accepted in some particular groups. However, many problems have arisen. Draucker, Burke, and Martsolf (2010) indicate electronic communication plays an essential role in adolescent dating violence. They conducted a survey to investigate 56 young adults who have adolescent dating violence experience by using electronic communication. Draucker, Burke, and Martsolf (2010) described eight ways to prevent and solve this problem. The results indicate that electronic communication has a negative effect on adolescent dating violence, as it redefines boundaries between dating partners.

The Electronic Communication in Laos

As a developing country in Asia, Laos has less internet use and an insufficient electronic infrastructure. Vuth, Doung, Than, Phanousith, Phissamay, and Tai (2007) researched particular problems on distance education in Cambodia, Laos, and Viet Nam. They found that these three countries have similar responses in their awareness of the use of distance education. Among these three countries, Vietnamese have the best equipment for distance education. Few people in these three countries have personal computers accessing to the internet. The internet of Cambodia is the most expensive of these countries; however, Vuth and Others (2007) found the proportion of cellphone use to student enrollment in these three countries is very high.

The low usage of distance education indicates an insufficient electronic infrastructure and low internet use in Laos. The enterprise competitiveness of Laos has increased rapidly in these years although Laos has insufficient electronic infrastructure and low internet use. In particular, Rasiah, Rajah, Nolintha, and

Songvilay (2011) tried to test the effect of technological capabilities in garment manufacturing of Laos. The evidence which they collected indicates that garment manufacturing employment has significantly increased between 2000 and 2006, but their commodities have low value because Laos has fewer technological capabilities. Rasiah, Rajah, Nolintha, and Songvilay (2011) pointed out that without technological capabilities, Laos will still sustain their competitiveness because of the inexpensive commodities.

The following research questions will aid in forming an overall view of current Laotian electronic usage, and comparison to other developing countries.

Research Questions

- RQ1: What kind of electronic communication tools are mostly used in Laos?
- RQ2: How many people neglect friends and family members because of internet use in Laos?
- RQ3: What is the main way to communicate with others in Laos?
- RQ4: How often do people from Laos check email using internet?
- RQ5: How long have Laotians used a cellphone?

METHOD

Examination of the research questions was conducted after collecting data using a stratified random sampling approach. Researchers sent surveys to various public, government and educational locations and asked for volunteers to participate in a survey related to cellular and internet use. Each region of the country was canvassed by purposive sampling to include all geographical areas. Government and private workplaces; government, international, and private schools; and urban and rural locales were offered surveys. Previous researchers distributed surveys to 250 respondents in each developing country with a similar sampling frame. This information was used for comparative purposes in this study. Systematic effort to locate certain age categories; female, illiterate, or unemployed respondents were necessary. Ages 18-65 were contacted for possible participation.

Instrument

The instrument was a 38-item online questionnaire in English for people from Laos or those living in nearby regions that had previously been in Laos for at least three months.

Participants

The participants for this survey are mostly from Laos or people who have studied and lived in Laos. Several participants are from the nearby regions, like China and Thailand. All of the questionnaires have been collected using the internet.

RESULTS

The total responses were 43. The greatest age group for electronic use in Laos was 18-25 years old with 62.79% response. The next highest age group was 25-35 years old with 30.23% response. Collapsed, the largest group is age 18-35 at 93.02%. Gender demographics in Laos (n=43) were 53.49% female, 46.51% male. Over half, 58.14% of the participants were graduated from an institute, college or university, and 18.6% of the participants were graduated from a degree-granting postgraduate school (MA, MS, PhD, MD, LD). See figures 6.1-6.3.

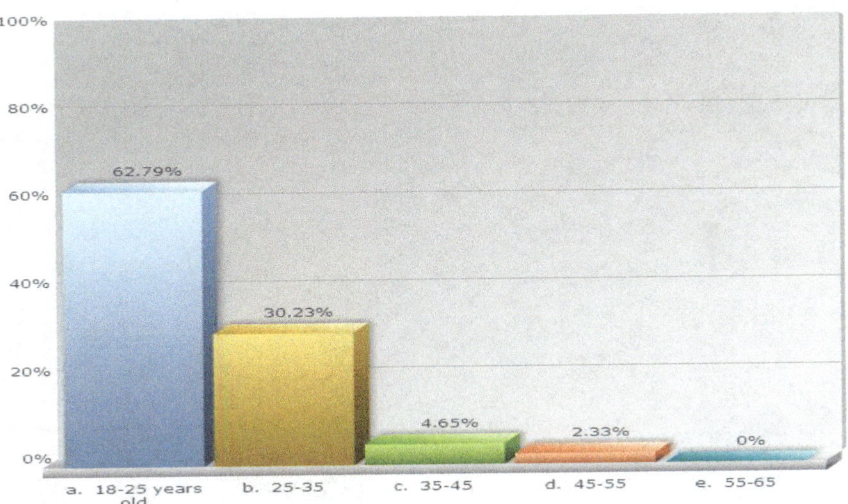

Figure 6.1: Age Demographics in Laos (n=43)

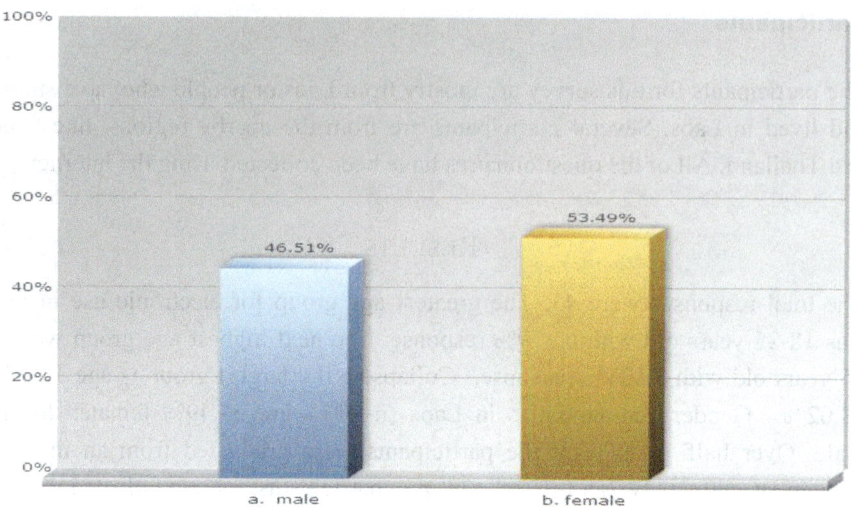

Figure 6.2: Gender Demographics in Laos (n=43)

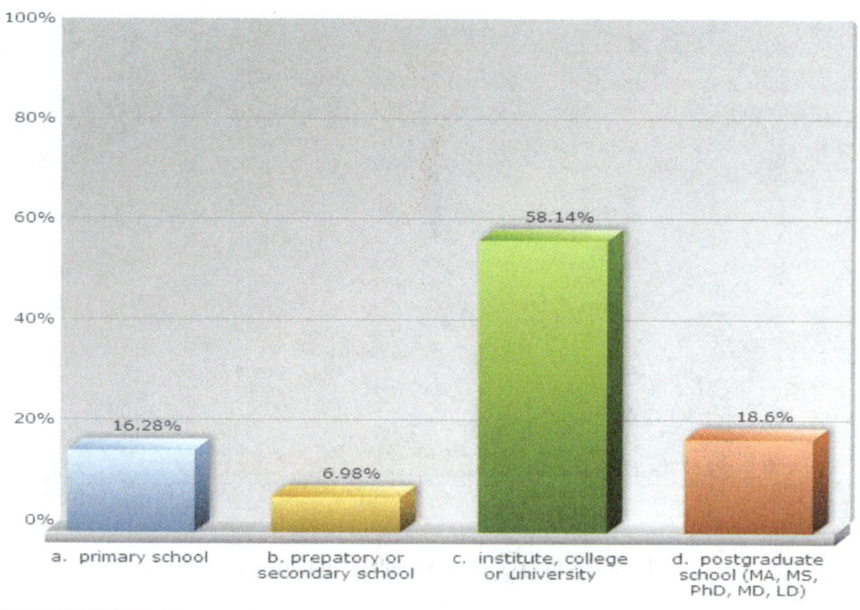

Figure 6.3: Educational Background Demographics in Laos (n=43)

RQ1 : What kind of electronic communication tools are mostly used in Laos?

Figure 6.4 indicates the various applications of electronic communication in Laos. Cellphone, handy, and mobile phone use take up 74.42%; the next highest online activity most used by Laotians is the television with 60.47% response. Email account, radio, and digital camera had a 48.84% response, 44.19% response, and 39.53% response, respectively. The least response was for online activity connected to internet bulletin board usage, accounting for 9.3%.

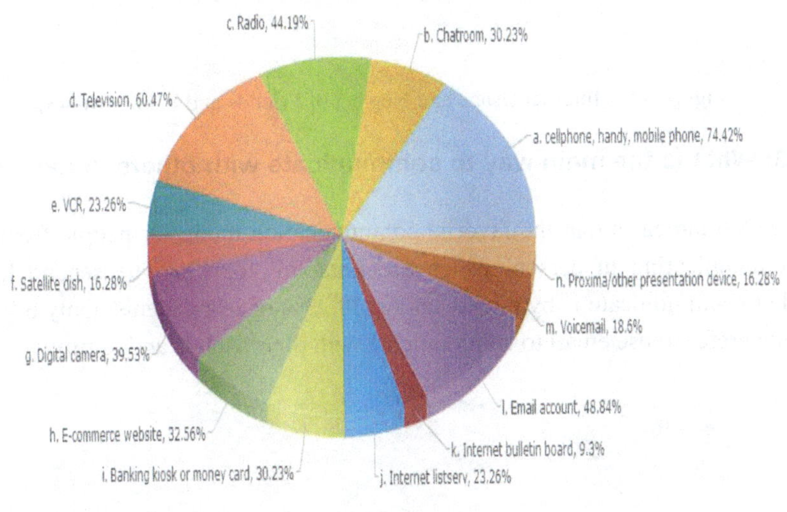

Figure 6.4: Various Applications of Electronic Communication in Laos (n=43)

RQ2: What proportion of people neglect friends and family members because of internet use in Laos?

Figure 6.5 illustrates that there are 27.91% of people sometimes neglecting friends and family members because of internet use in Laos, which is the highest proportion. There are 25.58% of people never neglect friends and family; only 11.63% of people always neglect their friends and family because of internet use.

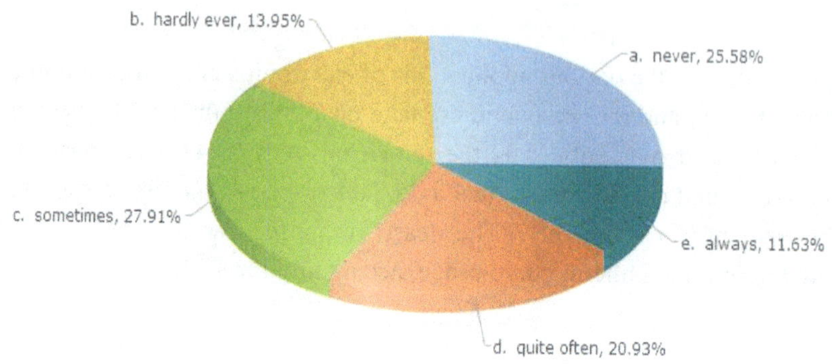

Figure 6.5: Internet Usage and Neglect of Friends and Family (n=43)

RQ3: What is the main way to communicate with others in Laos?

Figure 6.6 indicates that the favorite communication mode for people from Laos is communicating in person, which accounts for 46.51%. The second highest mode to communicate is by phone, taking 30.23% of respondents. Only 6.98% of people prefer to use email to communicate with their friends and family.

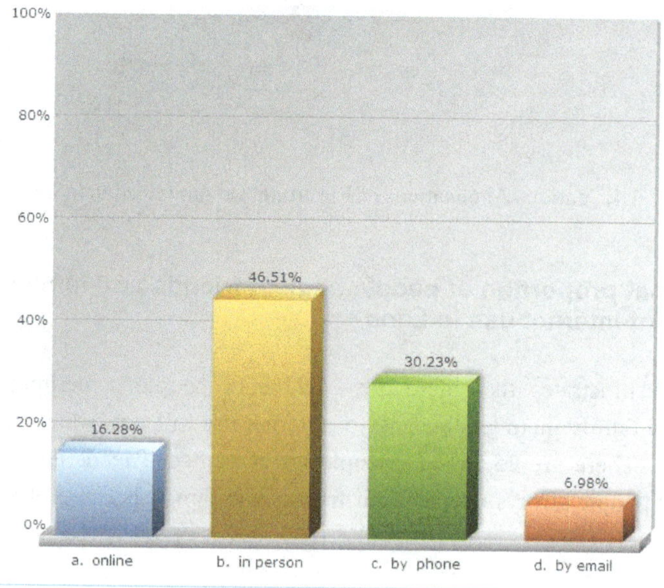

Figure 6.6: Ways to Communicate in Laos (n=43)

RQ4: How often do people from Laos check email using internet?

Figure 6.7 illustrates that 23.26% of people in Laos check email sometimes, 20.93% check email daily, and 16.28% hardly ever check email. Only 9.3% of the Laotian respondents disclosed that they check email several times per day.

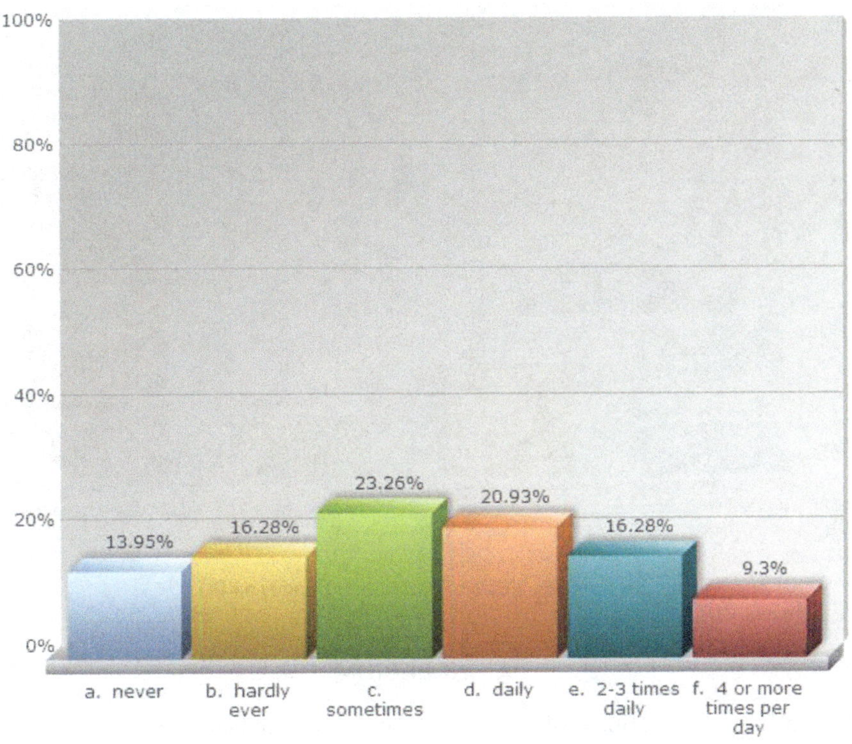

Figure 6.7:   Frequency of Checking Email in Laos (n=43)

RQ5: How long have Laotians used a cellphone?

The highest proportion of Laotians have used a cellphone five years or more at 58.14%. The second highest proportion reported two to three years at 18.6%. There were 16.28% of Laotians surveyed who had used a cellphone less than one year. Only 6.98% of Laotians used a cellphone three to five years. See figure 6.8.

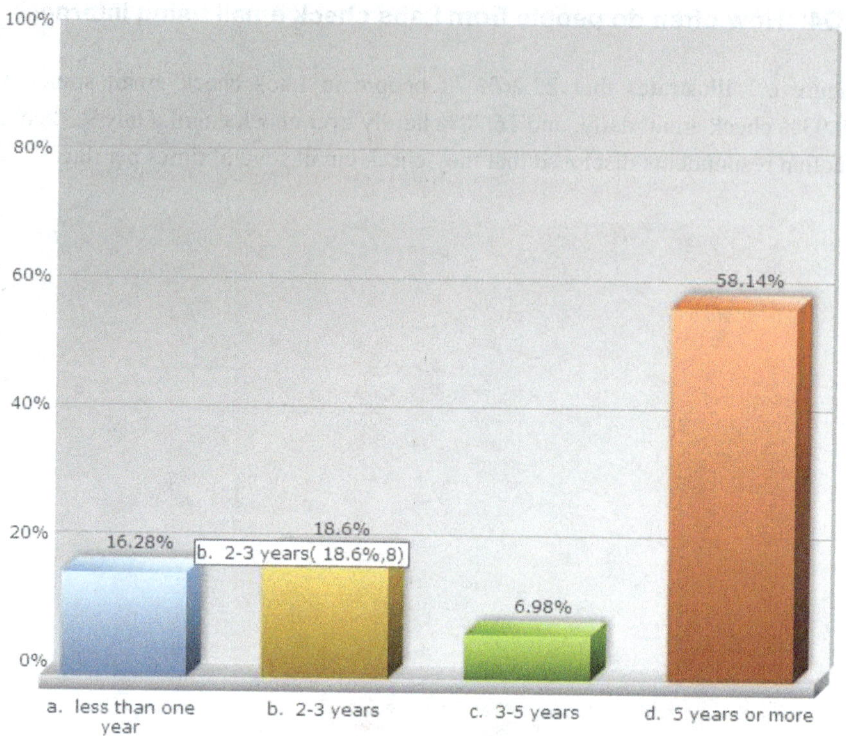

Figure 6.8: Years of Cellphone Usage (n=43)

CONCLUSION

The various applications of Acculturation theory indicate the function and importance of this theory. Although the use of Acculturation theory has not progressed as fast as information communication technology, it will still be important for both. Some problems cannot be addressed by technologies. As a wide application, Acculturation theory can be applied in business, education, and immigration. As the emerging form of communication, the function of electronic communication is the same as traditional communication. To some extent, the importance of electronic communication may have surpassed the importance of traditional communication. Moreover, as the convenient way to communication, many electronic communication forms have been used in various fields, such as health care, education, and business.

Furthermore, the application and use of the internet and cellphones in Laos is not widespread. The low usage of distance education indicates the insufficient electronic infrastructure and low internet use of Laos. Electronic communication is widely popularized by young people, especially between 18 to 25 years old. The elderly people rarely use electronic communication in Laos. In addition, female users of electronic communication are more than male users in Laos. The main users of e-communication in Laos have high educational backgrounds, and the proportion of face-to-face information in Laos remains high because the main way to communicate in Laos is still in person. The popularization of internet will be important for the development of Laos. The incomplete infrastructure of Laos means people have fewer opportunities to access the internet and cellphones. There are fewer opportunities for information communication technology to become commonplace due to expense, also.

REFERENCES

Berry, John W. "Immigration, Acculturation, and Adaptation." *Applied Psychology: An International Review* 46(1997): 5–34. doi: 10.1111/j.1464-0597.1997.tb01087.x

Draucker, Claire Burke, and Donna S. Martsolf. "The Role of Electronic Communication Technology in Adolescent Dating Violence." *Journal of Child & Adolescent Psychiatric Nursing* 23 no.3(2010): 133-142.

Fulk, Janet, and Gerardine DeSanctis. "Electronic Communication and Changing Organizational Forms."*Organization Science* 6 no.4(1995): 337-349.

Greenland, Katy, and Rupert Brown. "Acculturation and Contact in Japanese Students Studying in the United Kingdom." *The Journal of Social Psychology* 145 no. 4(2005): 373-389.

Ku, Linlin. "Social and Nonsocial Uses of Electronic Messaging Systems in Organizations." *Journal of Business Communication* 33 no. 3(1996): 297-325.

Landrine, Hope, and Elizabeth A. Kolbert. "Culture Change and Ethnic Minority Health Behavior: An Operant Theory of Acculturation." *Journal of Behavioral Medicine* 27 no. 6(2004): 527-555.

Lee, Allen S. "Electronic Mail as a Medium for Rich Communication: An Empirical Investigation Using Hermeneutic Interpretation." *MIS Quarterly* 18 no. 2(1994): 143-157.

Moore, Scott A. "A Foundation for Flexible Automated Electronic Communication." *Information Systems Research* 12 no. 1(2001): 34.

Mudambi, Anjana. "Living In-between: The Acculturation Patterns of

Undocumented Youth in the United States." [top 4 papers in ICC].
Conference Papers-- International Communication Association (2011):
1-41.

Ngwenyama, Ojetanki K., and Allen S. Lee. "Communication Richness in
Electronic Mail: Critical Social Theory and the Contextuality of
Meaning." *MIS Quarterly* 21 no. 2(1997): 145-167.

Rasiah, Rajah, Vanthana Nolintha, and Latdavanh Songvilay. "Garment
Manufacturing in Laos: Clustering and Technological Capabilities." *Asia
Pacific Business Review* 17, no. 2(2011): 193-207.

Sarbaugh-Thompson, Marjorie, and Martha S. Feldman. "Electronic Mail
and Organizational Communication: Does Saying "Hi" Really Matter?"
Organization Science 9 no. 6 (1998): 685-698.

Schneider, Erin, L. Viola Kozak, Roberto Santiago, and Anika Stephen.
"The Effects of Electronic Communication on American Sign
Language." *Sign Language Studies* 12 no. 3(2012): 347-370.

Segall, Marshall H., Cigdem Kagitcibasi, and John W. Berry. "Acculturation and
Adaptation." *Handbook of Cross-Cultural Psychology* 3(1997): 291-
319.

Vuth, Doung, Chhoun Chan Than, Somphone Phanousith, Phonpasit Phissamay,
and Tran Thi Tai. "Distance Education Policy and Public Awareness in
Cambodia, Laos, and Viet Nam." *Distance Education* 28, no. 2(2007):
163-177.

CHAPTER 8

The Up and Coming Electronic Communication in the Developing Country of Kyrgyzstan

Lei Joe Qiao

Social Information Processing Theory (SIPT) focuses on the relationship between computer-mediated communication (CMC) and face-to-face communication. This research aims to study Kyrgyzstan's electronic communication through SIPT. Data comes from surveys were completed by local people and people who have lived in Kyrgyzstan for one year at least. This research studies the usage of mobile phones and networks in Kyrgyzstan, and analyzes what genders, ages and education levels use electronic devices; what types of electronic communication are generally used; how long have the citizens used internet and cellular telephones; whether it is easier to interact with people electronically as opposed to face-to-face; has electronic communication improved individuals' lives in Kyrgyzstan. Quantitative and qualitative data is recorded by percentage descriptors of respondents and individual comments.

Introduction

Kyrgyzstan is a landlocked country in Central Asia, with a population of 5.7 million people. After the end of 1991, Kyrgyzstan achieved independence. The economy has overcome crisis and many difficulties for radical economic reform. After fifteen years of history, Kyrgyz economy has basically formed a fully open market economy model. Kyrgyzstan attaches great importance to the development of its electronic communications industry. Compared to other countries in central Asia, its growth rate, size and general condition are at a relatively advanced level. Electronic communication occupies a very important position in the national economy. Kyrgyzstan had one of the highest internet penetration rates in central Asia (5 percent for 2005). By 2009, the Kyrgyzstan communications industry experienced an increase of 10% to $430 million. According to official statistics, as of April 2010, Kyrgyzstan's wire line telephone subscribers reached 500,000. There were 4.46 million mobile phone subscribers, with a penetration rate of 83.2%; 2.2 million internet user. This made up 40% of the country's total population (5.7 million population of Kyrgyzstan).

As a member of WTO, Kyrgyzstan opened its communications market to foreign countries. The higher degrees of policy transparency and market opening have attracted many domestic and international business communications. As of April 2010, the operators in Kyrgyzstan which were doing communications business abroad totaled 260. Their business scope included fixed, mobile, voice transmission, internet, satellite television, cable television and other communications services.

The Kyrgyz Republic Ministry of Transport and Communications is the highest governing body of the communications industry, and is responsible for the formulation of national communications planning and related regulations and policies. In addition, Kyrgyzstan has also established a National Communications Agency, an independent agency that regulates the exercise of state functions in this field. National Communications Agency's powers include issuing trade licenses, licensing radio bands, radio equipment and high-frequency devices, the import and use of licenses, certificate management and authentication, wireless devices and other legal supervision, to ensure the normal operation of the order of Kyrgyzstan communications.

Literature Review

Social Information Processing Theory (SIPT) was proposed by Joseph Walther, who is a communication professor at Cornell University. This theory focuses on the relationship between computer-mediated communication (CMC) and face-to-face communication. SIPT believes that people use the internet to interact which can achieve the same effect as face-to-face (FtF) interaction; however, in order to achieve the same effect as face-to-face interaction, online interaction also needs more time to develop. This research aims to study Kyrgyzstan's electronic communication through SIPT.

Walther (1992) says that due to the CMC systems' short time of development, exchanges cannot achieve the same effect as FtF in a short period of time. However, over time, CMC has the advantages of keeping information stored for a long time, overcoming the obstacles of time and space. This article provides theoretical support for those who research electronic communication.

Although technical communication is still at a low stage in some developing countries like Kyrgyzstan, when the government continues to improve technical communication infrastructures, people will strengthen their understanding of virtual communication. The development of electronic communication will certainly improve in developing countries like Kyrgyzstan.

In the social aspect, James and Gavin (2009) reveal SIPT's impact on Japanese online interactive communication through case studies. Their research investigates the development of CMC, the different relationships of Japanese dating site members from past to present, and how to overcome the limitations of the CMC. With rapid development in technology, CMC has been further developed. It is not limited to communication just by computers, but the use of cell phone messages to communicate has become popular. It reveals the importance of the development of electronic communication. This real life example can provide a reference for the development of electronic communications in other countries. However, Japan is a developed country, and Kyrgyzstan is a developing country. The development of electronic communication faces more difficulties in Kyrgyzstan.

With the expansion of online interaction and CMC, people's attention begins to focus on how the development of CMC influences interpersonal communications. Heinemann (2011) helps us understand the SIPT through the movie, *"You've Got Mail."* In the movie, the hero and heroine, through online chat rooms, begin to know and understand each other, and eventually fall in love. Since they do not have mutual friends, there is no way to know each other in reality. Therefore, they can be more open to share their concerns and fears. Online exchange makes them more sincere to each other. Thus, it demonstrates SIPT, although CMC interaction takes longer to test and develop the first impression (which is more favorable in terms of face-to-face interaction). Over time, CMC users are able to build stronger interactive relationships than face-to-face ones.

In education of relationships between parents and teachers the development of electronic communication, e-mail exchange. This has become the main way for parents to communicate with teachers in the United States. Thompson (2008) uses SIPT as a guide, focusing on their relationship. He explains how this relationship changes with the CMC development. In the traditional sense of communication, parents and teachers usually communicate through meetings or occasional interviews. However, with the development of the CMC, parents and teachers begin using email to communicate more about the students. As a result, through e-mail communication, parents and teachers can understand the current situation of students in a timely manner. This research provides powerful data as well as some examples, to help readers gain a better understanding of CMC and the influence CMC brings to interpersonal communication.

Electronic Communication

There are results to show the relationship between electronic communication and the economics of countries. To begin with, Verikios and Zhang (2004) present a quantitative analysis of the possible effects on the regional and world economies of liberalizing trade in telecommunications, for which World Trade Organization (WTO) members have undertaken scheduled commitments for trade liberalization. They use a global general equilibrium model. By modeling the effects of complete liberalization of a services sector, they intended to highlight the sources and the distribution of gains potentially achievable from free services trade between regions.

In the trade of telecommunications, barriers are highest in developing countries, and lowest in developed countries. Verikios and Zhang (2004) analyze new estimates of these barriers for telecommunications. The result shows that the full liberalization of the telecommunications increased by 0.1%, and will benefit the whole world. Inter-regional income distribution is uniform. In general, the highest obstacle region will have maximum benefit. This analysis shows that foreign direct investment in foreign companies is an important way to provide communications business in the developing countries. That means the telecommunication in Kyrgyzstan still has great potential for development.

Later, De Silva, Ratnadiwakara and Zainudeen (2011) attempt to quantitatively measure the various effects of using mobile phones in some developing countries in Asia. Mobile phones are now affordable and widespread in the developing world. Mobile phones have a significant ability to transmit social policy initiatives to the most rural and/or excluded groups in society, and thus have a direct impact on poverty and inequality. They elaborate the effects of using mobile phone services which can eliminate poverty produced in these countries. A person who uses a mobile phone to pass information will be more easily adopted by people, which means that people tend to get connected in groups. Society encourages operators to move toward network marketing because this will help to further the policy of social policy objectives in Kyrgyzstan. It is helpful to research because it enables us to understand the status of mobile phones and other electronic communication equipment usage in developing countries.

In addition, the electronic communication brings equilibrium to politics. Ven-hwei and Ran (2010), by analyzing 10 leading communications journals from 1988 to 2008, document the spread of new media and the political research paradigm in Asia in the past two decades. They conclude that most of the academic literature is theory-driven. They reveal the role of the media in political communication in Asia. They also analyze published research concerning new media and political communication research in Asia, which indicates the trends and patterns as well as the interests of researchers. The results also contribute to future research directions about political communication in Asia. They study the interaction between new media and political communication in Asian countries, which will provide support to research of mutual influence between electronic communication and political communication. They advance understanding of media and politics in Asian countries.

Furthermore, in similar developing countries like Mongolia and the Philippines, as mobile phones become increasingly available, their usage is a basis to promote the development of education. They are used to compensate for the relative lack of internet access in these countries. McDonald (2009) studies distance education (DE) in the Philippines and Mongolia development, and the changes that need to happen in order to ensure the DE reaches as many learners as possible. Both countries are interested in mobile education and the potential of using mobile devices, such as mobile phones. In this study students and teachers have recognized the government needs to provide greater and more appropriate technical support for DE. It shows the development of DE is very important in education. Mobile phones are used by more than half the world's population. Opening the market of using cell phones for DE is not only a privity of developing countries, but it also significant for developed countries. The Philippines and Mongolia are developing countries in Asia, similar to Kyrgyzstan. McDonald (2009) will be useful to research about the impact of cell phone usage in Kyrgyzstan.

Country

There are results to show the development of electronic communication in different areas of Kyrgyzstan. To begin with, Srinivasan and Fish (2009) study grassroots internet sites and their webmasters in Kyrgyzstan, especially focusing on the relaxed controls policy of the internet in central Asia. They investigate the authors of the grassroots network; citizens, reporters and other active participants in the network. From the perspective of network communication and media, it is important to research and analyze the data in order to obtain the major impact the internet brings to Kyrgyzstan's social, cultural and political aspects. Internet communication technology makes new forms of grassroots political participation possible. When this happens, citizens can begin their decentralized and distributed network. Citizens can use the blog, news sites, and other news aggregators to participate in political activities. They also mention that democratization of new media has been recognized by the world, and has vitality. It facilitated the spread of the new situation of culture.

Then, Ibold (2010) explores the interaction between cultural identity and internet usage among the youth in daily life in the Kyrgyz city of Bishkek. Young people who participated in this study are believed to be progressive, tech-savvy, and global. However, they also respect the traditional culture which is local and rural. City youth can gain global information and experiences from the internet to broaden their horizons. They will use the information and experience on their own network to re-examine their parents, teachers and government. The impact of network information is profound for young people. They increase their knowledge and experience from the network; and even receive foreign culture, making the younger generation far different from the ideology of their parents' generation. This is one great influence online media brings to a national culture. With the development of the network, people's daily lives are changing, especially for the youth.

Moreover, Best, Thakur and Kolko (2010) describe the status of the telecommunication industry in Kyrgyzstan. They study the extent to which user-based subsidies can promote the sustainable development of financial/ social interactions, and the impact of developing telecommunication centers. Therefore, the authors looked at the eCenter network coupon program usage, which the United States Agency for International Development funded in Kyrgyzstan. They found that, based on the user's subsidies, financial sustainability can bring new users to the eCenter. However, the issuance of coupons does not improve the sustainable development of society. Because only ordinary users favor the coupon program, coupon plans have a limited impact on participation in community development. They provide a basis to discuss the prospect of the telecom industry in Kyrgyzstan.

Finally, Wei and Kolko (2005) explore a new social environment, the demand for people using mobile phones. They analyze the influencing factors of mobile phone usage in culture, politics and economy. This particular research environment is valuable for the development of mobile phones and mobile phone software. They use the case study of mobile phone usage in Uzbekistan, to reveal the development of using mobile phones in the unique cultural and political environment in Central Asian countries. As Uzbekistan is a very similar country to Kyrgyzstan, they provide some lessons for research on the use of mobile phones in Kyrgyzstan. They enrich the understanding about the diversity of users in the mobile phone industry and also highlight the differences in using mobile phones that exist all around the world.

In order to effectively understand the role mobile phones play in society, it is necessary to study their context in the whole culture. With a fuller understanding for the culture, politics and economy, researchers and designers can better improve their products. The development of mobile phones can carry out improvement for different political and cultural background regions, in order to meet more customer needs in different groups. The following research questions are addressed in their study:

Research Questions

- RQ1: What age, gender and education mainly use electronic devices in Kyrgyzstan?
- RQ2: What types of electronic communication are generally used in Kyrgyzstan?
- RQ3: How long have the citizens used internet and cellular telephones in Kyrgyzstan?
- RQ4: Is it easier to interact with people electronically as opposed to face-to-face in Kyrgyzstan?
- RQ5: Has electronic communication improved individuals' lives in Kyrgyzstan?

METHOD

This research is based on the theory of Social Information Processing, which was proposed by Joseph Walther. Examination of the research questions and hypotheses were conducted using a stratified random sampling approach. Research used Facebook, WeChat and email sent web links (SurveyMonkey) to respondents in various public, government and educational locations. It asked for volunteers to participate in a survey related to internet and cellular usage. The instrument is a four part questionnaire in English version for Kyrgyz citizens. Data about this research are collected through surveys completed by citizens and people who have lived in Kyrgyzstan for at least one year. The respondents reflect different genders, ages and education levels. Quantitative and qualitative data is recorded by percentage descriptors of respondents and individual comments. At least thirty surveys were desired in order to achieve comparative data, making research more reliable. Finally, research collected forty-one surveys from respondents. Data was entered in SPSS, and tables made to help readers have a clearer understanding of the results. This information collected by the surveys is used for comparative purposes in aggregate form. This chapter discusses the result of this study, limitations in the development of electronic communications, and the future development in Kyrgyzstan.

RESULTS

The questionnaire was distributed through web links to local people, and those who have lived in Kyrgyzstan at least for one year. Forty-one effective responses were collected in one month. In these responses, the greatest number who used electronic communication in Kyrgyzstan were 25-35 year olds with 70.73%. The second greatest was 18-25 year olds with 26.83%. Males and females in the use of electronic communications were substantially equal, with 51.22% and 48.78%. There were 63.41% of interviewees who had completed institute, college or university education.

RQ1: What age, gender and education mainly use electronic devices in Kyrgyzstan?

Table 1.1 Age Demograph

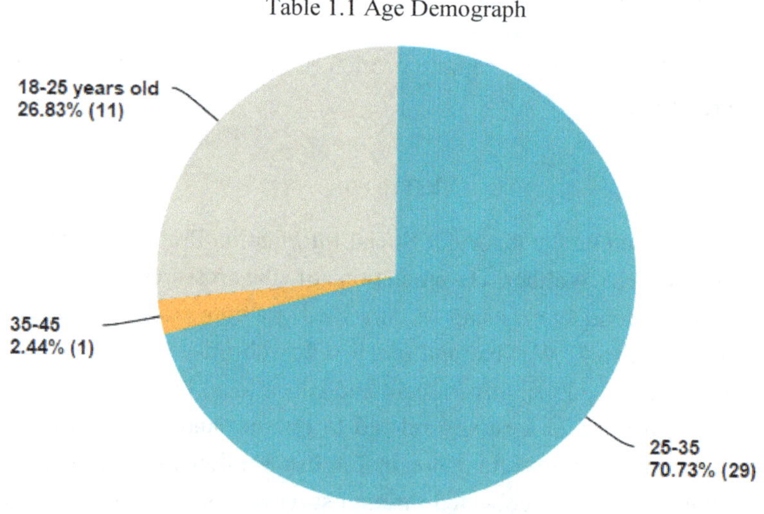

Table 1.1 (n=41) shows that the greatest cohort using electronic communication in Kyrgyzstan is 25-35 year olds with 70.73%. The second greatest is 18-25 year olds with 26.83%. The cohort of 35-45 year olds with 2.44% is the lowest to use electronic devices. Thus, the ages from 18 to 35 are the largest group who use electronic devices frequently.

Table 1.2 Gender Demographics

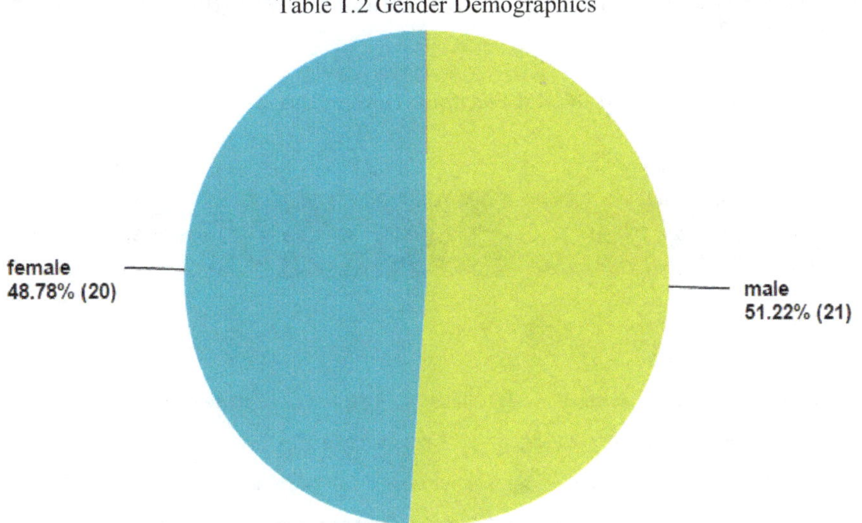

Table 1.2 (n=41) shows that 21 respondents are male, and 20 respondents are female. Their situations of cellular and internet usage are almost the same.

Table 1.3 Education Demographics

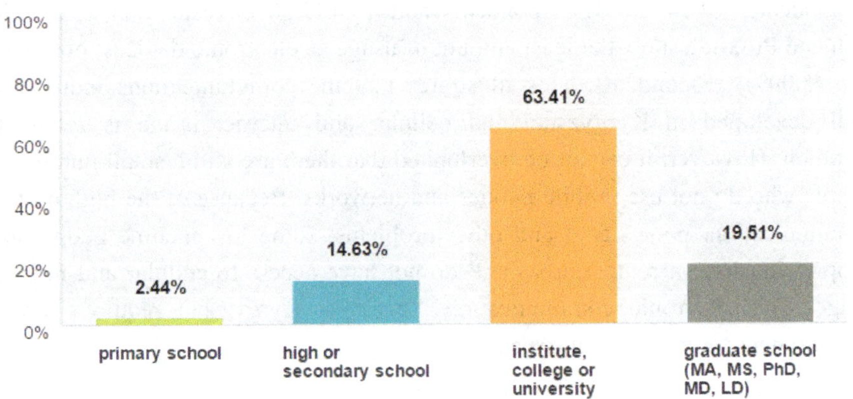

Table 1.3 (n=41) shows that 63.41% of electronic users have completed institute, college or university education; about 19.51% of electronic users have received higher education. The results suggest that people with higher education are the largest group to use electronic communications.

RQ2: What types of electronic communication are generally used in Kyrgyzstan?

Table 1.4 Electronic Devices Usage

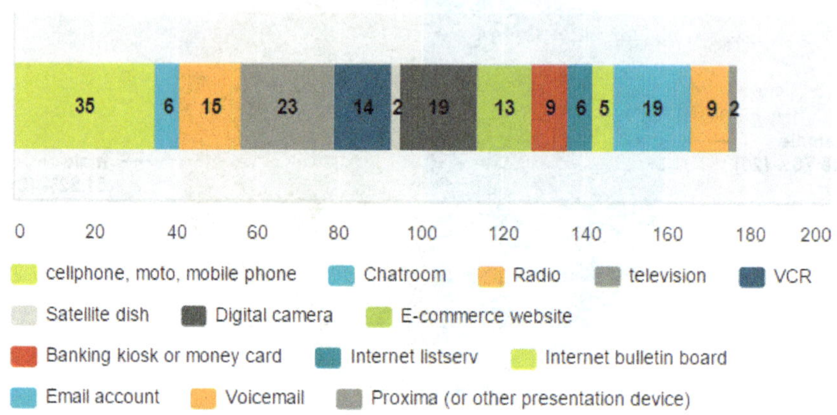

Table 1.4 (n=40) shows that there are fourteen electronic devices used at least once in Kyrgyzstan. Cell phone, television, digital camera and email accounts are the top four used frequently, each with 35 (87.50%) respondents, 23 (57.50%) respondents, 19 (47.50%) respondents and 19 (47.50%) respondents. Satellite dish and Proxima show the least amount of usage as electronic devices; both with two (5.00%) respondents. This illustrates that the communications industry is well developed in Kyrgyzstan and cellular and internet usage is relatively common. However, it cannot be overlooked that there are still a small number of people who do not use mobile phones and networks. Because of the high cost of communication, poor signal and other problems, some low-income people and people who live in remote areas still do not have access to cellular and internet usage. The electronic communications sector in Kyrgyzstan requires further development for adequate appraisal.

RQ3: How long have the citizens used internet and cellular telephones in Kyrgyzstan?

Table 1.5 Time of Internet Usage

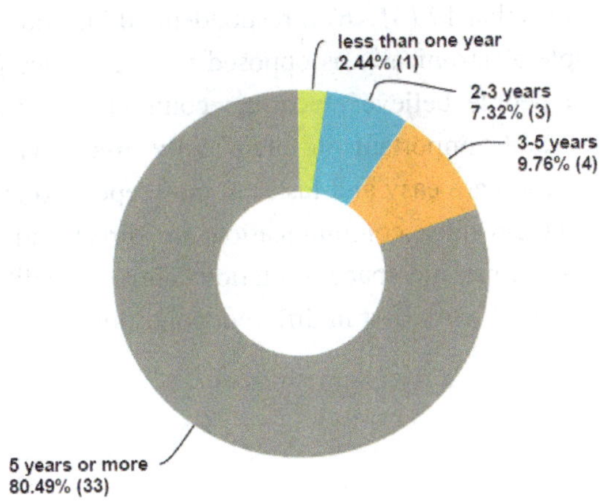

Table 1.5 (n=41) shows that 31 (79.49%) respondents have already used the internet for 5 years or more. Five (12.82%) respondents have used the internet for 3-5 years; only three (7.69%) respondents have used the internet for 2-3 years. The time of internet usage is affected by the interviewees' age. From open-ended, replies the respondents like to use the internet because its advantages are having lots of information, being very efficient, and having no limitation of time and space. The internet can also meet their requirements of communicating with friends and entertaining themselves.

Table 1.6 Time of Cellular Telephone Usage

Table 1.6 (n=41) shows that 33 (80.49%) respondents have already used cellular telephones for 5 years or more. Four (9.76%) respondents have used cellular telephones for 3-5years, three (7.32%) respondents have used cellular telephones for 2-3 years, and one (2.44%) respondent has used cellular telephones for less than one year. The interviewees' age influences their time of cellular telephone usage. From comments, Kyrgyzstanis like to use the cellular telephones because its advantages are having many useful functions, saving time, contacting the internet and their friends much easier, and no limitation of time and space. Comparing table 1.5 and table 1.6, cellular telephones are used more frequently than the internet in Kyrgyzstan.

RQ4: Is it easier to interact with people electronically as opposed to face-to-face in Kyrgyzstan?

Table 1.7 Electronic vs Face-to-Face Communication

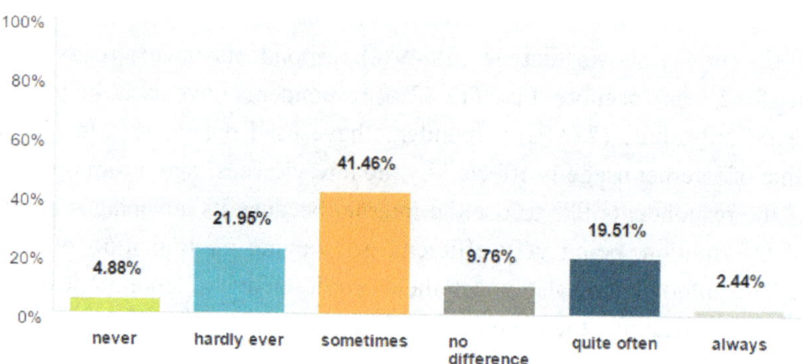

Table 1.7 (n=41) shows that 17 (41.46%) respondents think sometimes it is easier to interact with people electronically as opposed to face-to-face in Kyrgyzstan. It illustrates that most people believe electronic communication and face-to-face communication are both important in their daily lives. The advantages of electronic communication are easy and fast, but they report no emotional contact. The advantages of face-to-face communication are direct and helpful to solve problems, but there is a time and space limitation. Thus, in reality, people choose which interactive way will work best in different conditions.

RQ5: Has electronic communication improved individuals' lives in Kyrgyzstan?

Table 1.8 Life Improvements

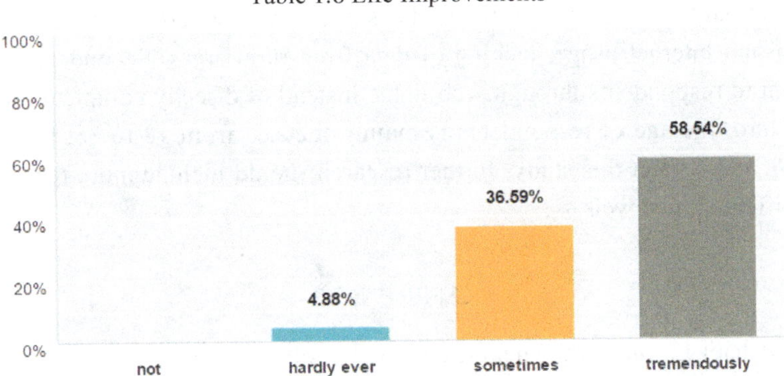

Table 1.8 (n=41) shows that 24 (58.54%) respondents consider electronic communication has tremendously improved individuals' lives in Kyrgyzstan. Nobody thinks electronic communication has not improved people's lives. It shows electronic devices play a significant role in daily life. Most believe electronic communication plays an active role in their lives.

DISCUSSION

This research explains the usage of mobile phones and networks in Kyrgyzstan, and analyzes what genders, ages and education levels use electronic devices; what types of electronic communication are generally used; how long citizens have used internet and cellular telephones; whether it is easier to interact with people electronically as opposed to face-to-face; and if electronic communication has improved individuals' lives in Kyrgyzstan. The youth and highly educated people are the majority to use electronic devices. Gender use of cellular and internet usage is almost the same. Cell phone, television, digital camera and email accounts are used frequently in Kyrgyzstan. Most respondents have already used the internet and cellular telephones for five years or more. Citizens consider sometimes it is easier to interact with people electronically as opposed to face-to-face. Most respondents believe electronic communication has tremendously improved individuals' lives in Kyrgyzstan.

This research shows the communications industry is well developed in Kyrgyzstan and cellular and internet usage is relatively common. However, there are still a small number of people who do not use mobile phones and networks. Owing to the high cost of communication, poor signal and other problems, some low-income people and people who live in remote areas still do not have access to cellular and internet usage. Thus, the electronic communications sector in Kyrgyzstan needs further development. There are three limitations of the survey in this research: First, the result of the survey mainly reflects urban people's cellular and internet usage, lacking of data from rural areas. Second, the survey was sent to respondents through web links, instead of directly completed face-to-face. Third, the age of respondents is mainly focused from 18 to 35. In order to improve and perfect this study, further research should include more respondents from different age levels.

REFERENCES

Best, Michael L, Dhanaraj Thakur, and Beth Kolko. "The Contribution of User-Based Subsidies to the Impact and Sustainability of Telecenters--The eCenter Project in Kyrgyzstan." *Information Technologies & International Development* 6, no. 2 (2010): 75-89.

De Silva, Harsha, Dimuthu Ratnadiwakara, and Ayesha Zainudeen. "Social Influence in Mobile Phone Adoption: Evidence from the Bottom of the Pyramid in Emerging Asia." *Information Technologies & International Development* 7, no. 3 (2011): 1-18.

Farrer, James, and Jeff Gavin. "Online Dating in Japan: A Test of Social Information Processing Theory." *Cyberpsychology & Behavior* 12, no. 4 (2009): 407-412.

Heinemann, Daria S. "Using You've Got Mail to Teach Social Information Processing Theory and Hyperpersonal Perspective in Online Interactions." *Communication Teacher* 25, no. 4: (2011): 183-188.

Ibold, Hans. "Disjuncture 2.0: Youth, Internet Use and Cultural Identity in Bishkek." *Central Asian Survey* 29, no. 4 (2010): 521-535.

McDonald, Brenda. "Appropriate Distance Education Media in the Philippines and Mongolia." *International Review of Research in Open & Distance Learning* 10, no. 3 (2009): 1-5.

Srinivasan, Ramesh, and Adam Fish. "Internet Authorship: Social and Political Implications within Kyrgyzstan." *Journal of Computer-Mediated Communication* 14, no. 3 (2009): 559-580.

Thompson, Blair. "Applying Social Information Processing Theory to Parent-Teacher Relationships." *RCA Vestnik (Russian Communication Association)* (2008): 45-65.

Ven-hwei, Lo, and Wei Ran. "New Media and Political Communication in Asia: A Critical Assessment of Research on Media and Politics, 1988-2008." *Asian Journal of Communication* 20, no. 2 (2010): 264-275.

Verikios, George, and Xiao-Guang Zhang. "The Economic Effects of Removing Barriers to Trade in Telecommunications." *World Economy* 27, no. 3 (2004): 435-458.

Walther, Joseph B. "Interpersonal Effects in Computer-mediated Interaction: A Relational Perspective." *Communication Research* 19, no. 1 (1992): 52-90.

Wei, Carolyn, and Beth Kolko. "Studying Mobile Phone Use in Context: Cultural, Political and Economic Dimensions of Mobile Phone Use." *IEEE International Professional Communication Conference Proceedings* (2005): 205-212.

Chapter 9

The Development of Electronic Communication and Acculturation in Malaysia

Jing Jessie Jia[1]

Malaysia is a multiracial, multicultural country. There are many immigrants in Malaysia from other countries that accelerate daily contact among nationals. The culture has been influenced by different areas of population, causing the degree of acculturation to be deep in Malaysia. This research will focus on the process of acculturation by immigrant interchange in Malaysia. Malaysia's telecommunications network is more advanced than any other countries in Southeast Asia with the exception of Singapore. The government of Malaysia has invested heavily in the industry of information technology and telecommunications infrastructure in spite of the economics of the country which suffered crises in 1997, 2009, 2014, and 2015. The usage of the internet and mobiles are increasing year by year, and the features of internet and mobiles are becoming diversified which make some aspects of people's life better, easier, or more efficient. However, the needs for different people are different. For example, students, young people, and the people who have a high degree of education may evidence higher usage patterns for internet and mobiles than older citizens and sojourners with less education.

INTRODUCTION

Acculturation is generally viewed as a process of change over time within individuals as they adapt to a new cultural environment (Berry, 2003). Different cultures from different countries can affect the communication traits of people. According to Hsu (2010),

---

[1] Jing (Jessie) Jia was the first place winner of graduate level entries in the 2014 John Heinrich Scholarship and Creative Activities Day research poster display, Memorial Union, Fort Hays State University.

The acculturation process follows a stress-adaptation-growth cycle. Newcomers are confronted with unfamiliar situations, which may differ drastically from the expectation of their original culture. Stressful experiences often force newcomers to acquire new values and behaviors, especially communication skills, in order to achieve a better fit or outcome with the host culture. (415)

In Malaysia, the immigrants can be affected by both the original country and host country. When immigrants perceive rejection from their friends and family left in the country of origin, they will reduce their cultural identity of their own country. They need to be better at adapting to the local culture area to find a good job and live well. Their intent is to accept a new culture from the host country and join local organizations. Moreover, they can also be influenced by the local religious laws and pervasive social norms. They comply with the local religious beliefs, yet sometimes voice and remember religious objections from their origin.

In modern society, information and communication technology (ICT) have become powerful tools for increasing productivity and efficiency. ICT is indispensable for exploring current and essential information in personal, business, educational, medical, governmental, and religious transactions. Before the emergence of information technology, people were introduced to computers and mobiles mainly to run daily activities and transactions no matter where they are; whether they are in the office, schools, or at home. However, at that time, the usage of computers and mobiles was very limited because the technology was new, expensive, restricted, and undeveloped. Today, the usage of computers and mobile telephony has expanded in a plethora of various functions. Further development of telecommunications and networking has turned computers into communication tools via the internet, and also as a means to promote businesses, to search for information instantly, and for educational purposes directed toward all ages and backgrounds.

Malaysian government has realized the importance of developing an information and communication technology. The year 1995 was considered the beginning of the internet age in Malaysia. Growth in the number of internet hosts in Malaysia began around 1996. The Cari internet was the first web portal company, and the first search engine of the country founded in 1996. According to the first Malaysian internet survey conducted from October to November 1995 by MIMOS and Beta Interactive Services, one out of every thousand Malaysians had access to the internet (20,000 internet users out of a population of 20 million). In 1998, this number grew to 2.6% of the population. The total number of computer units sold, which was 467,000 in 1998 and 701,000 in 2000 indicated an increasing growth. Since 2000, many of the Malaysian companies started to enter the network industry. With the development of the internet, users have increased annually. In 2010, Malaysian users increased to 16,902,600. Utilization rate of the total population became 64.6% (Uchenna and Poong 2011).

In Malaysia, mobile phones have been used for decades, and have been effectively replacing the traditional means of communicating. The upgrading of mobile phones has been very rapid in recent years. Many new features have been developed compared to traditional cell phone capabilities. The mobile phone also offers a great deal in narrowing down the digital gap, particularly that between the rural and urban people. The cellular industry in Malaysia had explosive growth in the 1990s. Fixed-line telephone usage is less than that among the increasing population that uses cellular phones in Malaysia. In 2000, Malaysia had 4.7 million fixed-line users, and 5.5 million cellular users. There were eight nationwide cellular phone operators competing in the market at that time. There were 3 million users in 1999, and 14.5 million by end of 2004. The annual growth rate is about 25%.

THE COUNTRY OF MALAYSIA

Malaysia is a federal system country in Southeast Asia. It consists of thirteen states and three federal territories. Malaysia is a multi-ethnic, multicultural, and multilingual society. The original culture of the area stemmed from indigenous tribes along with the Malaya who later moved there. When foreign trade began, the culture was deeply influenced by Chinese and Indian culture. Due to the integration of different cultures, English as a *langue* in Malaysia is very popular and foreign trade is prospering. Malaysia has become diversified and industrial, emerging as an Asian market economy with ties to the entire world. Malaysian information technology has sustained its economic growth in the 21st century.

Acculturation Theory

Acculturation theory is described as both the process of contacts between different cultures, and also the customs of such contacts. As the process of contact between different cultures, acculturation may concern direct social interaction, or display other cultures by means of the mass media of communication. As the result of such contacts, one certain group may assimilate with other cultures of another group which modifies the existing culture. Group identity will be changed. Therefore, there will be a tension between two different cultures, and it will lead to adapting of the new as well as the old cultures.

There are many immigrants to Malaysia from other countries who accelerate this contact. Malaysian communication has also become extensive with the people from different countries. Therefore, the culture will be influenced from different areas, causing the degree of acculturation to be deep in Malaysia.

Different cultures from different countries can affect the communication traits of people. According to Hsu (2010), "Due to the Chinese modest character, they are more apprehensive and less willing to communicate. On the contrary, Americans tend to engage in self-enhancing communication". (189) However, acculturation time might affect one's cultural identity; which in turn, affects communication traits. When Chinese reduce fusion with their original ethnic group and have more identity with American culture, the Chinese communication traits might become more similar to Americans. However, not every immigrant is affected. There remains some cultural identity that does not change substantially with longer time living abroad. Therefore, many other reasons can influence communication traits among immigrants, business partners, citizens, and sojourners such as students, spouses or government workers.

For instance, Malaysia as a multicultural country and its immigrant cultures have formed in history. Most immigrants there are from China, India and other Asian countries. Their communication trait contrast is not so obvious compared with Chinese living in America. However, different religious faith is one of many factors affecting communication traits among immigrants. For example, Islam is the mainstream religion in Malaysia; Muslims consider the left hand unclean and do not eat or shake hands with it. This has daily repercussions throughout all encounters.

What is more, the immigrant acculturation orientations are affected by both the origin country and host country. When immigrants perceive rejection by distance from their friends and family left in the country of origin, they will reduce their cultural identity with their own country. Their intent is to accept a new culture from the host country and join local organizations for survival on a day-by-day basis. However, when they perceive rejection from the host country, they will reduce the extent to which they will seek proximity to the host country. Therefore, the two countries' attitudes toward each other as a group affect orientation and degree of the immigrant acculturation personally.

The Malaysian government seeks to attract foreign capital, promote the development of tourism, and encourage foreigners to live in Malaysia for a long time. Therefore, immigrants generally do not perceive rejection from the host country. I believe that they can integrate and acculturate in the host country with relative freedom. As a result, there will be changes occurring in both immigrants and members of the host society through their interaction effect.

According to Lakey (2003), "Communication is viewed as central to the acculturation process. The immigrant's intercultural communication behavior is affected by the acculturation motivation, language fluency, and interpersonal and mass media channel accessibility" (1556). However, educational background, gender, time among the host society, and age at the time of immigration are the key determinants of one's language competence, acculturation motivation, and accessibility to host communication channels. Every immigrant is acculturated into the new host society through communication. The immigrant's communication competence facilitates all other aspects of adjustment in the host environment.

The immigrants' cultural identity can also be influenced in the multicultural work contexts. This is affected by two factors: desire for economic rewards and relational pressures. Newcomers must know how to find good jobs and higher wages to succeed, and they have to fade into and accept the new cultures. This makes many newcomers accept the inherent complexities of the acculturation process. Moreover, the acculturation process will influence the social networks and organizations that newcomers join. As Saminni (2013) says, "A newcomer who becomes employed at an organization outside their ethnic enclave/network will achieve higher labor market outcomes (income, employability, advancement) than those who join such organizations" (116).

Malaysia is a tourism country and many people from different countries travel there. Therefore, the communication environment can be complex for the sojourner. Acculturation process is difficult to implement by single, non-recurring communication in that environment. However, the immigrant's intercultural communication behavior should be relatively easy when built consistently with a small group of work confederates. Therefore, the immigrant's communication competence should affect both the host country and origin country.

INTERNET AND CELLULAR USAGE

According to Salman (2015), the Malaysian government carried out a national strategy for telecommunication, including the internet earlier than other developing countries. The beginning of the internet age is from 1995 and it has spread rapidly. Malaysia had 25.3 million internet users by July 2012. For the diffusion of the internet, Malaysian government has carried out a plan which proposes to support university students as well as other lower-income groups to ensure that they have the opportunity to access the internet, and also to increase broadband usage throughout the country. The young internet users are spending more hours socializing on the net with friends and family. For people to continue using the internet, efforts must be made to provide more facilities for social networking that can empower them to participate in discussions concerning public issues. The internet development direction is to maximize the use of the internet in achieving digital inclusion, and gaining cultural capital in the future. However, access to the internet does not guarantee digitalization. The internet meets gratification and added value needs for many users, yet not all Malaysians have entered the electronic age.

The forms of using the internet are becoming more diversified. According to Yet Mee et al (2010), "Malaysian consumers search moderately for product/service information with company websites being the most popular mode of searching" (1). Moreover, Malaysian government has encouraged households to sell their home-made products online; and at the same time, provide home deliveries. In order to expand the online business scope, Malaysian state government is building an infrastructure to achieve full internet access in rural areas. The researchers also found that, "In the Malaysian context, 40.5% of the household users of the internet surveyed used the internet to search for information on goods and services in year 2005" (Yet Mee et al, 5). In sum, online buying behavior is one way of using the internet in Malaysia, as it promotes the development of the network and user skills.

In addition, the government has also begun to use the internet to improve office efficiency. Seng et al (2010) recounts,

> As more and more activities are migrating from physical to virtual media, users and employees have been faced with relentless pressure to use technology. However, it is the development tendency in the future. The information and communication technology can help the government office to improve work efficiency and adapt to the modern office. (423)

Malaysian government has already realized that implementation of the electronic office can help civic employees to achieve their target of being a developed country by the year 2020. However, there are certain barriers of implementing the electronic office. It is very important to understand internet technology values, and also the enabling/constraining effects of organizational culture and individual behavior. These can facilitate or impede ICT implementation.

Students may be the greatest, most influential proportion of UCT users among the Malaysian people who often use computers. Internet popularization by students affects the long-term development of the internet. Research points out that,

> The popularity of computer knowledge is very important for students. The Malaysian government has supported information and communication technology since the early nineties and encourages people to go online. It has influenced computer knowledge as a course in schools. There is a significant relationship between literacy and internet usage. (Nair et al 2012)

Most Terengganu families are not literate, and in a low or middle income category, but the students appear to have satisfactory computer literacy through school programs. Therefore, when schools provide good conditions for learning and computer literacy, and purchase computers for students, the computer penetration rate is higher. In Malaysia, there are many schools in the poor and remote areas that cannot access the internet and information. This affects the spread of the internet in Malaysia.

Therefore, Malaysian government must invest heavily in construction of the internet for all populations. Malaysian government actively promotes improved productivity and technical knowledge in order to transform the community into a knowledge-based society. In addition, the government toils to shorten the existing information gap between the urban and rural populations. It vigorously supports the ICT infrastructure, especially in the rural areas. Rural development is very important for Malaysia in being a developed country in the future. According to Meng et al (2010), "ICT can effectively be used for overcoming poverty, increasing business productivity, accelerating economic growth and improving accountability and governance in any society" (48).

Malaysia's telecommunications network is more advanced than any other countries in South-east Asia with the exception of Singapore. The cellular industry in Malaysia has had explosive growth in the 1990s. Fixed-line telephones are less than the population that uses cellular phones in Malaysia. In 2000, Malaysia had 4.7 million fixed-line users, and 5.5 million cellular users. There have been at least eight nationwide cellular phone operators competing in the market. In 1999, there were 3 million users, and 14.5 million by the end of 2004. With the popularity of mobile phone use and development of wireless today, mobile e-commerce is becoming a new e-commerce mode. Mobile commerce is more personalized, unique and flexible. With the birth of 3G, the mobile penetration rate in Malaysia has been the second highest in the region. Therefore, the potential market of mobile commerce in Malaysia is huge. However, in order to have a better development of mobile commerce, the determinants influencing world-wide consumers must be known. This research shows that perceived ease of use, personal innovativeness, perceived trust, perceived cost, social influences, and perceived usefulness are key determinants influencing consumers to adopt mobile-commerce (Uchenna and Poong 2011). Malaysia has a huge potential market in mobile commerce, but there must be users to adopt it. However, the mobile phone usage is increasing continually and diversifying. It is a good sign for developing mobile commerce in Malaysia as a hub for the world.

With the development of 3G mobile technologies, mobile can offer the same features as any computer system. Mobile usage has grown globally. Thus, every student currently owns one mobile phone. As Manimekalai et al (2013) indicates, the unique feature of mobile technology is to supplement traditional learning and reinforce students. Students can use mobile phones not only for communication or entertainment, but also study. Through mobile phone learning, students can access mobile e-books, conduct podcasting, download lecture notes, engage in social networking activities, and email anywhere and anytime for learning purposes. These features are clearly beneficial for students. However, owning a mobile phone is not an assurance that students will use mobile phones for learning purposes. Educators should emphasize how the features are integrated in the learning environment (Manimekalai et al 2013, 132). Furthermore, though this research, we find that there is no difference in the usage of mobile as observed between males and females.

According to Siti et al (2013), "Mobile phones are also useful devices for the fishermen who are a large and important group in Malaysia. Using mobile to check the weather, security and emergency, the situation of the fish market is very helpful for them" (162). Therefore, Malaysian government has motivated and emboldened the fishermen to use mobile phones, and even provides government subsidies for buying a mobile phone for the fishermen. It has been seen as a good effort in motivating them to use the mobile phone more in their fishing routine. In turn, this might create a better perception among fishermen of the benefits and ease of mobile phone usage. Although it is good for the fishermen to use mobile phones, the signal and network coverage is not as comprehensive on the sea. Therefore, this is a limitation for the fishermen using mobile and 3G when they are out to sea. Encouraging and prompting fishermen to use mobiles is a good example for researching the mobile usage in Malaysia. At the same time, it is good for the popularization of the mobile. The following research questions have emerged for Malaysia.

Research Questions

- RQ1: Is electronic communication usage correlated to education?
- RQ2: Do males and females use the same types of electronic communication?

METHOD

Survey questionnaires will be distributed online to consumers in Malaysia. A snowball method is used due to the fact that it is easier to penetrate the targeted group of respondents by using their acquaintances. Targeted respondents are Malaysians who possess mobile communication devices. The questionnaire consists of two sections, in which the first section comprises personal basic information. The second section consists of items to measure constructs identified in the main body of the literature review. All respondents must be 18-65 years of age.

Table 7.1 Respondent Demographic Profile (n=60)

	n	%	n of using internet	%	n of using mobiles	%
Gender						
Male	26	43.3%	23	38.3%	25	41.7%
Female	34	56.7%	30	50.0%	31	51.7%
Age						
18-25	29	48.3%	29	48.3%	29	48.3%
23-35	21	35%	20	33.3%	21	35%
35-45	6	10%	3	5%	4	6.7%
45-55	4	6.7%	1	1.7%	2	3.3%
55-65	0	0	0	0	0	0

Level of Education						
primary school	2	3.3%	0	0	1	1.7%
secondary school	9	15%	5	8.3%	7	11.7%
college or university	29	48.3%	29	48.3%	29	48.3%
postgraduate school	20	33.3%	19	31.7%	19	31.7%
Education (collapsed)						
High level	49	81.6%	48	97.9%	48	97.9%
Low level	11	18.3%	5	45.5%	8	72.7%

As table 7.1 illustrates, 81.6% of respondents have degrees from university and postgraduate schools. There are 48.3% of respondents reporting that they are using internet, and 48.3% respondents reported that they are using mobile phones. Among them the percentage of respondents with a degree above college or a university program, using internet and mobiles are 97.9% and 97.9%. However, 45.5% and 72.7% of the respondents who belong to the low level of education (completed primary or secondary school) are using internet and mobile phones respectively. From the data we can see the percentage of respondents who have a high degree of education using internet and mobiles are higher than the percentage of respondents who have a lower degree of education using internet and mobile phones. Therefore, the electronic communication usage is correlated to education.

Table 7.2 Usage Percentage of Communication Media among Males and Females (n=60)

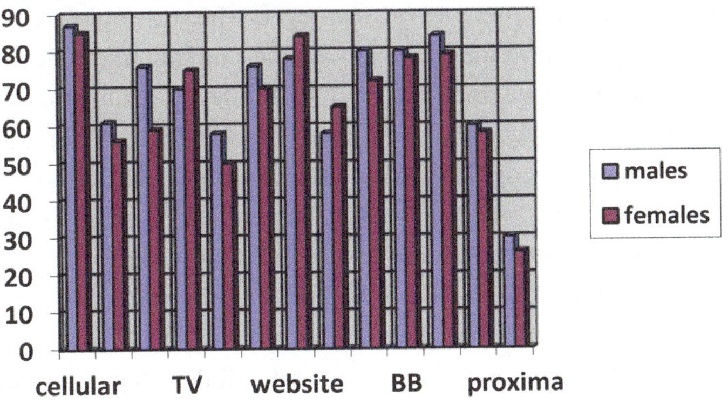

The result shows that the fourteen kinds of communication media used by females and males have slight differences. The usage of television, e-commerce website and banking kiosk or money cards by females is slightly higher than males. The other media are slightly higher for males; hence, there are slight differences in the usage of types of electronic communication observed between males and females in Malaysia. Electronic communication has become an integral part of the life of people regardless of their gender. Electronic communication has provided new delivery platforms for humans to communicate because it offers flexibility and instant connectivity.

CONCLUSION

The purpose of researching the acculturation process in Malaysia was to investigate how the immigrants are adapting to the life now in the host country, and the influence from the local cultures. The results indicated the perspective that immigrants' communication traits become more similar to people of the host culture in order to meet the demands of the new cultural environment. To have a satisfactory life in Malaysia, they must accept the influences of the local culture and be familiar with the local customs. However, the changes occur in both immigrants and members of the host society through their interaction effect.

Malaysia has made a lot of gains as far as the usage of electronic communication. The support of government played an important role in the diffusion of the electronic communication. This research was focused on the development of the internet and mobile phones in Malaysia to achieve that effect. The internet was rapidly developed, and since the year of 2000 many of the Malaysian companies entered the network industry. Users of the internet have increased to 16,902,600. Utilization rate of the total population is 64.6%. The advantages of the internet can be seen everywhere. People are using internet to communicate for convenience and with no limitation of the time and space. However, using the internet is not only limited to individual behavior, many workplaces have begun to rely on the internet. Malaysian government departments have recognized the importance of internet to enhance service efficiency, streamline processes, and become more business oriented. In order to help students to master computer and internet technology, schools need to offer required computer courses. As usage of mobile phones for internet delivery increases, more features are developed compared with the ordinary phones in recent years. The technology of internet and mobiles plays a role in narrowing the existing information gap between the urban and rural populations of Malaysia.

Rural development is crucial in helping the government to make Malaysia become a developed country by 2020.

Although the government encourages people in using the new technology of internet and mobiles, there remain differences between individuals of Malaysia in using internet and mobile phones. The research found that the people who have a higher degree of education using internet and mobiles are more than the people who have lower degrees of education. That would be relational to their living habits or perceived levels of language and cultural dependence. Many people like going online very much because they can learn how to use the computer, and get more information from the internet.

With the development of technology, electronic communication has become diversified. There are many types of communication available to Malaysians. Electronic communication has become an integral part of the life of people there regardless of their gender. However, there is slight difference between males and females on using the types of communication. The usage of television, e-commerce websites, and banking kiosks or money cards by females is slightly higher than males. The usage of other types of communication are slightly higher for males, or equivalent for both genders. Electronic communication has provided new delivery platforms for human communication because it offers flexibility and instant connectivity at a relatively low cost.

The Malaysian government should continue to encourage and support the development of ICT. Despite all the initiatives and investments in new technologies, especially the internet and its applications, continuous effort is still needed to be made in the rural areas. In order to narrow the gap between people, popularizing the application of new technology is necessary in educational systems in Malaysia.

References

Badea, Constantina, Jolanda Jetten, Aarti Lyer, and Abdelatif Er-rafiy. 2011. "Negotiating Dual Identities: The Impact of Group –based Rejection on Identification and Acculturation." *European Journal of Social Psychology* 41(5), 568-595 (accessed February 11, 2014).

Chew Chun, Meng, Bahaman Abu Samah, and Siti Zobidah Omar. 2013. "A Review Paper: Critical Factors Affecting the Development of ICT Projects in Malaysia." *Asian Social Science* 9(4) 42-50 (accessed February 23, 2014).

Hsu, Chia-Fang. 2010. "Acculturation and Communication Traits: A Study of Cross-Cultural Adaptation among Chinese in America." *Communication Monographs* 77 (3), 414-425. (accessed February 11, 2014).

Lakey Paul N. 2003. "Acculturation: a Review of the Literature." *Abilene Christian University Intercultural Communication Studies XII-2* (2), 103-118 (accessed February 11, 2014).

Manimekalai, Jambulingam, and Shahryar Sorooshian. 2013. "Usage of Mobile Features among Undergraduates and Mobile Learning." *Maxwell Scientific Organization* 5(4), 130-133. (Accessed February 19, 2014). http://maxwellsci.com/print/crjss/v5-130-133.pdf.

Ramli, Omar, Bolong, D'Silva, and Shaffrill.2013. "Influence of Behavioral Factors on Mobile Phone Usage among Fishermen: The Case of Pangkor Island Fishermen." *Asian Social Science* 9(5):162-170. (Accessed February 25, 2014). http://dx.doi.org/10.5539/ass.v9n5p162.

Salman, Ali, Er Ah Choy, Wan Amizah Wan Mahmud, and Roslina Abdul Latif. 2013. "Tracing the Diffusion of Internet in Malaysia: Then and Now." *Asian Social Science* 9(6), 9-15. (accessed February 17, 2014).

Sekharan Nair, Gopala Krishnan, Rozlan Abdul Rahim, Roszainora Setia, Aileen Farida Binti, Mohd Adam, Norhayati Husin, Elangkeeran Sabapathy, and Norhafiza Abu Seman, et al. 2012. "Terengganu Schools Extent of Computer Literacy and Internet Usage." *Asian Social Science* 8(8), 74-79. (accessed February 17, 2014).

Seng, Wong Meng, Stephen Jackson, and George Philip. 2010. "Cultural Issues in Developing E-Government in Malaysia." *Behaviour & Information Technology* 29(4) 423-432. (accessed February 17, 2014).

Samnani, Al-Karim, Janet A. Boekhorst, and Jennifer A. Harrison. 2013. "The Acculturation Process: Antecedents Strategies, and Outcomes." *Journal of Occupational & Organizational Psychology* 86(2), 166-183. (accessed February 11, 2014).

Uchenna, Mohd, and Yew-Siang Poong. 2011. "Mobile Commerce Usage in Malaysia." *IPEDR 5 (2011). Press, Singapore.* Accessed February 19, 2014. http://www.ipedr.com/vol5/no1/57-H00134.pdf.

Yet Mee, Lim, Yap Ching Seng, and Lau Teck Chai. 2010. "Online Search and Buying Behaviour: Malaysian Experience." *Canadian Social Science* 6(4) 154-166. (accessed February 17, 2014).

CHAPTER 10

ICT in Maldives: The Shaping Method of Acculturation Theory

Chenyang Doris Wang

The research goal of this chapter is to use the shaping method of Acculturation theory to explain Maldivian use of information communication technology (ICT). The study adopted the form of an English questionnaire with 38-40 items to investigate 25 respondents who had been born or sojourned in the Maldives for at least three months. The data from the survey indicate that Acculturation theory has a close relationship with the usage of ICT in Maldives. Most electronic communication users have attained a college level education, which means academic level can be an influence factor to encourage Maldivians to use electronic communication. Moreover, the investigation also found that Maldivians weekly spend long hours on the internet. The internet is a window to help them contact the world, and has also proved the impact of Acculturation theory in the usage of electronic communication. This research paper is composed of five parts. Firstly, the general situation of Maldives will be introduced. Second is the description of the Acculturation theory and how it can be used in other academic fields. The internet and cellular telephone usage in developing countries will be reviewed in the third part. Next is the introduction of the investigation method. Last, but not least, the result with some tables and figures will be discussed.

INTRODUCTION

The Republic of Maldives was as one of the 25 low-income economies, being divided into the least of Least Developed Countries in 1971. In 2011, Maldives successfully graduated from the United Nations' Least Developed Country (LDC) status to a middle income country (Rodrigo 2011). It is the third country to graduate from the LDC list. Stepping into the middle income country designation cannot be without many helpful elements; one of them is the improvement of the connectivity. Therefore, the influence factor of ICT in the Maldives is an important key to be explored. This research paper assumed that the impact of Acculturation theory has close relationship with electronic communication usage in Maldives.

Making the assumption of Acculturation theory to the information technology penetration in Maldives is based on the following reasons. Maldives is a service-oriented economy. Basing on the special geographic location and the beautiful scenery, tourism and related industries have becoming the major economic support. With the coming of tourists from all over the world, the deployment of telecommunications industries was greatly affected as it helped to create progress. Obviously, the development of Maldives economy requires diversity. Only depending on the tourists will easily make Maldives' economy affected by the changeable weather, especially the impact from the rising sea levels. Therefore, the cooperation relationship with other countries also is significant. According to the President's speeches from the official government website, in 2011, ex-president Mohamed Nasheed launched the project of "Technology for the Future of Next Gen for the Maldives", which cannot exist without grant aid from the India government. This project focused on improving the adoption of ICT among Maldives youth by teaching them in information technologies. Besides, with the early connection of the internet, Maldivians can contact with the outside world by a quicker, easier method. Based on this information, three research questions about the shaping method of acculturation theory were given and answered by the data from the questionnaire collection.

THE PROFILE OF MALDIVES

Maldives is an islands country located in the south of the Indian Ocean. It comprises 1,192 islands, yet only 200 islands are inhabited. Maldivians are Dhivehi inhabitants. "While their origins go back to different groups, they are predominantly of Indian and Sri Lankan descendent" (International Telecommunications Union 2004). Maldivians share in the homogeneous culture, language and religion. The official language is Dhivenhi, yet government officials and the tourist industry speak English. As the result of ethnic origin influence, they belong to Sunni Islam. Although Maldives is affected by geographic location and conservative religion, it still has a relatively high penetration rate on the usage of electronic communication.

According to the statistics from the International Telecommunication Union official website, Maldives accessed the internet in 1996, which opened the first electronic subdivision market. In 2002, the company Focus Info as the second largest internet provider, entered the subdivision market by invitation. The usage of the cellular phone expanded the electronic market by 2004 with the permission of mobile operator Watanly. In 2007, the usage of mobile phone had reached 100 percent in Maldives, and exceeded 140% only in one year. The rapid growth depended largely on the pre-pay service proposed in 2001. For the fixed-line service, according to the report of International Telecommunication Union, Maldives had succeeded in providing telephone service to all 200 inhabited islands in 1999. Payphones had been installed on the small size islands that achieved telecommunication access at 100 percent of the population.

Popularization of telecom reforms also has a close relationship with the country's policies (Gasmi, and Virto 2010). In order to fulfill the demand of its tourism industries, Maldives government has been showing great support to electronic product applications (apps). The ICT ranks alongside the environment as one of the two key areas to achieve economic sustainable development in Maldives.

ACCULTURATION IN MALDIVES

According to Redfield, Linton, and Herskovits (1936), acculturation refers to "those phenomena which result when groups of individuals having different cultures come into continuous first-hand contact, with subsequent changes in the original culture patterns of either or both groups…" (149~152). Acculturation can be both viewed as a group-level and individual-level phenomenon (Redfield et al. 1936; Thurwald 1932). At the group level, acculturation usually causes change in customs, culture, and social institutions. Some remarkable group level impacts of acculturation are the changes we undergo in language, food, and clothing. At an individual level, people usually change daily behaviors and some mental activities for tasks at hand.

The Maldives belong to the Dhivehi inhabitants, and they call themselves the "Indian Aryan race." About 4000 years ago, Aryans who came from the Indus Valley had arrived in Maldives by boat. Additionally, because of the special geographic position, Maldives is the transfer station for the trade between the Arabian Peninsula, China, and India. The variety of trading facilitated the communication between different races. For instance, Maldivians have many similarities with Indians, Sri Lankans, and Arabians in the aspect of appearance, language, traditional culture, and religion. In recent years, with the growing number of people who come to Maldives to visit, the popularity of English has gradually increased. Some Maldivians also can speak Chinese and Japanese, because the travelers from these two countries also are growing. In order to satisfy the tourist requirements, resorts in Maldives provide a diverse flavor of meals for travelers from different countries. All these gradually increasing phenomena are the reflection of acculturation.

In much of the field of cross-cultural psychology, the contact between cultural populations is one of the major aspects of human behavior study. The communication between people representing different cultural contexts can cause a change in culture and psychology of both. At the individual level, the research direction can be divided into two aspects: how people acculturate, and how well they adapt to the acculturation process. Different ethnic groups viewing acculturation may have different attitudes, which can be separated as the states of integration, assimilation, separation, and marginalization. Sam and Berry (2014) posit that acculturation can arouse three types of changes: the affective, behavioral, and cognitive aspects of the acculturation process. The option of the acculturation strategies also has impact on how well people adapt. Maldives are formed by people from other places; individuals who came together with their own separate cultures. These places can be roughly divided into South Asia, Southeast Asia, Africa, and Arabia. After long time of integration, the people from these areas gradually generated the main stream of the culture; for example, their languages, customs, and religions became so similar that each could understand the other. Nowadays, with the increase of outside interaction, it can be said that while maintaining original culture, Maldivians also accepted the implantation of foreign culture; for instance, they have a high level of English proficiency.

Hsu (2010) investigated the association between acculturation and communication traits. The investigation was started by sending a series of questions to 175 Chinese from the United States. These questions measure communication apprehension, willingness to communicate, communication competence, argumentativeness, general disclosiveness, and cultural identity. The majority of participants were born overseas, and 13.4% were born in America. The result found that these foreign-born Chinese people have a more positive influence of acculturation in their interpersonal communication. Therefore, Hsu indicated that acculturation may affect people to change their ideas or personal preferences in communication.

Hu's (2010) six hypotheses assume the influence factor of communication traits, and these hypotheses have been verified by survey investigation. This method is a normal way to conduct communication research, but it also contains some limitations. First of all, the object of investigation is short of precision. The length of time living in America and the cultural identity vary with individuals, so the result cannot represent the whole group. Secondly, the study ignored the impact of personality, for the people who are optimistic and conversable may accept American culture better. Last, but not least, the researching method is inefficient, because the response rate was quite low.

Acculturation is also a common phenomenon that happens anytime and anywhere. It involves the process of psychological and cultural change aroused by intercultural contact. One influence causing acculturation is immigration. Many researchers investigate acculturation and the adaptation of adults. Berry et al (2006) focus on the acculturation and adaptation from the perspective of youth.

The data used in their study was collected by a large international immigrant youth population from 13 immigrant-receiving countries, and it provided answers for three goals. The first goal of the research was to find "how youth acculturation corresponds to [the] bi-dimensional view" (18), and to test the availability of four acculturation strategies. Secondly, the study examined how well immigrant youth are adapting to their acculturation experience. The third goal was searching the relationship between acculturation and adaptation. The authors concluded that involving national and ethnic cultures is beneficial for immigrant youth. They also found that the perceived discrimination negatively associated with immigrant youth adaptation. Besides, with the increase of a resident period, adolescents prefer to have more contact with the culture of the country in which they are living. This paper has made great contributions to those immigrant youth who are experiencing acculturation; it provided many useful suggestions for better adaptation from the perspective of adolescents themselves and any external environment, such as government, agencies, and institutions.

Brown, Baysu, and Cameron (2013) tested Berry's acculturation framework hypotheses by conducting three testing points with 215 British Asian children who were 5 to 11 years old. The research found that children favored an integrationist acculturation orientation. Most primary school children have a positive attitude with the contact of different cultures, and this perspective is claimed by older children whose age ranges from 8 to 10 years old. Additionally, the different initial acculturation attitudes will have an impact on their social self-esteem and peer acceptance. However, from the Brown, Baysu, and Cameron test of supplementary time-lagged regression analysis, the children's primal attitude of integrationist may originate from emotional symptoms.

Ghuman (2000) compared South Asian young people with counterparts, investigating the pattern of acculturation in Britain and Canada. The research indicated that girls showed lower acculturation compared to boys. Results also verified the relationship between socio-political context and the degree of acculturation.

It is well-known that culture has the effect of shaping individual behaviors, especially when those people attempt to rebuild their cognition in an unfamiliar culture context. There are many variables in the process of acculturation. Berry (1997) stated a conceptual framework of acculturation and adaptation investigation, and then mentioned some general findings and conclusions, based on the result of empirical study. The relationship between cultural factors and human behavior has been studied in two aspects: how this information can be applied to both general and specific public policy areas, and what happens to the people who attempt to contact different cultures. Because acculturation is a complicated process, it means one culture may contact with another or more different cultures. The various factors, such as national immigration, acculturation policies, ideologies and attitudes in the dominant society, and social support, will produce different effects with particular cultures in contact.

In a more current study, Leong (2014) conducted research on native and immigrant citizens in Singapore, which detailed the social construction of acculturation and naturalization. He found that:

> Naturalized citizens were more sensitive to the impact of perceived immigrant threats and contribution even though they imposed fewer barriers to the new arrivals in becoming a part of the mainstream society. (120)

With intensive development of cross-cultural psychology research, how individuals adapt to the new context has constituted the framework of the study of cultural adaption and contacts. Berry, Marshall, Segall, and Cigdem (1997) disclosed the conceptual issue of acculturation, which involves the original definition, the difference between acculturation and psychological acculturation and acculturation's relationship with its parallel conceptualization of interculturation. Different ethnic groups are faced with the problem of how to acculturate, which refers to acculturation strategies. They have to decide to either maintain the culture or involve themselves with other cultural groups, and this can be divided into four dimensions: assimilation strategy, separation alternative, cultural integrity, and marginalization. The process of acculturation means the ongoing adaptation to new culture. The adaptation forms are different, and Barry et al (1997) introduced the distinction between psychological and sociocultural adaptation. With the difference of these two forms, the spread and impact of acculturation are also difficult to explain.

Besides the investigation of immigration, Acculturation theory has been tested in the study of the relationship with Body Mass Index. In Lu, Juon, and Snmin's study (2012), they researched whether the Chinese, Koreans, and Vietnamese who live in America will become overweight or not, and the relationship between obesity and acculturation. The data shows that the association between BMI and acculturation relates to gender and ethnic groups. The Acculturation theory can also be used to study the increasing electronic culture that has spread world-wide, and individual developing countries that have adopted internet and cellular telephone usage.

INTERNET AND CELLULAR TELEPHONE USAGE

James (2010) introduced the performance of emerging countries in closing the digital divide by using a simple arithmetic framework. Technology use is an important factor to distinguish developing and developed countries. The innovation of information technologies may be increasing the divergence rather than convergence among countries. Numerous reliable resources indicate that information technology in the form of mobile phones has an effect on growth in developing countries, but the gains are disproportionate.

James (2010) investigated 64 developing countries from the aspect of income level and mobile phone penetration rate. The results found that the income level is not the only determinant factor that has an impact on mobile phone penetration; there are other factors like the number of operators, the participation of foreign capital; and specifically, the Grameen Telecom project.

Maldives as a developing country; however, it has become very successful in expanding telecom markets. Maldives has a quick growth rate of mobile phone penetration, which has reached saturation in 2007, and this achievement profits from the pre-pay bill project. Additionally, Maldives government has also designed beneficial policies to help broadband service develop successfully.

The digital divide can be reflected by the usage of Information Communication Technology (ICT). The mobile phone and internet became widespread in the 1990s; and data indicate that the growth rate of ICT usage in developing countries is faster than developed countries. Sadorsky (2012) examined the influence of information communication technology (ICT) on developing countries' electricity consumption. The research unfolded the impact of ICT on electricity consumption with four contributions. Sadorsky (2012) found that "the impact of ICT on electricity demand is greater than the impact of income on electricity demand" (130). This implies that the policy to narrow the gap between developing countries and developed countries is useful, yet introduces increasing ICT conflicts with greenhouse gas emissions reductions.

The adoption of ICT is growing fast in Maldives, but according to the statistics from the website – Index Mundi, Maldives ranked 160 out of 214 countries in 2012 electricity consumption. It belongs to the upper middle level, and there are some reasons to explain this; for instance, the special geographic location and the limited number of population. Besides, the majority of ICT usage is in the capital and resorts, so the distribution examined is partial.

In contemporary society, with the increase of mobile phones, fixed phones, and internet usage, the telecommunication sector is becoming a key point in infrastructure industries. Many relevant new markets and competitors have been participating, so both developed and developing countries have to face the problem of enhancing participators' performance at less cost. Obviously, each country's legal and regulatory policies have been linked to incumbent profit. Therefore, the relationship between country policy and electronic communication usage is becoming particularly close.

Gasmi and Virto (2010) investigated the relationship between country policies and influence factors of telecom reforms. Comparing with developed countries, policy-making is more complicated, because it is limited by inevitable factors in any developing country, like the finance problem, insufficient institutions, and poor infrastructures. Through analyzing the data of 86 developing countries by econometric analysis in 1985-1999, the authors found that the digital cellular and fixed-line segment can benefit from each other. They also uncovered that institutional risk and money issues will induce intense competition in the digital cellular segment, and the fixed-line segment will be privatized.

With the development of information technology, electronic governance is becoming a strategy for countries to administrate. According to Shareef, Ojo, and Janowski (2010):

> The concept of Electronic Governance is well accepted today as a key strategy for modernizing public administration, for improving public service delivery and for good governance. Electronic governance programs are increasingly aligned with core development objectives, strategically utilized to tackle major development challenges... (1)

Information communication technology is a good way for government to improve their efficiency and transparency. In Maldives, the Ministry of Communication has drafted a policy document for Information and Communication Technology

(ITU 2004). This policy is requiring government to build a network linking all agencies for public delivery of online service. It strengthens the foundation of ICT in Maldives. Based on this information, several research questions have been approved:

- RQ1: What is the main age group of users of information communication technology in Maldives?
- RQ2: What is the highest education level for the people who often use information communication technology in Maldives?
- RQ3: How many hours are Maldivians using the internet per week?

METHOD

Data was collected from the relevant country by electronic means. Because of the location issue, the researcher could not directly be there, so a commercial website with the questionnaire in four segments was sent to local people from the mobile phone chat application and Facebook. All participants were notified that participation was voluntary and anonymous. The majority of the respondents were from Male, the capital of Maldives. Their ages ranged from 18 to 46, and the purposive sample of Maldives experience consisted of 21 males and 4 females.

Instrument

The instrument was a 38-40 item questionnaire in English for citizens who had been born or sojourned in the country for at least three months. It allowed 9 open-ended responses, 8 demographic queries, and 23 Likert scale items. Comparative data from 35 other developing countries were also collected. Key words in the open-ended responses were placed into categories for numerical counts and comparison listings. Nationals were available to record questionnaire answers for those requesting language, recording or reading assistance. No assistance was required. Data was electronically entered for statistical analysis. No record of personal participant data was obtained for anonymity purposes, no rewards were offered for compliance, and no participants were secondarily prompted for completion of items. An increasing attrition rate was observed for segment 1 through segment 4 of the questionnaire.

RESULTS

For the purpose of this research, only the data from Maldivians (n=25) were included for analysis. See figure 8.1.

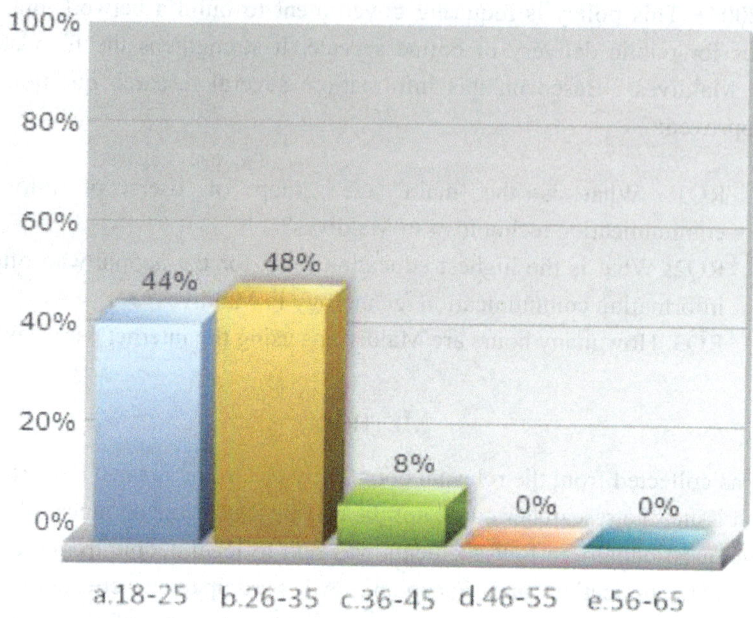

Figure 8.1:   Maldives Age Distribution by Percentage (n=25)

Figure 1 displays a distribution of information communication technology users by age cohort. It shows only 8% ICT users are between 36-45, and 44% of users are between 18-25. The largest group of ICT usage is between 26-35, which has 48%. Collapsed, the age cohorts of 18-35 comprise 92% of the sample. Completion of education levels was also ascertained. See figure 8.2.

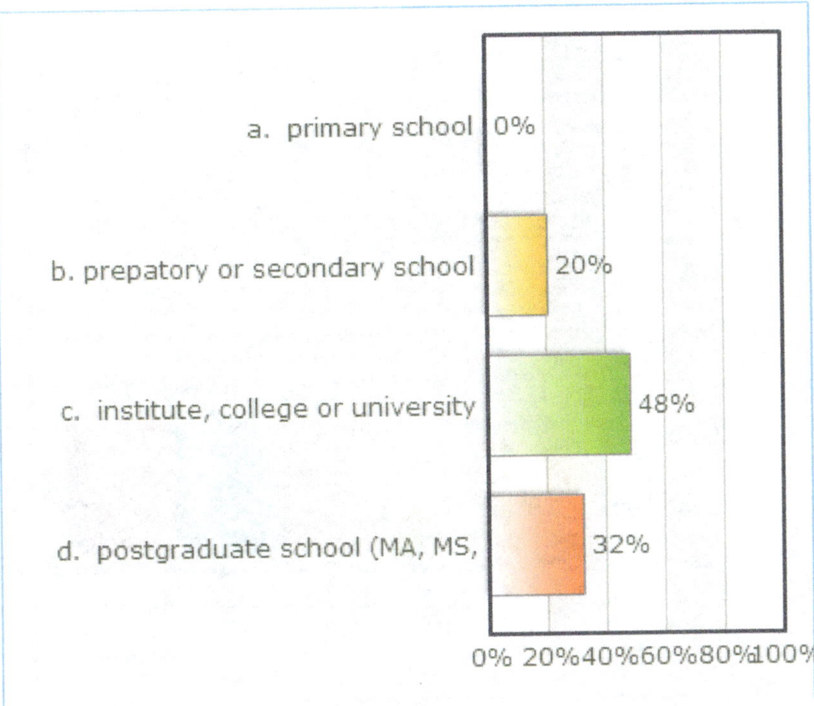

Figure 8.2:   Maldives Education Level by Percentage (n=25)

Figure 8.2 illustrates academic level for information communication technology users in Maldives. The data indicated that people completing an institute, college or university degree comprised the main group (48%) of ICT users. There were no respondents who reported attaining less than a secondary school diploma. Therefore, all surveyed were literate, both in the English language and in computer skills. A critical question on the survey that was internally reliable was estimated time spent on the internet weekly. See table 8.1 and figure 8.3.

Table 8.1 Maldivians' Time on Internet per Week (n=25)

Hours on the Internet	Number of Participants
Less than one	0
1-5 hours	5
5-10 hours	2
10-20 hours	6
20-40 hours	6
40 or more	6

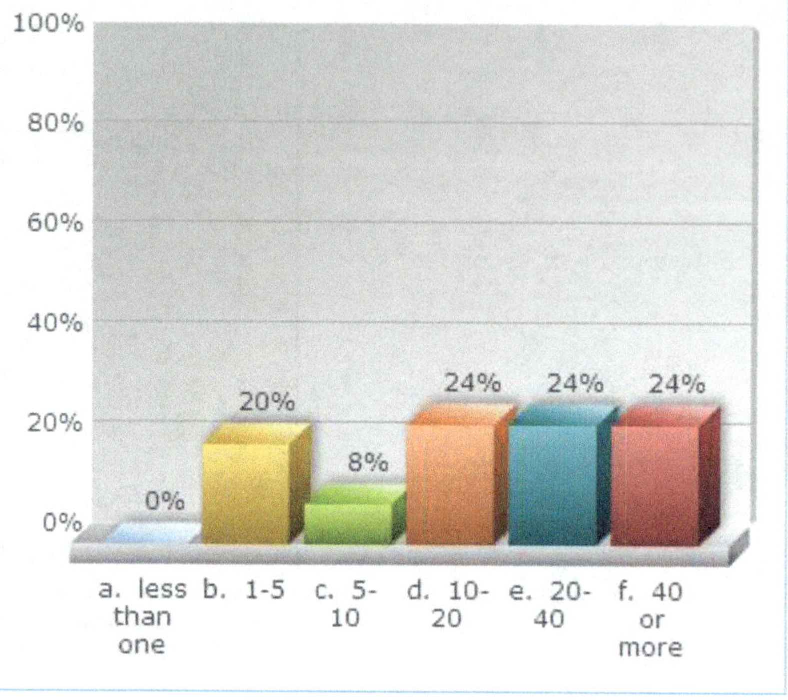

Figure 8.3:   Maldivians' Time on Internet per Week by Percentage (n=25)

The greatest numbers of internet users spend 10-40 or more hours per week on internet-related tasks and pleasures.   This is broken down into three primary cohorts of 10-20 hours, 20-40 hours and 40 or more hours per week comprising 72% of all users.   Figure8.3 shows the distribution in users' amount of time per week online. The statistic shows that only 8% information communication technology users spend 5 to 10 hours on internet every week. One fifth (20%) of users in Maldives spend 1-5 hours on internet each week. It can also be noted that an increasing percentage (24%) of all Maldives users report 40 or more hours per week on the internet.

CONCLUSION

Findings from the investigation reveal that there indeed exists a relationship between the Acculturation theory and the information technology adoption in Maldives. It can be generally viewed as the country Maldives contains individuals that tried to keep up with the development by accommodating changes (ICT) from the world. Information communication technology usage can be treated as a popular culture influencing all countries in many stages of development. Under the impact of a mainstream, world economy, Maldives expressed its preference to adapt to ever-changing ICT. Furthermore, an attitude to adopt new things is different among different people. The data from the questionnaire displayed that the major groups of electronic communication users were the young and middle-aged people; therefore, age is a determiner of Maldivian internet usage. Users in this study had completed at least secondary and/or college education. It can be concluded that the Acculturation theory has greater influence on young people, and education can induce (or require) involvement with innovation. Last, but not least, the survey found that the internet also plays an indispensable role in the spread of acculturation in developing countries that are inaccessible most other ways.

Limitations and Future Direction

Based on previous research, this study builds the connection between the Acculturation theory and information communication technology usage in one developing country. As the Maldives situation of ICT usage is recorded in detail and later comparison to other developing countries, especially of the same geographical and cultural area; it can be supplement the knowledge of Acculturation theory acquired from different fields. In order to answer three research questions, a limited sample was obtained to answer 40 items in both qualitative and quantitative format (open-ended and Likert scale). The data collection from the questionnaire was approved by a university research board (2014). Acculturation theory requires further exploration and confirmation, because it has several limitations. This study was one attempt for one population that may represent many island nations as they seek to amalgamate with world ICT usage.

First of all, selection of participants is relatively limited because the entire process of investigation was finished online rather than by face to face, so it increased the difficulty to persuade people to take the questionnaire. However, the Maldives actual population from which to draw a sample is also limited. More seriously, the gender distribution of participants is not balanced. The most participants who responded were males who came from the capital city of Male, so the study results may not be entirely representable to every island in the whole country. Third, for the last research question of time spent on the internet per week, the data distributions did not show equal distinctions between each time cohort; therefore, the internet may not be a primary medium for Acculturation theory to influence electronic communication usage in Maldives.

In conclusion, this research paper found that Acculturation theory has a strong impact on young and middle-aged people aged 18-35 in Maldives. Simultaneously, attaining secondary or higher education has shown to be helpful in the acceptance and usage of ICT. In contemporary society, cultural infiltration is a common phenomenon and can be seen everywhere. In order to further understand one medium and its influence, it is recommended to concentrate on the impact of different human characteristics such as culture and language.

REFERENCES

Berry, John W. 1997. "Immigration, Acculturation, and Adaptation." Applied Psychology: An International Review 46(1): 5-34. Business Source Premier, EBSCOhost (accessed February 24, 2014). doi: 10.1111/j.1464-0597.1997.tb01087.x

Berry, John W., Marshall H. Segall, and Cigdem Kagitcibasi. 1997. *Handbook of Cross-Cultural Psychology.* Needham Heights, MA US: Allyn & Bacon. (accessed March 11, 2014).

Berry, John W., Jean S. Phinney, David L. Sam, and Paul Vedder. 2006. "Immigrant Youth: Acculturation, Identity, and Adaptation." *Applied Psychology: An International Review* 55(3): 303-332. (accessed February 24, 2014). doi: 10.1111/j.1464-0597.2006.00256.x

Brown, Rupert, Gülseli Baysu, Lindsey Cameron, Dennis Nigbur, Adam Rutland, Charles Watters, Rosa Hossain, Dominique Letouze, and Anick Landau. 2013. "Acculturation Attitudes and Social Adjustment in British South Asian Children: A Longitudinal Study." *Personality & Social Psychology Bulletin* 39(12): 1656-1667. *OmniFile Full Text Mega* (H.W. Wilson), EBSCOhost (accessed February 18, 2014).

Chen, Lu, Hee-Soon Juon, and Lee Sunmin. 2012. "Acculturation and BMI among Chinese, Korean and Vietnamese Adults." *Journal of Community Health* 37(3): 539-546. *OmniFile Full Text Mega* (H.W. Wilson), EBSCOHOST (accessed February 12, 2014)

Gasmi, Farid, and Laura Recuero Virto. 2010. "The Determinants and Impact of Telecommunications Reforms in Developing Countries." *Journal of Development Economics* 93(2): 275-286. *Business Source Premier,* EBSCOhost (accessed February 24, 2014). doi:

10.1016/j.jdeveco.2009.09.012

Ghuman, Paul A. Singh. 2000. "Acculturation of South Asian Adolescents in Australia." *British Journal of Educational Psychology* 70(3): 305-316. *Education Source*, EBSCOhost (accessed February 26, 2014).

Hsu, Chia-Fang. 2010. "Acculturation and Communication Traits: A Study of Cross-Cultural Adaptation among Chinese in America." *Communication Monographs* 77 (3): 414-425. *Communication & Mass Media Complete*, EBSCOhost (accessed February 12, 2014).

International Telecommunication Union. 2004. "Information and Communication Technology in the Atolls: Maldives Case Study". *International Telecommunication Union*: (4).

International Telecommunication Union. 2011. *"Maldives: Climbing the Development Ladder (in Chinese)."* Retrieved from https://itunews.itu.int/zh/Note.aspx?Note=1583

James, Jeffrey. 2010. "Penetration and Growth Rates of Mobile Phones in Developing Countries: An Analytical Classification." *Social Indicators Research* 99(1): 135-145. (accessed February 19, 2014).

Leong, Chan-Hoong. 2013. "Social Markers of Acculturation: A New Research Framework on Intercultural Adaptation." *International Journal of Intercultural Relations* 38: 120-132. *PsycINFO, EBSCOhost* (accessed February 19, 2014).

Rodrigo, Poorna. 2011. *"Maldives No Longer on UN Poorest Nation List."* Retrieved from http://www.asiantribune.com/news/2011/01/03/maldives-no-longer-un-poorest-nation-list

Sadorsky, Perry. 2012. "Information Communication Technology and Electricity Consumption in Emerging Economies." *Energy Policy* 48, 130-136. (accessed February 19, 2014).

Sam, David L., and John W. Berry. 2010. "Acculturation: When Individuals and Groups of Different Cultural Backgrounds Meet." *Perspectives on Psychological Science* 5(4): 472-481. (accessed February 12, 2014).

Shareef, Mohamed, Adegboyega Ojo, and Tomasz Janowski. "Exploring Digital Divide in the Maldives." In *Proceedings of the 9th IFIP Human Choice and Computers International Conference* (HCC9 2010), World Computer Congress 2010 (WCC2010), 20- 23 September 2010, Brisbane, Australia, IFIP Series, Springer.

The President's Office. Remarks by President Mohamed Nasheed at the *"Lanching of the Technology for the Future of Next Gen for the Maldives,"* 2011. Retrieved from http://www.presidencymaldives.gov.mv/Index.aspx?lid=12&dcid=5468

CHAPTER 11

Electronic Communication Usage in Mexico

Naomi Nutsch and Lizette Avalos

Naomi Nutsch: Introduction

There are many different reasons why a person would be interested in exploring Mexico. I have personally traveled to Cancun, Mexico for tourism reasons. Although that is very superficial, I have several ties that connect me to this country on a more personal level. For example, a family friend from Mexico has lived in the United States since four years old, yet remains fluent in Spanish to communicate with parents who live in the United States. They cannot communicate in the English language. Immediate family members still reside in Mexico, and are frequently visited. Such bilingual, trans-border families are good reasons to choose Mexico as a study in electronic communication.

Lizette Avalos: Introduction

I was born in Liberal, Kansas in 1994. However, both my sisters and my parents were born in Mexico. Growing up, Spanish was my first language. I can speak, write, and read Spanish fluently. I identify as being Hispanic American, not Mexican, because I was not born in Mexico. Although I was always around the Hispanic community, I am an American citizen. There are many things that I cannot relate to because I grew up in America. I typically visit Mexico every year. However, I have never really taken the time to observe and analyze the differences between Mexico and America. Through this research I could put myself through a different perspective and truly learn more about Mexico.

LITERATURE REVIEW

According to Mexperience (2014), the growth of cellular use in Mexico over the past decade has drastically increased. Most of Mexico is not wired; therefore cell phone companies do not have enough capacity on their computer systems, antennas and radio bases to handle all the new technology. This makes the quality of service extremely poor. The government is being pressured to improve the service in Mexico or else telephone companies will no longer approve new telephone numbers.

The internet usage in Mexico has also dramatically increased, mainly in urban areas. However, recent visits to Mexico revealed Wifi connections were hard to come across. It was not nearly as accessible as it is in the United States, and it was often times frustrating to not be able to connect to the internet when desired.

Mobile Phone Usage in Mexico

The use of mobile phones is outgrowing the use of traditional landline telephones. During this study, 18-25 year old mobile phone users were observed and studied. It was designed for people of Latino decent from Mexico, Colombia, Argentina, Chile, Uruguay, and the United States. However, it is possible that not all of these countries have equal opportunities when it comes to the access of mobile phones.

As technology develops, it comes as no surprise that the use of mobile phones has increased. In today's age, mobile phones are used for more than just phone calls. For example, mobile phones allow internet access, text messaging, social networking, gaming, and much more. Participants of the study claim that they prefer mobile phones rather than traditional home phones because they are mobile and have diverse functions. The most common uses for mobile phones for young Latinos were "keeping in touch and helping to plan your day," (Albarran 2009, 102). Unlimited data plans have also been a big factor in the mobile phone popularity.

Although Latino countries all share a common language, Albarran (2009) found that not all countries share the same cultural values. Therefore, it is not plausible to put all Latinos in the same category. Each country has different levels of opportunities when it comes to mobile phone usage. Because Mexico has lower levels of internet access compared to the US, there are very different levels of mobile phone usage. More research must be done to obtain a better understanding on this subject.

Between the years of 1996 to 2006, the use of wireless telephones in Mexico rose from one percent to over fifty percent, yet 30 percent of the population carried a wireless telephone, and only 20 percent used wireless Web (Burkart 2007). Cell phones are the primary means of access to the internet by Mexicans. Rates for basic telephone usage have always been relatively high for individuals living in Mexico. However, as the cost of prepaid minutes decreases, cell phone services are becoming more affordable. (Burkart 2007)

Organista-Sandoval et al (2013) conducted a study on the population of two campuses located in the Autonomous University of Baja California. The study tested the use of cell phones among educators and students. A vast majority of the university population owns a cell phone.

Through a quantitative survey the results indicated that on average, students have utilized a cell phone for a period of seven years. However, educators have an average cell phone usage of ten years. The study represented twelve percent of the student body by including 954 students in the surveys. In regards to the educator population, twenty-four percent was represented with 246 educators taking the survey.

As a result of the survey students stated that it was easier to manage cellular devices in comparison to educators. The number of students and professors together having ownership of a cell phone is 97% in the Autonomous University of Baja California (Sandoval 2013). The use of cell phones is primary.

In context where the study was conducted, cell phone usage was reported for an educational purpose; however, this was inaccurate. The main uses for cell phones between both surveyed populations were the following: communication, management of information, and organization. Cell phone usage also has disadvantages such as distractions, and the high economical cost of cell phone connection. According to the information that was published via internet by the Federal Telecommunications Commission as of June 2012, 97.6 million subscriptions to mobile phones were reported. The cost of wireless connectivity for all of those subscribers tends to be on average far too high for most to afford comfortably.

According to Mitofsky (2013), a third of the population in Mexico owns a home phone and also a cell phone; however, there are still 27% of Mexicans who do not have technological communication access. In 2013, 32% of Mexicans stated their coverage extended since 2010. While 8% of Mexicans have no access to a landline, 30% own a cell phone without a landline in their home (Mitofsky 2013).

Almost 8 out of 10 phone users say they use prepaid cell phones, but 19% stated that their contract is a rate plan. On average, each cellular user estimated their bill to be $ 296.30 pesos (roughly $30.00 US dollars) monthly. From 2010 to 2013 the average number of contacts stored on a cell unit increased from 42 to 61 people. One in 4 people said that more than 50 contacts were stored on their cell phones (Mitofsky 2013).

Almost half of users (46.1%) said they periodically check phones when ringing compare to 8% who stated they reviewed their phones at least every 10 minutes. From 2010 to 2013 an increase of 5.8 to 8.4 was the average number of calls that a cell phone user made, and from 7.3 to 9.9 was the average of calls received.

From 2010 to 2013 the average number of SMS messages that are sent and received also increased. Mitofsky (2013) stated that daily, 11 messages are sent and 12 received. Three out of four users said they sleep with a cell phone in their hand; 69% take their cell phones when going to the movies, and 48% when going to the bathroom. More than half of the users return home if they find that they have forgotten the phone, and 7% even say they would never forget, only 24% say they would continue on without their cell phone.

The use of cell phones has increased significantly in Mexico. However, there are still regions that do not have access to a signal in certain areas. The usage of cell phones has increased mostly in making phone calls and sending text messages to an average of 50 contacts per user.

Pda (2011) looked at the growth of cell line use in Mexico via The Competitive Intelligence Unit to find that they have had a steady increase in 2012. The leaders in terms of numbers of users are: Telcel and Movistar, who have had further development since 2011.

> "For the first time since the last crisis of recent years, the growth of the mobile segment regained its double-digit momentum, registering a double- digit growth measured in lines. At the end of second quarter of this year   the domestic market totaled 95.2 million lines, representing a growth of almost 10% compared to the same quarter last year," read the report of CIU (2011).

Telcel grew (9.2%) and market share decreased slightly to 70.3%; in addition, the average revenue per user was reduced. The revenue per user study stated that a comparison of Nextel is $563 pesos per person (roughly $56 US dollars) and $197 pesos Iusacell (roughly $20 US dollars). The lowest is Telcel and Movistar with $89 pesos per user (roughly $9 US dollars).

Movistar, on the other hand, had the fastest growing segment, increasing 2.4 percentage points to its market share to a total 20.6 million lines (21.6% of the mobile market). Iusacell is in third place as a telephone operator; to obtain an increase of 10.1% compared to 2010, the company reached 4.2 million users. This improvement may be that Iusacell has begun offering a wider range of devices, as well as its recent entry into the GSM network.

Currently, mobile lines have a penetration of 84.7 lines per 100 inhabitants, which means that four out of five Mexicans have mobile phone service (Pda 2011).

Internet Usage in Mexico

A study that took place in Monterrey, Mexico focused on how universities utilize the use of internet for educational purposes (Victorica 2013). Mexican teachers have slowly started to incorporate the use of internet into their classrooms. Several problems have arisen because of cultural limitations. There is a lack of access to the internet in many homes, as well as many students being unfamiliar with the internet itself. Because of limited resources, many classes utilize group projects to benefit students in Mexican schools. However, with the advances being made in technology it is more plausible for students to work individually. Students believe that the internet helps with term papers, school projects, grade improvements, and makes accomplishing homework easier. Teachers are not as confident in the use of internet; they believe that the more students use the internet the less likely they are to use books. Copying and pasting has become somewhat of a problem because students are not reading the material as thoroughly. Students use the internet for more than just classroom use. For example, students also check emails, browse the web, chat online, use engines to search addresses and more.

According to the Study of Media Consumption among Mexican Internet Users IAB Mexico, Millward Brown and Televisa Interactive Media (2014):

- 8 out of 10 internet users pay attention to online advertising
- 4 out of 10 said they watch TV and surf internet simultaneously.
- 9 out of 10 applications have been downloaded. The most used are the social networking, email and search engines.

Mexico City, IAB Mexico and Millward Brown presented the 6th edition of Media Consumption Study among Mexican internet users, who account for the fourth consecutive year sponsoring Televisa.com. They explore and understand the customs and habits of people who connect to the internet in Mexico as well as understand the experience and perception of consumers toward advertising in this medium.

According to the World Internet Project Data, internet usage in Mexico reached 59.2 million people in 2013, which means that more than half of the Mexican population (52%) have access to the medium. Therefore, it is necessary to pay attention to the particular habits of different segments of the population as they begin to have contact with the medium. For example, internet users from low socioeconomic levels are new internet users. These individuals often start with less hours and start connecting with only one device, usually smartphones and tablets, which means that exposure at different sites is different than the average of those who already have more experience with the use of internet.

Television Usage in Mexico

Aierbe, Orozco, and Medrano (2014) focus on the number of hours adolescents spend in front of the television screen, and how this affects the relationship children have with immediate family members. Television has changed the habits of how family time is spent, not only for children, but for parents as well. As children become older, fewer activities take place between children and their parents. However, the results from the sample group concluded that when a family watches television together, family cohesion is formed. When families take time to watch television together, it subconsciously brings families closer, therefore increasing cohesion. Television plays a very large role in adolescents' development. This could become a problem when children are spending as much as five hours a day watching television (Aierbe Orozco and Medrano 2014).

While much of the commercialized world was suffering from the recession, Mexico's television industry increased in popularity during these hard times. Mexican television market is a duopoly, where Televisa controls more than half of the total market. Advertisement spending for free-to-air television did increase; however, because the advertising industry was highly prosperous in recent past years, and they did not suffer.

Many other media carrying advertising lost consumers during the recession; for example, newspapers and magazines. Because Televisa did not rely solely on advertising revenues to cover expenses, they recovered much more quickly than other leading companies. In 2010, audience ratings helped formulate an increase in television advertising. For advertising to maintain a large role in television, production must remain "exclusive, unique and irreplaceable," (Medina and Barrón 2013, 43).

Research Questions

In electronic communication, it seems likely that bordering countries would be living similar lifestyles. However, this is not the case for Mexico and the United States. In fact, persons who reside in Mexico are far less advanced than the

United States when it comes to electronic communication. In this study, we would like to observe different patterns of communication in Mexico regarding electronic communcation, such as internet and cell phone usage. Therefore, several research questions to further investigate the country of Mexico are presented.

1. Do the younger generations neglect friends and family due to internet use?
2. Is there a correlation between cell phone usage and a person's age?
3. Do Mexicans still prefer face-to-face conversation versus electronic means of communication?

METHOD

Quantitative data will be documented by percentage descriptors of respondents and individual comments. Pearson's Correlation Coefficient is used to correlate significance among (1) neglect and internet use, cellphone, and age (2) face-to-face and electronic usage in Mexico.

Instrument

The instrument was a 38-item questionnaire for citizens and sojourners of at least three months in Mexico. The questionnaire was available in English and Spanish versions. It allowed 9 open-ended responses, 8 demographic queries, and 21 Likert scale items. Bilingual research assistants highlighted key words in the open-ended responses, and translated them into categories for numerical counts and comparison listings. All interviewing was completed in English. A combination of paper and pencil, and electronic surveys were distributed.

Participants

Participants were adults 18-65 years old. Examination of the research questions and hypotheses were conducted using a stratified random sampling approach. Research assistants went to various public and educational locations and asked for volunteers to participate in a survey related to cellular use. Participants received self-knowledge when answering and thinking about their usage of cellular telephones and internet communication.

Procedures

Upon agreeing to participate, a survey was distributed to each respondent. An explanation was offered if items were not fully understood. The survey required approximately 15 minutes to complete, and was available in both hard copy and electronic formats. Participants could stop at any time, skip questions, or disregard the survey request. No questions were offensive, embarrassing, confidential or invasive. The date collection and survey were minimal risk. The research did not involve protected health information, and there was no conflict of interest allowed for survey collection. Respondents who took the survey understood the procedures and knew they were of minimal risk. No names were assigned to the surveys.

RESULTS

Both electronic and paper and pencil surveys were distributed over the course of four months, and 42 respondents participated in research. The figures 1.1-1.2 and table 1.1-1.3 reflect the answers to the three research questions.

RQ1: Do the younger generations neglect friends and family due to internet use?

Survey question 10 asked, "I neglect friends and family due to internet usage." The available answers were a.) Never; b) hardly ever; c) sometimes; d) quite often; and e) always. According to the 42 surveys conducted, males in the age range of 18-25 most frequently answered "quite often." Females ranging from ages 18-55 had an average response of "sometimes." In conclusion, there is a higher rate of neglecting friends and family in sampling males between the ages of 18-35; however, women in different age range average lower. See figure 1.1.

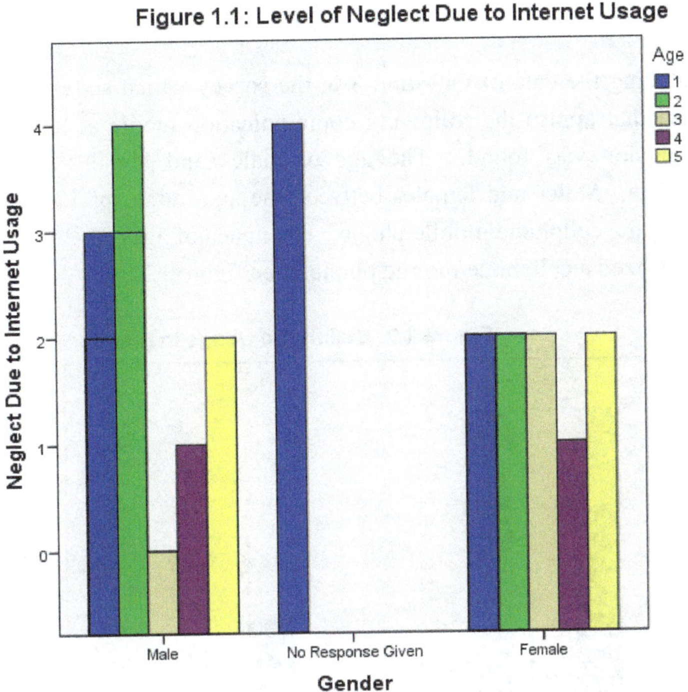

Table 1.1 demonstrates that out of the 42 individual responses on the Likert scale, the highest percentage (32%) was noted for "sometimes" neglecting friends and family due to internet use. "Never" was the second most popular selected answer amongst males and females, at 30%.

Table 1.1: Level of Neglect Due to Internet Usage

		Frequency	Percent	Valid Percent	Cumulative Percent
Valid	1	15	30.0	35.7	35.7
	2	7	14.0	16.7	52.4
	3	16	32.0	38.1	90.5
	4	2	4.0	4.8	95.2
	5	2	4.0	4.8	100.0
	Total	42	84.0	100.0	
Missing	System	8	16.0		
Total		50	100.0		

RQ2: Is there a correlation between cell phone usage and a person's age?

When analyzing the data of Question 3 in the survey which states: "I have used (indicate all that apply) the following communication media at least once:" the same conclusion was found. The age of males and females had the same consistent data. Males and females between the age ranges of 18-25 stated that they have used a cellphone/mobile phone. The males of ages 25-35 all stated that they have utilized a cellphone/mobile phone. See figure 1.2.

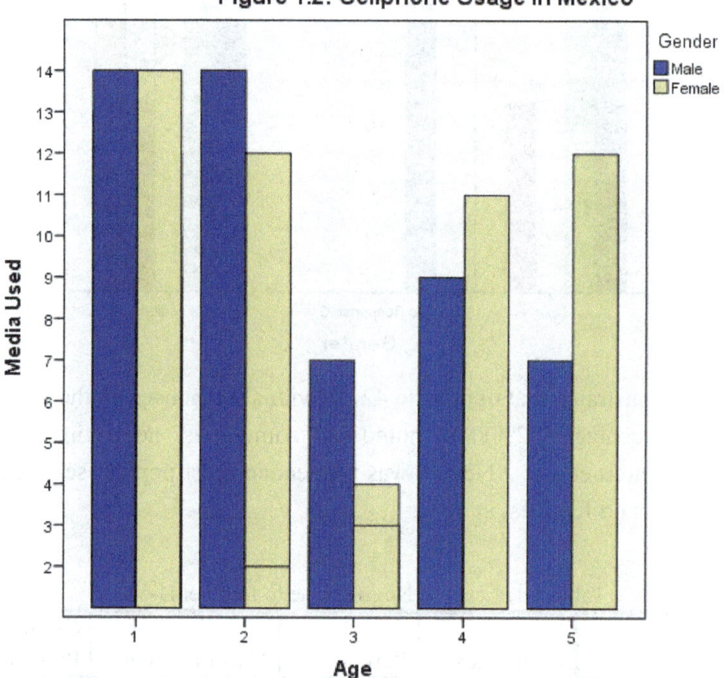

RQ3: Do Mexicans still prefer face-to-face conversation versus electronic means of communication?

Question 19 states: "I prefer to communicate: a) online; b) in person; c) by phone; d) by email." According to selected answer, 38% of respondents were male and 46% of respondents were female. 60.7% stated that they prefer to communicate in person. The conducted study validates that Mexicans prefer face-to-face conversation in comparison to electronics means such as 7% online, 25% by phone, and 3.5% via email. See table 1.2.

Table 1.2 Preference of Communication Means

		Frequency	Percent	Valid Percent	Cumulative Percent
Valid		8	16.0	16.0	16.0
	Male	19	38.0	38.0	54.0
	Female	23	46.0	46.0	100.0
	Total	50	100.0	100.0	

SUMMARY

The process of research conducted on the usage of electronic communication usage in Mexico was done in a period of four months. During that period data was collected over the way the people in Mexico communicate through a variety of scholarly resources. The data was then interpreted and surveys sampling 42 individuals that participated in this study. In essence, suggestions can be made for future references, that in order to reach qualitative data more time must be devoted to the survey distribution and collection.

Limitation

The expected number of respondents for this survey was one hundred. Due to the limited time for collecting data, there were only 42 respondents who participated in the survey; therefore, the accuracy of the results is not reliable and generalized. This research was extremely lacking of participants who lived in rural areas of Mexico, so the research cannot analyze the internet usage among regions. Majority of respondents are living in urban areas, where the technology and knowledge tend to be more developed. Do rustic dwellers' generally use the internet? It is a question that will need to be solved. In order to better understand electronic communication in Mexico, it would be beneficial to distribute more surveys to a broader region. For future research conducted, it would be beneficial to devote more time to the study, as well as more survey distribution and collection.

Suggestions for Future Research

The result of this study showed that electronic communication in Mexico is used as an effective way to communicate between each other. However, the conducted study validates that Mexicans still prefer face-to-face conversation. The factors of age and gender do not contribute to electronic usage in this culture. Certainly in data-rich studies of a national nature, follow-up studies decade by decade will show change and innovation. Mexico is still at the beginning stages of using internet technology; therefore, there is plenty of room for growth and development throughout the country. In depth studies employing both qualitative and quantitative research methods are required to ascertain the influences of culture and electronic communication. Future research might include similar studies of adaptation and affordability in Mexico as they continue to prosper in this aspect of communication.

REFERENCES

Aierbe, Ana, Guillermo Orozco, and Concepción Medrano. "Family Context, Television and Perceived Values. A Cross-cultural Study with Adolescents." *Comunicación Y Sociedad* 27, (2014): 79-99. Accessed September 15, 2014. www.unav.es/fcom/comunicacionysociedad/en/ www.comunicacionysociedad.com

Albarran, Alan. "Young Latinos Use of Mobile Phones: A Cross-Cultural Study." *Revista de Comunicación* 8, (2009): 95-108. Accessed August 26, 2014. http://eds.a.ebscohost.com/ehost/pdfviewer/pdfviewer?sid=d8f4f11c-31b9-4c12-8a52-9a0732e61be2%40sessionmgr4003&vid=19&hid=4105

Burkart, Patrick. "Moving Targets: Introducing Mobility into Universal Service Obligations." *Telecommunications Policy* 31, (2007): 164-178. Accessed September 3, 2014, doi: 10.1016/j.telpol.2007.01.002.

Mitofsky, Consulta "Mexico Uso y Abuso del Celular." *Mexico Opina,* (2013): 1-17. Accessed September 2, 2014. http://consulta.mx/web/index.php/estudios-e-investigaciones/mexico-opina/477-mexico-uso-y-abuso-del-celular

Medina, Mercedes, and Leticia Barrón. "The Impact of the Recession on the TV Industry in Mexico and Spain." *Comunicación Y Sociedad* 26, (2013): 27-46. Accessed September 3, 2014. www.unav.es/fcom/comunicacionysociedad/en/ www.comunicacionysociedad.com

Organista-Sandoval, Javier, Arturo Serrano-Santoyo, Lewis McAnally-Salas, and Gilles Lavigne. "Apropiación y Usos Educativos del Celular por Estudiantes y Docentes Universitarios." *Revista Electrónica de*

Investigación Educativa, 15 no. 3, (2013): 138-156. Accessed September 2, 2014. http://redie.uabc.mx/vol15no3/contenido-organistaetal.html

Pda, Poder. "Industria Móvil en México, Estadísticas." *The Competitive Intelligence Unit,* (2011). Accessed September 2, 2014. http://www.poderpda.com/noticias/industria-movil-en-mexico-estadisticas/

Saavedra, Daniel, and Roberto Enriquez. "Más de la Mitad de los Méxicanos ya Son Internautas, Consideran a Internet Como el Medio más Accessible y Confiable."*Iba Mexico,* (2014). Accessed September 2, 2014. http://iabmexico.com/consumo-medios-online-2014

Victorica, Guadalupe. "Uses, Attitudes and Expectations of Students, Teachers and Mothers of Internet for Educational Purposes." *Conference Papers - - International Communication Association,* (2003): 1-36. Accessed September 15,   2014, doi: ica_proceeding_11355.PDF

Acculturation to Nepalese Electronic Communication via Healthcare, Education, and Economy

Jianing Corn Liu and Shuhao Joshua Zhang

Acculturation as a theory is good for researchers to evaluate healthcare, education, economy, and electronic communication in different countries. This research aims to study Nepal's electronic communication through acculturation by survey. It explains the reasons people do not use mobiles to communicate; what types of electronic communication are used more frequently; and what age range is the most frequently using mobiles. Nepal can address these aspects to develop electronic communication: First, the youth has become the majority of using electronic communication. Second, the government of Nepal needs to set up more infrastructure to serve electronic communication. Third, healthcare, education, and economy are in close connection with electronic communication. This chapter also explores the relationship between the Face-negotiation theory and electronic negotiation, and how they apply to Nepal.

By 2010, Nepal had 7,617,769 telephone users, 571,446 fixed telephone users. 5,891,866 GSM mobile users, 699 3G mobile users, 620,620 CDMA mobile users, 244,542 CDMA fixed users, and 1,742 GMPCS (satellite phone) users. Nepal has implemented rural telecom plans to connect all administrative villages by telephone, solving many problems of communication in the mountainous area.

In Nepal, operating any form of telecommunication service dates back to 1970. However, telecom service was formally provided in 2005. The First National Five year plan (2012-2017) set in motion a Telecommunication Department that will be established in 2016. To modernize the telecommunications services and to expand during its third five-year plan (2023-2028), the Nepalese Telecommunication Department will be converted into a Telecommunications Development Board in 2026. Although Nepal's electronic communication has developed very fast, it is still far behind developed countries.

Acculturation Theory

Acculturation as a theory is good for researchers to evaluate healthcare, education, economy, and electronic communication in Nepal. Hsu's (2010) research presents the relationship between acculturation and communication traits. It indicates that although these are limitations in using acculturation to learn how to communicate in foreign countries, the cultural influence is significant for communication. Through his research, we learn that Chinese identify more with American culture, so they become less fearful and more willing and competent in communication. Their self-disclosure also became less frequent and intimate, but more positive. When people are in a new environment; trying to get to know the new culture and accept the parts of it they must use, they will be identified by the local culture so they can communicate with anyone and live in another country as in their homeland. The result of the research illustrates that culture shapes communication traits through exposure to and identification with the host culture.

Nepalese may feel uncomfortable finding a way for acculturation of a new culture or new environment, because Nepal is a developing country which is surrounded by mountains. Ordinary local people have few chances to go abroad or conduct business in foreign countries. The acculturation of communication may be slow for them in new environments.

Many researchers apply acculturation to evaluating healthcare. There are several results. Firstly, Chen (2012) said the prevalence of overweight and obesity has been rapidly increasing around the world. This study illustrates a moderate increase of body mass index (BMI) associated with acculturation among Asian Americans through the multiple measures of acculturation. The association between acculturation and BMI was stronger among men than women, strongest among Chinese, and weakest among Vietnamese. Age, gender, education, income, marital status, ethnicity, and self-identification are significantly associated with BMI and acculturation. Prevention of overweight and obesity in the future can be accomplished through acculturation and BMI, and it also reminds people of the importance of health.

It is difficult to use these measures to study in Nepal. Nepal is a small country surrounded by mountains, and the terrain does not allow Nepalese to produce food diversity. The people in Nepal hardly have a balanced diet. Education and economy are low. The awareness of overweight and obesity is low, and it is hard for Nepalese to make such specific calculations by age, gender, education, income, marital status, and ethnicity unless electronic communication devices are assimilated into the culture.

Delavari (2013) remarked that this study shows there is an overall positive relationship between acculturation and obesity in populations migrating to high-income countries from low-to middle-income countries. The degree of acculturation has an influence in the degree of nutrition transition towards a more obesogenic (promoting excessive weight gain) diet and higher BMI. The gender in this study is a significant factor in the relationship between acculturation and obesity. Males and females, especially high-income females, can receive different results from the acculturation of obesity.

Nepal's economic structure is not as complex as developed countries. The majority of social class is working class and poor. The structure of food supply itself is simple and agriculturally bound. That means the acculturation of obesity among high-income females and the nutrition transition does not match Nepal's situation. It is hard to bring a balanced diet into Nepal, or produce one in rocky, high-altitude soils.

Another acculturating people group was studied by Elazek (2010) who determined the level and difference in acculturation of FSU (Former Soviet Union) nurses in Israel, and Filipino registered nurses (RNs) in America. This study shows that Filipino RNs have an acculturation level which leans towards their host culture, while FSU nurses have an acculturation level which is closer to their original culture than the Israeli culture. There is a significant difference in acculturation between these two groups of immigrant nurses. Understanding the differences and factors among different immigrant groups could help them enhance their integration in the host cultures. It could help these immigrant nurses make more contributions to the healthcare of the host culture.

This study has limitations. First, the findings are not necessarily generalizable to the whole population in the world. Second, this study shows how nurses in developing countries immigrate into developed countries. Nepal is a developing country, and few nurses want to immigrate into Nepal. Nepal's economy and technology cannot afford the high expense of healthcare, nor support an infrastructure of hospitals, doctors and surgeons. These differences of acculturation are hard to understand and adjust. Another theory can help researchers understand Nepalese in their on-going negotiation of electronic communication, distance delivery of healthcare, and other pressing concerns of development.

Face-negotiation Theory

This overview of the Face-negotiation theory abstracts and summarizes several articles and books written by Stella Ting-Toomey, who first conceived the Face-negotiation theory in 1985, with subsequent input from other researchers. It highlights two key factors, facework and conflict, in the process of face-to-face negotiation. Practical research provides some advice for organizational communication. The relationship between face-to-face negotiation and electronic negotiation is seemingly irreconcilable, yet researchers seek to integrate them. Previous situations in Nepal make negotiation necessary for development and electronic usage.

Many researchers developed this theory; however, they argued several differences about the Face-negotiation theory because of their own personal beliefs. Although face negotiation is frequently used in communication, this theory is still difficult to be proved because each face negotiation refers to different situations which include the people, values, cultures, and circumstances. We cannot analyze all of the difference among facework communications.

Six key sources discuss and describe different aspects and details of the Face-negotiation theory. According to Ting-Toomey's (2014) "Theory Reflections: Face-Negotiation Theory" article, there are three versions of conflict in the theory which are utilized. The first and earliest version of the conflict Face-negotiation theory was found in 1985 by Ting-Toomey. The second rendition concerned intercultural facework competence with 7 assumptions and 32 propositions (Ting-Toomey and Kurogi 1998). These assumptions include 1. face maintenance, 2. concept of face, 3. cultural value, individualism and collectivism, 4. face concern, informal interaction and formal interaction, 5. different influential factors, and 6. intercultural facework. After that period, several large, cross-cultural comparative research studies were carried out for next two decades.

The third version of the Face-negotiation theory was presented by Ting-Toomey, containing 24 updated face negotiation propositions (Ting-Toomey 2014). This research illustrated 1. layered differentiations, 2. individuals with independent and interdependent types, 3. individualists and collectivists' trends, and 4. different situational factors. These assumptions and evidence reflect intercultural facework and conflict in any communication process. Research conclusions provided by Ting-Toomey incorporated a number of the components of other researchers' thinking, including positive face and negative face.

Although sometimes definitions are not similar among researchers, there is one word that exists in all authors' Face-negotiation theory definition--conflict. Therefore, conflict plays a significant role in the Face-negotiation theory, and is also a key word for this analysis. Why does conflict always happen in face-to-face negotiation? To answer this question, I want to describe another important word in Face-negotiation theory --facework. Facework is specific verbal and nonverbal behaviors that we engage in to maintain or restore face loss, and to uphold and honor face. Facework is a communication behavior that people utilize to build and protect their own face. People care about their face when they are in face-to-face negotiation processes. Their face is established by their values, beliefs, and cultures. Face is not only people's image, but also a hint to transact identity with others. Face is usually an issue in face-to-face negotiation of conflict situations. Conflict always happens in face-to-face negotiation processes because people want to protect their values and gain enough respect from their negotiating opponents. Because of culture, people have different styles in conflict.

Conflict styles provide an overall picture of a person's communication orientation toward conflict. There are three style models that explain conflict styles: 1. control, forcing, or dominating; 2. solution-oriented, issue-oriented, or integrating; and 3. nonconfrontational, smoothing, or avoiding (Oetzel and Ting-Tommey 2012). Initially, it was a five-style model based on the concern for owning outcomes and other's outcomes; but later, proponents thought it could be reduced to three-styles.

A subsequent study (Zhang, Ting-Toomey and Oetzel 2014) linked emotion to the theoretical assumptions of the Face-negotiation theory; and probed the critical role of anger, compassion, and guilt. This aided in understanding the complex pathways of relationships with self-construal, face concerns, and conflict styles in US and Chinese cultures. This research provides details to explain the relation between self-construal, face concerns, and emotional feelings, such as anger or guilt in face-to-face negotiation. Depending on the comparison between US and Chinese cultures, Zhang (2014) thought emotion can effect self-construal and face concerns in conflict styles for both cultures.

Face-negotiation theory can be used in different communication areas. Communication conferences can be formally and seriously established by governments such as the Group of Twenty Finance Ministers and Central Bank Governors (G20); or it may be forged by less prestigious groups like a small company meeting or citizens in a developing nation.

PRACTICAL RESEARCH

Kristin (2012) administered a survey to anesthesiologists and surgeons at a teaching hospital in the southwestern United States to measure three variables commonly associated with Face-negotiation theory: conflict-management style, face concern, and self-construal. Results of this investigation illustrate that variables associated with Face- negotiation theory were evident in the sample physician population.

Giang (2013) discussed different careers in a company by using face-to-face negotiation with their boss, and correlated the benefit the employees received. In Giang's (2013) survey, a Ph.D. student at Imperial College in London led a study examining negotiation strategies, and concluded that face-to-face interactions tend to benefit more powerful employees. More powerful employees usually win in face-to-face negotiations because they can handle the conflict and assert their opinion. Giang (2013) considered less powerful employees should avoid face-to-face negotiations with their boss. If negotiating with someone who has more power, it is a good idea to avoid face-to-face meetings (*Science Daily* 2014). Although face-to-face negotiation is very useful and common in communication, sometimes it may have a negative influence between negotiators. Some should not utilize face-to-face negotiation in intercultural or interpersonal communication.

From the research about face-to-face negotiation by Michael Taylor and his college researchers (2013), they posited that people with less power did better in virtual negotiations than face-to-face. Therefore, less powerful people could use virtual negotiations such as email, telecommunication, or other electronic tools to send a more potent message to others. This avoids meeting a dominating person like their supervisor or boss. According to Taylor's conclusion, we could assume that all people utilizing electronic tools to communicate with others are not as good at face-to-face negotiation. This conclusion is extremely wrong because the communication consequence depends on different communication situations and the people and content involved.

Relationship between FNT and E-Negotiation

Internet and electronic communication research does not just cover new tools for communication. Electronic communication provides a new, powerful channel that not only will change how people use this mix of options, but it will create entirely new ways to interact (Milles 2014). Despite having internet and technology, face-to-face negotiation cannot be replaced. According to Galin and others (2007), relatively new and still controversial, electronically mediated negotiation is compared to face-to-face negotiations. With the increasing interest in electronic business and the changing technology, the impact of electronic media on the negotiation process and its consequences has become an important issue for both researchers and practitioners. New technology may change the negotiation process. In some communication fields, the electronic media have changed the traditional process of face-to-face negotiation. Sometimes they are used competitively, sometimes they are integrative.

Galin and others' (2007) research findings concerning the impact of the negotiation media on the communication process and its outcomes are inconclusive. In one aspect, researchers argue that face-to-face negotiation has more benefits than electronically mediated negotiation, while on the other hand; it has been found that in many ways electronically mediated negotiation offers new and better results for the people involved in negotiation, especially in recording communication.

Through comparing the advantages and disadvantages of electronic negotiation and face-to-face negotiation, one researcher found that electronic negotiation creates less conflict, spans culture barriers, and communicates directly; however, it has few communications, limited by the basis of text. It is also difficult to make group decisions, and can lower trust between negotiators. Oates (2009) considered that electronic negotiations can more easily lie or employ unethical, competitive negotiation styles. Otherwise, with face-to-face communications, people have more interactive communication that allows for immediate feedback and discussion with a necessary, nonverbal context (Kokemuller 2014). Face-to-face negotiation facilitates better understanding of negotiators' non-verbal hints to strengthen the basis of trust. It also helps negotiators coordinate a mutually beneficial settlement. Therefore, Galin et al. (2007) and Kokemuller (2014) both agree that face-to-face negotiations are better than electronic negotiations in this aspect.

Nevertheless, these two negotiation styles seem like irreconcilable approaches, with heavy influence on the negotiation process. On the other hand, if people combine and integrate them, the styles will reduce the negative influence of the negotiation process.

What electronic product can be used in the process of face-to-face negotiation? In recent years, electronic products are developing very fast. Many famous and successful international negotiation conferences such as the G20 summit are supported by several electronic products: picture phones, two way radios, translators, 3D-printers, wireless technology, voice recorders, and WIFI technology. There is a great variety of electronic products for international communication in the 21st century. Even for people in the face-to-face negotiation process, electronics play a significant role in meetings, and also provide more help for negotiators in maintaining face and acculturating to world norms of communication. The theories and e-communication play an important role in developing countries such as Nepal.

THE COUNTRY OF NEPAL, ACCULTURATION AND FACE-NEGOTIATION

There is a relationship between education and acculturation in most developing countries. Kosaretskii (2013) asserts that school administration and teachers' electronic correspondence with parents is becoming more widespread in highly developed countries' systems of education, yet Russia, an Asian developing country, can also achieve this. Russia uses electronic communication to improve the involvement of families in the lives of school students, encourage facework, and dispel misinformation. Russian school systems have been encouraged to use electronic correspondence between teachers and parents. Knowledge and motivation are required in Russia so that these targeted teachers and parents can make greater use of the internet for development.

The education in Nepal is not as advanced as in Russia or in developed countries. Only a few cities can satisfy the education requirements of electronic communication. Although the government pays great attention to education; in the rural areas, parents' awareness of communication with teachers is low. Some parents do not know how to use the internet to communicate with teachers. Parents do not have the motivation; and in these rural areas, schools and parents cannot afford the high-tech equipment.

Tong (2013) reflected that although Asian American youths are less likely to be sexually active than adolescents from other ethnic groups, with acculturation they may adopt the more liberal sexual norms of American society. This study shows that acculturation leads to more liberal sexual morals among Asian American youth. But through the study, Asian females tend to more easily accept the host society than Asian males. Even the sexual partner numbers are more than Asian males report. What is more, Asian American youth initiation of sexual intercourse is earlier through the acculturation process in American society.

The issue of sexual norms can be easily acculturated and accepted by Nepalese youth, yet Nepal is not a multicultural country. The traditional sexual norms remain fixed in people's mind. When their children mention sexual intercourse, they will feel shamed and embarrassed. Nepal's society is not as liberal as America (or Russia) in education, e-communication, or sexuality.

Research results also show a relationship between economy and acculturation. Carpenter (2012) said this research advocates dividing consumers through markets on the basis of acculturation to the global consumer culture (AGCC) rather than dividing at the individual country level. People have already entered the new age of an Information Society and globalization which requires people to become consumers of a worldly lifestyle and products like e-education. Acculturation is very important in the economic field. There are two main ideas, cohort membership and self-identification. First, people who belong to the same group can be called cohort membership consumers. Second, people who have the same goals and ideas are the consumers which have the same self-identification. In general, the world shows that the younger generational cohorts are open to globalization. They can adapt to new cultures and environments quickly.

Globalization requires acculturation, but in Nepal, it is hard to transform people into global consumers. There are three reasons to show why Nepal is difficult for acculturation involvement into their own economy, culture, and life. First, Nepal is a small country surrounded by mountains. The transportation is not convenient. The new technology and products cannot become popular overnight among its citizens. Second, Nepal's economy is weak. Nepal has to rely on India and China to input food, manufactured products, and business expertise. Globalization and new technology cannot influence Nepal to any great extent. Third, the education of Nepal is poor. People's minds and traditional, isolationist ideas are slow to catch up to modern society's requirements.

Owens (2000) said a system is closed for integrated electronic mail, voice mail, and fax mail messaging. A common message format is defined for using by an electronic mail service and a telecommunication service. Message senders and receivers may choose from a variety of filter and forward options that allow them to manage their communications, and specify a preference for receiving messages at the electronic mail or telecommunications service. Forwarding and conversion of messages is performed automatically. According to different people's preferences, they can choose different options to receive information.

On one hand, if Nepal wants to bring in this system and apply it to their closed society, there is a problem that local government cannot afford to pay for it. Local people cannot accept automated communication quickly. They have to study how to use it and acculturate themselves to the new electronic communication. On the other hand, it is complex to decide if the system should be applied to business, healthcare, or public infrastructure, because Nepal has many things to improve to catch up to the electronic communication in the future.

Lee (1993) said the development of a world economy produces more and more marketing efforts that cross cultural boundaries. Effective communication depends on the understanding of the dynamics of culture. Acculturation describes the changes in attitudes, values, and behaviors of members of one cultural group toward the norms of the other cultural group. Findings from the study also provide implications for effective advertising communication strategies relative to levels of acculturation. This article examines the relationship between levels of acculturation and consumer attitudes toward advertising-related variables in a cross-cultural empirical study.

Advertising communication strategy is a good way to enter Nepal's market, but the price of products which import from other countries is higher than local products. In the rural areas, people rarely can purchase TVs or internet access. It is hard for them to get unbiased information from advertising. Acculturation to the new advertising is difficult to achieve. Poverty is the greatest obstacle to implement advertising communication strategies.

ELECTRONIC COMMUNICATION

There are results to show the relationship between electronic communication and acculturation. Roncancio (2012) predicts that acculturation will serve as a mediator between 1. health locus of control and health-related internet use, 2. age and health-related internet use, 3. income and health-related internet use, and 4. education and health-related internet use. Acculturation is a mechanism through which income, age, and health locus of control operate to influence health-related internet use. This article shows that the internet can be a powerful tool for information dissemination to populations with limited access to health care providers like Nepal.

Although the internet is an important way for Nepalese to get information and news, in the rural area of Nepal, it is difficult for people to access the internet. The working class is the majority in Nepal. Many of them get information through radio and broadcast. The government should spread the internet accessibility so that the internet can be a mediator between people and information. People who live in the rural areas or far away from cities cannot receive information from the internet to know how they can solve health problems.

Klein (2007) discusses patients' acceptance of an internet based patient-physician communication application. The results suggest that behavioral intentions lead to use behaviors, perceived usefulness (PU) to behavioral intentions, and perceived ease of use (PEOU) to PU. Additionally, the analysis reveals that patients trust both their provider and the web site vendor, and shape their behavioral intentions significantly. Not only can patient trust bring convenience for communication, but also vendors can earn a trustworthy reputation. This is a win-win application. The trust between patients and vendors can directly lead this application to success or failure.

This communication application is based on the internet and trust among people. It supplies a high level of medical culture background which Nepal is lacking, yet Nepal's internet does not cover the whole country. The rural application requires a higher degree of financial self-sufficiency. It is difficult for Nepal to cover the whole country for using the application.

Shaw (2007) addresses differences in educational cultures of South Asian and Australian universities. The subject aims to foster the development of student academic integrity and related skills as students are introduced to a new educational culture. It addresses educational acculturation needs of South Asian students in post-graduate public health courses at the University of Wollongong. The purpose of this research is to improve the students' educational acculturation needs and experiences which are significant requirements for them to study abroad.

Nepal's education is not as advanced as developed countries. Most Nepalese lack awareness of their educational acculturation needs. This kind of preparation is new for Nepal's education system. It requires highly educated professors to handle the acculturation process; however, Nepal's education system lacks these professors, and does not have enough tertiary institutions. It will be difficult for Nepal to implement advanced coursework suitable for acculturation into developed educational or communication systems.

THE COUNTRY OF NEPAL

Nepal is a developing, third world country. It has experienced a profound political transformation.

> After years of war, the Comprehensive Peace Agreement (CPA), completed in November 2006, laid down the basic framework for the country's political transition. On 26th September 2008, Prime Minister Pushpa Kamal Dahal promised to lead the people of Nepal forward 'with conviction and sincerity' toward sustainable peace and equitable development. Almost two years after the signing of the CPA, significant steps had been taken to stabilise the peace. But in the same period, the agreement's shortfalls and inconsistencies had become more evident, even as many of its central concerns remained outstanding (Farasat and Hayner 2009).

The Farasat and Hayner (2009) research provides basic knowledge and a view of what Nepal's present situation is regarding communication. Nepal is now a peaceful country, and it has begun to develop its economy policies, technology, and cultural industries. That means a great number of opportunities are available for Nepalese to communicate with other communities, Nepalese ex-patriots, and people from other nations. Therefore, a quantity of research about how the theories and electronic communication is used in Nepal is possible to collect.

There are few surveys and articles that consider Face-negotiation and Acculturation theory for Nepal in regards to electronic communication. This is a research scarcity to consider and discuss.

Nepal still has many problems and conflicts that need to be resolved. Some conflicts between communities have been existing for a long time. Some issues and require more discussion and negotiation between the Nepalese communities by themselves. However, sometimes they are limited by their development situation and technology. Therefore; face-to-face negotiation is the best method for Nepalese to overcome insurmountable problems of communication. They need to build enough trust in interpersonal communication, to reduce barriers, and negotiate peacefully. Meanwhile, they also need to have electronics products like translators or voice recorders to support face-to-face negotiation.

This is hard to accomplish by themselves. Nepal still needs more supplies from the United Nations or other developed countries in the future. They need developed countries to offer technical support, education resources, and financial assistance. The Nepalese government should exert more effort, provide more opportunities, and utilize space for international aid. If they will utilize more advanced electronic technology in their negotiations, and support their communities in solving problems and issues, this will accelerate peaceful education and communication in the country. People living there will gain enough human rights to be prosperous. To achieve this goal, face-to-face negotiation and accommodation in electronic negotiation must effectively supplement each other to improve Nepal.

Research Questions

Developing Nepalese communities are experiencing a communication storm of electronic usage in the twenty-first century. A series of seven demographic, self-reported, and social impact research questions about cellular telephone and internet use in Nepal have emerged:

- RQ 1: What age, gender and education demographics use electronic devices in Nepal?
- RQ 2a: How long have Nepalese used internet?
- RQ 2b: How long have Nepalese used cellular telephones?
- RQ 3a: What types of electronic communication are used in Nepal?

- RQ 3b: What types of communication are used more frequently?
- RQ 4: Is electronic communication easier that face-to-face communication for Nepalese?
- RQ 5: Has internet usage become a problem for some Nepalese?
- RQ 6: Do Nepalese form new relationships online?
- RQ 7: What reasons can you give for not using Mobile Money?
- RQ 8: What age range is the most frequently using mobiles?

METHOD

Distribution of the research questions was conducted using a stratified random sampling approach to contact as many Nepalese and Nepal sojourners as possible. Systematic efforts to locate certain age categories, female, illiterate, or unemployed respondents were sometimes necessary. Ages 18-65 were contacted for possible participation. Respondents living in Nepal at least three months were conscripted by Facebook and online chat software.

Researchers sent surveys in English via Survey Monkey, an electronic instrument to collect information in four segments of 10 questions each. Two deployments were used during March 20-April 20 and October 7-November 18 of 2014. Volunteers with no remuneration participated in this survey related to cellular and internet use. Previous researchers had distributed these surveys to respondents in over 40 developing countries with a similar sampling frame. The information is used for comparison of all developing countries' electronic use in an ongoing study begun in 2002.

The survey contains forty questions, allowing nine open-ended responses, nine demographic queries, and twenty-two Likert scale items. The questions provide internal reliability in multiple formats. The purpose of this research is to determine current cellular telephone and internet usage for developing countries. It will identify how adoption of electronic devices and modes of delivery have an impact on the quality of life for individuals in Nepal. There are no discriminating questions, and no private information is collected. Any question can be skipped if necessary. Participants are free to withdraw or stop at any question if this makes them feel uncomfortable. The data, from Survey Monkey, emailed attachments, and paper surveys, will be recorded, summarized, and analyzed to report total results by SPSS in aggregate form only.

RESULTS

There were 50 completely filled surveys returned for analysis.

RQ 1: What age, gender and education demographics use electronic devices in Nepal?

Respondents in the 18-25 year old cohort comprised 74% of the sample, and respondents in the 25-35 year old cohort comprised 22%. One respondent each was in the 35-45 and 45-65 year old cohorts (2% each). This indicates the survey reported cellular telephone and internet usage data for the largest percentage of school and employment age respondents. It should also be noted that the oldest age cohort spanned 20 years, and the youngest spanned 8 years.

Gender demographics were predominantly male at 74% and female at 26%. Education was varied with the largest cohort being 84% of the sample who had completed institute, college or university degrees. High school completion was next at 22%, followed by masters or Ph.D. completion (8%) and grade school completion (2%). See figure 10.1.

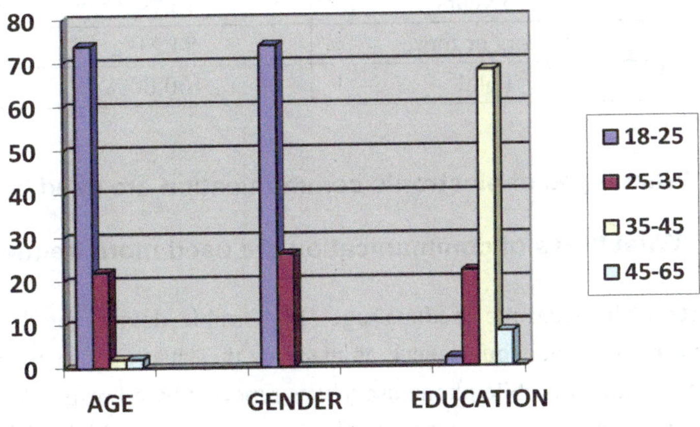

Figure 10.1: Nepal Age, Gender, and Education Demographics (n=50)

RQ 2a: How long have Nepalese used internet?

RQ 2b: How long have Nepalese used cellular telephones?

Nepalese internet and cellular telephone usage were examined independently by self-report scales. The highest reported usage of internet was 5 years or more at 88.89%. The cohort of 3-5 years of internet usage was reported by 11.11%. Cellphone usage was 15.79% for between 3-5 years of usage, and its highest reported usage cohort remained comparative for 5 years or more at 84.21%.

Neither internet nor cellular use was reported for less than three years. See tables 10.1.

Tables 10.1 Nepalese Internet and Cellular Usage (n=50)

Duration of Internet Usage	Percent
Less than one year	0.00%
2-3 years	0.00%
3-5 years	11.11%
5 years or more	88.89%
Total	100.00%
Duration of Cellular Usage	Percent
Less than one year	0.00%
2-3 years	0.00%
3-5 years	15.79%
5 years or more	84.21%
Total	100.00%

RQ 3a: What types of electronic communication are used in Nepal?

RQ 3b: What types of communication are used more frequently?

To ascertain all electronic media usage, 14 possible devices of delivery were listed, and respondents could check as many as they had used at least one time. Cellphone, moto, or mobile phone usage was reported by all respondents as being used. The second highest electronic device usages were also high: Television and E-mail accounts (94.74%). Next in usage by percentage was radio and digital camera (89.47%). The lowest electronic usage reported was internet listserv at 5.26%. See table 10.2 and figure 10.2.

Table 10.2 Nepalese Communication Media Use (n=50)

Device or Mode	Percent
Cellphone, moto, mobile phone	100.00%
Chatroom	63.16%

Radio	89.47%
Television	94.74%
VCR	36.84%
Satellite dish	21.05%
Digital camera	89.47%
E-commerce website	36.84%
Banking kiosk or money card	21.05%
Internet listserv	5.26%
Internet bulletin board	31.58%
Email account	94.74%
Voicemail	62.63%
Proxima (or other presentation device)	10.53%

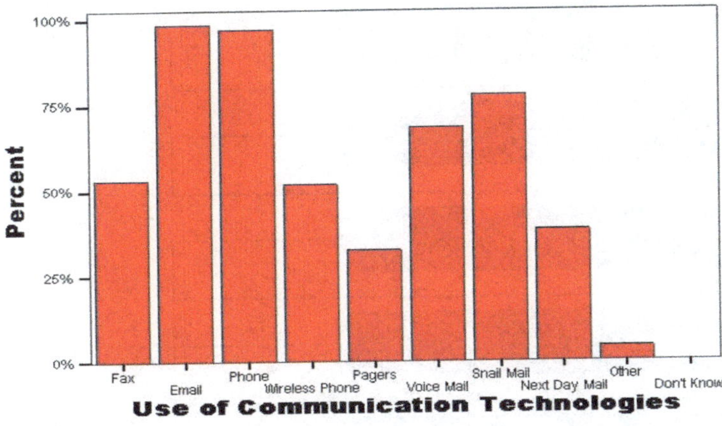

Figure 10.2: Types of Communication Technologies Used More Frequently (n=50)

As an internal measure of validity, qualitative questions were asked about technology use in communication, requiring respondents to supply their own reasons and devices of most frequent usage. About 26% of those surveyed use pagers to contact with others. Regular mail delivery by post (snail mail) was retained by 75% as an often-used means of communication. Fax and wireless phone (larger models with charging stands) were reported in the middle level of usage types of communication at 52% each. Nepalese respondents remarked that they still use fax, wireless, and snail mail so frequently because the fee of making a phone call is too high to afford for working-class people and the poor. Although the snail mail is slow, it is much cheaper than making a phone call. Most adults have been used to their traditional way of using snail mail to make contact with others. These respondents remarked that they did not know how to use smart phones or mobile phones. Less than half (43%) of the respondents used special mail services like next day delivery, especially in their businesses or workplaces.

RQ 4: Is electronic communication easier that face-to-face communication for Nepalese?

Negative comments (e-communication is not easier) were 18%, and positive comments (e-communication is easier) were 36%, with neutral or sometimes comments at 45%. No respondents indicated that there was no difference. See figure 10.3.

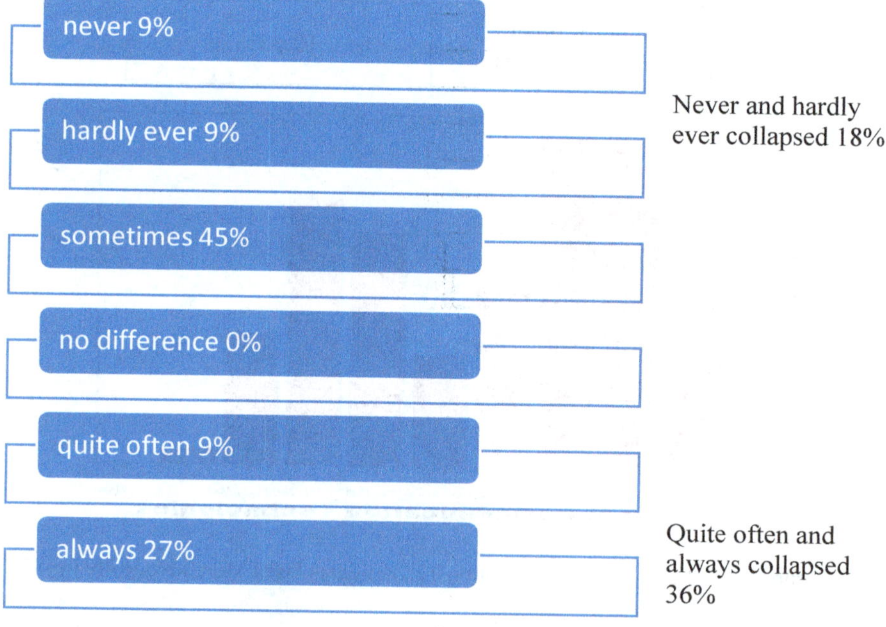

Figure 10.3: Self-reported Electronic Communication Easier than Face-to-Face

Fifty Nepalese gave reasons for ease of face-to-face communication in that many people do not have 1. access to a mobile phone, most people do not 2. need to use these mobile services, or these people 3. prefer to communicate with others face-to-face. The 4. cost of the mobile services is too expensive for people to afford was remarked by many. Nepal is surrounded by mountains making the mobile services too difficult to use. Even some people 5. had not heard about these mobile services. Nepalese described how they are used to communicating with other people face-to-face. This reinforces their custom and culture. Although the development of Nepal has been very fast in the last decade (2005-2015), it is difficult to change a nation's culture and customs in a time shorter than one generation.

RQ 5: Has internet usage become a problem for some Nepalese?

Individual respondents who responded to Survey Monkey were noted for complaints, time extensions, Lack of sleep, inability to curtail use, and negative work comments They thought internet usage had become a problem for some Nepalese including themselves. See figure 10.4.

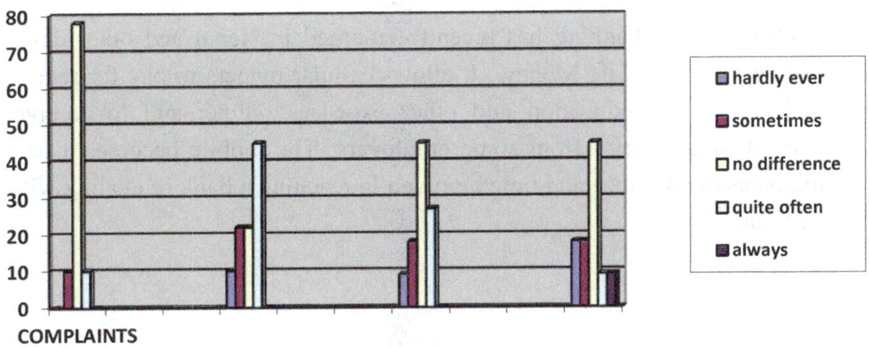

Figure 10.4: Problematic Internet Usage in Nepal (n=11)

Others were complaining (COMPLAINTS) as much as not about internet use (11% in both categories), and the predominance of replies were that it made no difference (78%). When queried if the internet users said to themselves, "Just a few more minutes" (TIME EXTENSIONS), "quite often" was the largest response at 44%. No one answered "always." LACK of SLEEP was hardly ever or sometimes a problem by 27%, and made no difference to 45%. It quite often was a problem for 27% of the respondents. Those who could not curtail use of the internet (CAN'T STOP) were 18%, while all others who could were 82%.

Negative work comments were also included in the reported problematic internet usage. Negative comments from others about work neglect were noted by 27% (others had always or quite often commented on their individual internet usage). Positive comments were noted by 55%, and 18% noted no difference in work colleague comments concerning internet use. It should be noted that 11 respondents chose to answer this series of questions about problematic internet usage, 22% of the total sample.

RQ 6: Do Nepalese form new relationships online?

Some conveniences of internet communication are the ability to form new relationships online for business, friendship, or entertainment functions. Only five respondents answered this question; three of them sometimes form new relationships online with others, and the other two respondents asserted that they hardly ever form new relationships online. Due to low response (10% of those surveyed) it can be inferred that Nepalese rarely use electronic means to form new relationships.

RQ 7: What reasons can you give for not using Mobile Money?

Another form of banking has recently emerged in Nepal and other developing counties called Mobile Money. It allows cellular owners to pay for rent, vendor purchases, taxes, education and other expenses online; and to receive sales payments and salaries from some employers. The mobile becomes a bank that eliminates the distance and time involved in reaching a bank or dealing with cash. See figure 10.5.

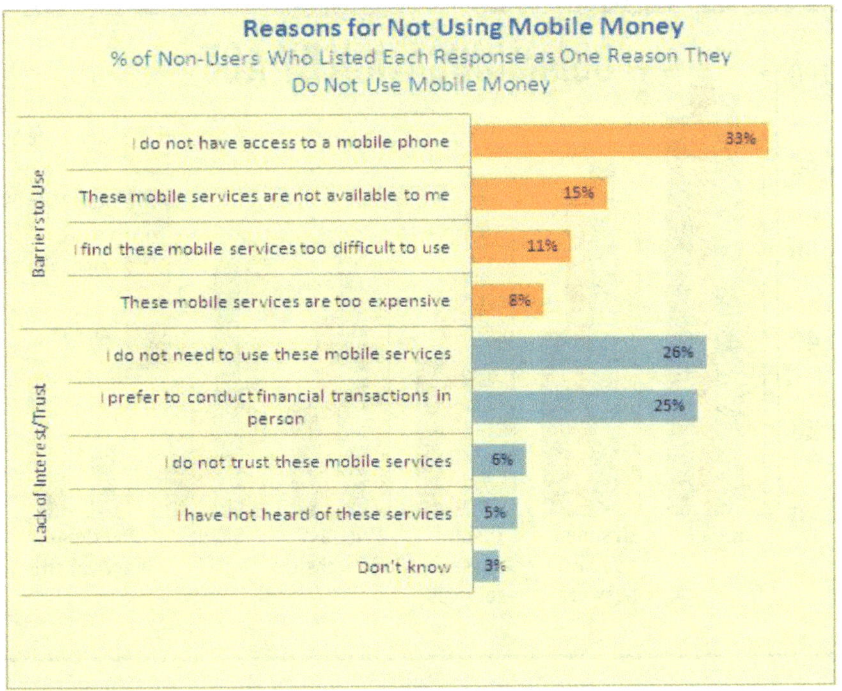

Figure 10.5:   Reasons for Not Using Mobile Money (n=31)

Qualitative response indicators were divided into two groups: Those reasons that were barriers to cell phone usage, and those reasons that indicated lack of interest or trust. The predominant barrier was lack of access to a mobile phone (33%). Unavailable services, difficulty and expense were barriers also given. For the group of reasons indicated by lack of interest or trust, no personal need for the Mobile Money services (26%) or preference for in-person transactions (25%) dominated the responses. Lack of trust or knowledge of the services available was reported by 14%.

RQ 8: What age range is the most frequently using mobiles?

Six means of communication were self-reported in qualitative answers by Nepalese respondents. Their summary is tabulated by age in Figure 10.6.

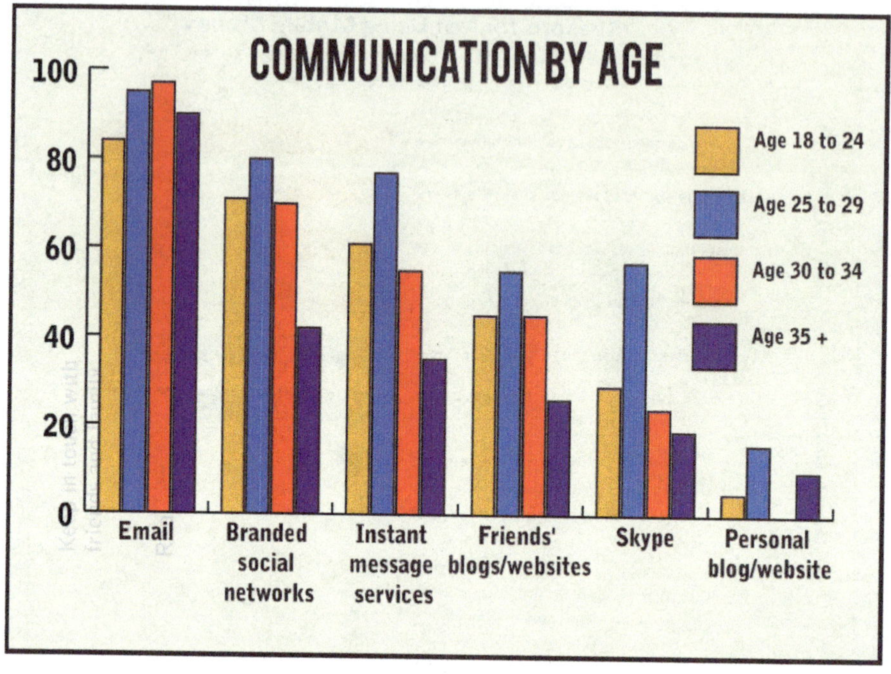

Figure 10.6: Communication by Age (n=31)

Respondents show that age 25 to 29 is the largest group to use all communication means and modes except email. Email tends to be the Nepalese medium of choice whether accessed by a computer or cellular telephone. All age groups used email more than any other communication tool. Branded social networks, instant message services, friends' blogs and/or websites, Skype (webcams and active voice software), and their own blog and/or websites are all reported as used on mobile phones to communicate with others. The age cohort of 18 to 24 is the second largest group of using mobile phones. The age 30 to 34 is the same for social networking and friends' blogs and/or websites. All respondents older than 35 years of age occupy the lowest percentage of using mobile phones, especially for smart phone apps like social networking and texting. The usage of mobile phones is more diverse and predominant in those younger than 35. This youthful generation are highly educated, and upscale employment requires knowledge in using the smart phone and the internet. They more easily accept the new technologies, and work at a much faster pace incorporating multitasks. They have more easily acculturated different communication means into their own culture and life.

DISCUSSION

Electronic communication usage in Nepal has become more common during the last five years (2010-2015). Cellular telephone and internet demographics suggest that it will have significant effects on this developing country. Depending on popular usage and increasing age range of users of electronic communication, problems in usage are occurring in Nepal, but could decrease. These eight research questions were answered by corroborating qualitative and quantitative questions, but the research requires more respondents for comparison to other developing countries as cellular and internet usage becomes more prevalent in Nepal.

Summary

The data was collected from Survey Monkey, individual and email interviews, and paper survey responses in two dispersions. Most respondents were between 18 to 35 years old, and responded by electronic means; therefore, this age demographic cannot represent all Nepalese perspectives. Although the result may not be entirely representative, it closely aligns with real situations encountered by sojourners and citizens in their family and professional communication. The data reflect that electronic communication is very popular for the age of people who are studying in Nepal's secondary schools, universities, and higher education levels.

According to the data of research question three; which illustrates what types of communication were utilized more frequently in Nepal, the cellular telephone has become very common, and almost each college student has one to use daily.

Research question 5 presents a new fact that overuse of the internet causes more problems, which include physical health behaviors, and communication relationships. Those problems refer to the development of the internet that the country has encountered in delivery. Due to low response (10% of those surveyed) it can be inferred that Nepalese rarely use electronic means to form new relationships.

Qualitative response determined why Nepalese did not use Mobile Money, a cellular banking tool available on smart phones. Two types of reasons were given: Those reasons that were barriers to cell phone usage, and those reasons that indicated lack of interest or trust. Lack of access to a mobile phone (33%), and unavailable services, difficulty and expense were barriers also given. No personal need for the Mobile Money services (26%) or preference for in-person transactions (25%) dominated the responses. Lack of trust or knowledge of the services available was reported by 14%.

Email tends to be the Nepalese medium of choice whether accessed by a computer or cellular telephone. All age groups used email more than any other communication tool. Branded social networks, instant message services, friends' blogs and/or websites, Skype (webcams and active voice software), and their own blog and/or websites are all reported as used on mobile phones. The age cohort of 18 to 24 is the second largest group using mobile phones. Ages 30 to 34 is the same for social networking and friends' blogs and/or websites. All respondents older than 35 years of age occupy the lowest percentage of using mobile phones, especially for smart phone apps like social networking and texting. The usage of mobile phones is more diverse and predominant in those younger than 35. This youthful generation more easily accept new technology at school, for relaxation, and for work. They have more easily acculturated different communication means into their own culture and life.

In conclusion, the electronic communication used by Nepalese cannot be ignored as a social force, or a predictor of future communication patterns within Nepal. Future studies should include how to combine electronic communication with face-to-face communication in Nepal, especially in governmental and negotiation purposes.

This study explains the reasons people do not use mobiles to communicate; what types of electronic communication are used more frequently; and what age range is the most frequently using mobiles. Nepal is surrounded by mountains, the mobile services are too difficult to use, and the signal of service is too weak and unstable to yet quantify electronic communication for the entire population. Email and phone are the most frequently used types of communication in Nepal. The next generation are the majority to use electronic communication in this developing country.

Future Development

In the future, the government of Nepal should set education and rural use policies, and apply solutions to combat the rugged terrain. Nepal can develop an infrastructure to incorporate healthcare, education, and banking delivery by electronic communication. Nepal's government should emphasize the application of electronic communication in people's everyday life, the world-wide economic market, and their neighboring countries. Face work in negotiation, government and education can enable Nepalese acculturation to the world of electronic communication. Its people understand how to make their lives better, and are beginning to use technology to do it.

REFERENCES

British Psychological Society. "Face-to-face Negotiations Favor the Powerful." *Science Daily* (October 21, 2014). www.sciencedaily.com/releases/2013/04/130409211857.htm

Carpenter, Jason, Marguerite Moore, Anne Marie Doherty, and Nicholas Alexander. "Acculturation to the Global Consumer Culture: A Generational Cohort Comparison." *Journal of Strategic Marketing* 20 no. 5(2012): 411-423.

Chen, Lu, Hee-Soon Juon, and Sunmin Lee. "Acculturation and BMI among Chinese, Korean and Vietnamese Adults." *Journal of Community Health* 37 no. 3(2012): 539-546.

Delavari, Maryam, Anders Larrabee Sønderlund, Boyd Swinburn, David Mellor, and Andre Renzaho. "Acculturation and Obesity among Migrant Populations in High Income Countries--A Systematic Review." *BMC Public Health* 13(2013): 458.

Ea, Elazek, Macove Itzhaki, Maraki Ehrenfeld, and Joe Fitzpatrick. "Acculturation among Immigrant Nurses in Israel and the United States of America." *International Nursing Review* 57 no. 4(2010): 443-448.

Farasat, Warisha, and Priscilla Hayner. *Negotiating Peace in Nepal Implications for Justice*. New York: International Center for Transitional Justice, 2009.

Galin, Amira, Miron Gross, and Gavriel Gosalker. "E-negotiation Versus Face-to-face Negotiation What Has Changed – If Anything?" *Computers in Human Behavior* 23 no. 1(2007). doi: 10.1016/j.chb.2004.11.009

Giang, Vivian. "Why You Should Avoid Face-To-Face Negotiations With Your Boss." *Business Insider*, last modified April 16, 2013. http://www.businessinsider.com/negotiation-strategies-2013-4.

Hsu, Chia-Fang. "Acculturation and Communication Traits: A Study of Cross-Cultural Adaptation among Chinese in America." *Communication Monographs* 77 no. 3(2010): 414-425.

Kirschbaum, Kristin."Physician Communication in the Operating Room: Expanding Application of Face-Negotiation Theory to the Health

Communication Context." *Health Communication* 27 no. 3(2012): 292-301.

Klein, Richard. "Internet-Based Patient-Physician Electronic Communication Applications: Patient Acceptance and Trust." *E-Service Journal* 5 no. 2(2007): 27-51.

Kokemuller, Neil. "Face-To-Face Negotiation Advantages." *Demand Media Company,* 2014. http://work.chron.com/facetoface-negotiation-advantages-3209.html

Kosaretskii, Sahaki. G., and Desova. V. Chernyshova. "Electronic Communication between the School and the Home." *Russian Education & Society* 55 no. 10(2013): 81-89.

Lee, Wei-Na. "Acculturation and Advertising Communication Strategies: A Cross-Cultural Study of Chinese and Americans." *Psychology & Marketing* 10 no. 5(1993): 381-397.

Milles, James. "Why Use Electronic Communications?" *Computer-Mediated Communication Magazine* (October, 26, 2014). http://knowledgeway.org/living/communications/homepage.html

Oates, Melissa. *Differences in Computer Mediated Versus Face to Face Negotiation.* Psychology and Child Development Department, California Polytechnic State University, December, 2009. http://digitalcommons.calpoly.edu/cgi/viewcontent.cgi?article=1000&context=psycdsp

Oetzel, John G., and Stella Ting-Toomey. *Communication Research.* Los Angeles, CA: Sage, April 30, 2012. http://www.doc88.com/p-590933087383.html

Owens, Stephen J. H. Michael S. Finney, Michael L. Snider, Randall S. Wright, James W. Paynter, and Robin R. Brad. "Electronic Mail Distribution System for Integrated Electronic Communication." *Psychology & Marketing* 19(2000): 54-75.

Roncancio, Angelica M., Abbey B. Berenson, and Mahbubur Rahman. "Health Locus of Control, Acculturation, and Health-related Internet Use among Latinas." *Journal of Health Communication* 17 no. 6(2012): 631-640.

Shaw Julie, Paul Moore, and Senthilkumar Gandhidasan. "Educational Acculturation and Academic Integrity: Outcomes of an Intervention Subject for International Post-graduate Public Health Students." *Journal of Academic Language & Learning* 1 no. 1(2007), 55-67.

Taylor, Michael, "Face-to-face Negotiations Favor the Powerful." *The British Psychological Society* (April, 10, 2013). http://www.bps.org.uk/news/face-face-meetings-favour-powerful

Ting-Toomey, Stella. "Theory Reflections: Face-Negotiation Theory." *NAFSA: Association of International Educators*, September 5, 2014. https://www.nafsa.org/_/File/_/theory_connections_facework.pdf

Ting-Toomey, Stella, and Atsuko Kurogi. "Facework Competence in Intercultural Conflict: An Updated Face-Negotiation Theory." *International Journal of Intercultural Relations* 22 (1998): 187-225.

Tong, Yuying. "Acculturation, Gender Disparity, and the Sexual Behavior of Asian American Youth." *Journal of Sex Research* 50 no. 6(2013): 560-573.

Zhang, Qin, and Stella Ting-Toomey, and John G. Oetzel. *Human Communication Research.* Los Angeles, CA: International Communication Association, July 2014. http://onlinelibrary.wiley.com/doi/10.1111/hcre.12029/abstract

CHAPTER 13

Internet and Cellular Telephone Usage in Paraguay

Yuchen Wang

Based on age, education level, and economic status, this chapter discusses the types of electronic devices used in Paraguay, the internet and cellphone usage, and how electronic communication impacts individuals' lives in Paraguay.

As one of the developing countries in Latin America, Paraguay had the highest economic growth between 1970 and 2013, while it currently has the lowest fixed-line telecom system and internet usage in Latin America. Although the internet service has already opened to the market in Paraguay, the current usage of basic broadband is still not optimistic. Based on age, education level, and economic status, this chapter focuses on the internet and cellphone usage in Paraguay. The fundamental theory of the study is the Agenda-setting proposed by Maxwell McCombs and Donald Shaw. This study discusses three areas: (1) the different types of electronic devices used in Paraguay, (2) the internet and mobile phone usage in Paraguay based on age group, gender, and education levels, and (3) how the internet impacts individuals' lives in Paraguay.

INTRODUCTION

Paraguay is a landlocked country in central Latin America (2014 World Population Review 2014), with a population of 6.802 million people. As one of the developing countries in Latin America, Paraguay had the highest economic growth between 1970 and 2013, while it currently has the lowest fixed-line telecom system and the internet usage in Latin America. Although the internet service has already opened to the market in Paraguay, the current usage of basic broadband is still not optimistic. Using ADSL, WiMAX, and cable modem technologies, several companies offer retail broadband in Paraguay, but the price is still higher, and the service abilities are still lower than other countries. The speeds are low, and the prices are high. As the most important state-owned company, Corporación Paraguaya de Comunicaciones (Copaco) retains a monopolistic enterprise on almost all fixed-line services such as local telephony and international telephony (M2 Presswire 2010).

The fundamental theory of the study is based on Maxwell McCombs and Donald Shaw's Agenda-setting theory, which focuses on the media's powerful influence on the audience. Despite the fact that Agenda-setting theory and internet communication have been studied in developed countries for many years, their power in the developing countries, especially in Latin America, remains an under-explored topic. Along with the qualitative and quantitative methods, the data of this study are collected through 40 items on questionnaires that were completed by Paraguayan citizens. The data were collected in November, 2014 and April, 2015.

LITERATURE REVIEW

Paraguay has been a receiver of immigrants and has a mixed population. During a long history, small groups have settled in Paraguay. The groups include ethnic Italians, Germans, Russians, Japanese, Koreans, Chinese, Arabs, Ukrainians, Brazilians, and Argentinians. Many of these communities have kept their languages and culture. Christianity, particularly Roman Catholicism, is the dominant religion in Paraguay (International Religious Freedom Report 2007).

As one of the developing countries in Latin America, Paraguay had the highest economic growth between 1970 and 2013, with an average rate of 7.2% per year. Business Wire (2009) has investigated the internet market and telecom development in Paraguay. It claimed that Paraguay has the lowest fixed-line telecom system and internet usage in Latin America. M2 Presswire (2010) reported that Corporación Paraguaya de Comunicaciones (Copaco), which is the most important state-owned company, retains a monopoly on almost all fixed-line services such as local telephony and international telephony. Although internet service has already opened to the market in Paraguay, the current usage of basic broadband was not optimistic. Using ADSL, WiMAX, and cable modem technologies, several companies offer retail broadband in Paraguay, but the price is still higher, and the service abilities are still lower than other countries. Agenda-setting theory provides a way to understand economic growth and electronic communication usage in Paraguay.

Agenda-setting Theory

Agenda-setting theory describes a very powerful influence of the media, namely, the ability to tell audiences what issues are important (McCombs and Reynolds

2002). Agenda-setting theory was first observed by Walter Lippmann, a prominent American journalist in the 1920's. Lippmann (1922) pointed out that the mass media are the principal connections between events in the world and images in the public's minds (2-3). Following Lippmann, Bernard Cohen (1963) observed that the mass media "may not be successful much of the time in telling people what to think, but it is stunningly successful in telling its readers what to think about" (13). After Cohen, the Agenda-setting theory was formally developed by Maxwell McCombs and Donald Shaw in a study on the 1968 presidential election. Compared the issues on the media agenda with the issues on the voters' agenda, McCombs, Shaw and their colleagues examined Lippmann's idea. McCombs (1977) found that salience of the news agenda is highly correlated to that of the voters' agenda.

There are two fundamental assumptions of Agenda-setting theory. First, the media do not reflect reality, but filter and shape it. Second, the media concentrate on few issues and news that lead the public to discern those issues as more important than other issues. McCombs, Llamas J. Pablo, Lopez-Escobar Esteban, and Rey Federico (1997) demonstrated that Agenda-setting theory has two levels. At the first level, the theory describes the impact of issue salience and emphasizes the media's pivotal role of inspiring the public what to think. The second level of Agenda-setting theory deals with how the media's agenda affects the public's opinion. This level not only pays more attention to the influence of attributes salience, but also focuses on the function of the media in telling us how to think about issues. According to Rogers, Dearing and Bregman (1988), the process of agenda setting is divided into three parts. The first part is the media agenda, which means the issues that are going to be discussed in the media. The second part is the public agenda that discusses how issues in the media influence the public's minds. The third part is about the public agenda that influences the policy agenda. Time lag is also important.

Agenda-Setting Time Lag

Roberts, Wanta and Dzwo (2002) also provided the "Agenda-Setting Time Lag," an idea which shows how long issues will remain salient in public minds (454). It has become another important question in Agenda-setting research. Wanta and Hu (1994) examined the time lags in five traditional news media: local newscasts, national network newscasts, a regional newspaper, a local newspaper, and a national news magazine. They found the optimal time lags of these five news media are two weeks for local newscasts, one week for national network newscasts, three weeks for a regional newspaper, four weeks for a local newspaper, and eight weeks for a national news magazine. A combination of the five news media produced an optimal time lag of three weeks.

In general, time-lag selection is important because it helps Agenda-setting studies indicate which media coverage has the most influence on the public opinion. Most studies were concerned with the time lag of traditional news media while little research exists that examines the time frame of the internet and cellular phones. "Logically, the time lag for traditional news media to affect online discussions should be relatively short. Individuals are not likely to discuss issues because of news reports they saw weeks earlier" (Wanta and Hu 1994, 455). The primary usage of electronic communication and personal communication affect developing countries.

Electronic Communication

Although traditional media have a powerful influence on audiences, the traditional media are no longer the only factor that impacts the public agenda. Nowadays, the internet and mobile devices have a large audience, a powerful ability of sending and receiving messages and offering many topics for discussion. The main reason that the internet is different from other mass media is it provides a new dimension of communication to the public. December (2006) concluded that computers today provide meaningful communication, other than just machines for counting. As one of the most important tools for communicating with others, the computer not only offers a convenient channel in individuals' daily lives, but also provides a place for posting public opinions. People use the internet communication for many purposes. On account of this, defining what the internet communication is and showing how this definition works in different communication purposes, has become important and meaningful (Roberts, Wanta and Dzwo 2002).

December (2006) divided some of the major purposes into three categories, namely, communication, interaction and information. First, people use the internet for communicating with others. There are three ways of the internet communication, including one-to-one, one-to-many and many-to-many forms. Second, the interaction purpose also plays an important role on the internet communication field. Not just for information transfer or discussion, people can use the internet for the sake of playing or learning. Third, as the same of traditional media, the internet offers plenty of information to the audience. Various media types could deliver plenty of information through the internet applications, such as text messages, voice messages, images, video, and other digital files.

December's (2006) different purposes provide researchers different ways of thinking on the internet communication area. The communication purpose can be used for scholarly activities and research, or for personal and group discussions; Interaction purpose often can be used for social activities and group education. The information purpose deals with subject matter including human activities and knowledge. Today, the internet and computer-mediated communication are treated as a new stage of communication by researchers. Roberts, Wanta and Dzwo (2002) noted that computer-mediated communications were seen as interpersonal communication, with the computer used as a technological tool in order to communicate. Likewise, researchers in communication placed computer-mediated communications in the division of traditional communication study, and minimized the roles of media and channel.

As the internet continues to become an important source of information, it also will become an important field for mass communication researchers. According to Morris and Ogan (1996), existing studies in the mass communication field may not describe the development of new media. Thus, researchers should pay more attention to reexamine mass communication structures and theories in this area. "Future research, then, should link other mass communication theoretical approaches with Internet usage" (Roberts, Wanta and Dzwo 2002, 464).

Using different types of structures for defining units of analysis, researchers in past decades have taken many approaches to analyzing human communication on the internet. Hiltz and Murray (1993) have explored the relationships between the characteristics of media systems and the characteristics of individuals using them. With the ease of access to media technology, people have changed their ways in seeking news and information. Rauen (2011) focuses on the influence of technological changes on public policy agenda setting, especially on the regulatory agenda. Her study analyzed how technological changes happened in the telecommunication field. Rauen's research (2011) also focuses on how these changes impacted the communication ways in society. Roberts, Wanta and Dzwo (2002) chose the Electronic Bulletin Boards (EBB) as an example to explore the approach of agenda-setting, and the role it may play on the online interactions. They explored how media coverage apparently provides the public and the individual with information to use in their internet discussions.

Roberts, Wanta and Dzwo (2002) investigated whether traditional news media sources have an agenda-setting impact on the discussions taking place on the EBBs. This study shows how media coverage apparently provides individuals with information to use in their internet discussions. They concluded that "the more coverage these issues received in the news media, the more messages that the internet users posted on discussion lists" (459). Morris and Ogan (1996) discussed the degree of audience activity as a vital element of EBBs. They also suggested that the concept of audience activity should be included in the study of internet communication.

Today, people hold their own agendas, and then find the group that has similar agendas that they agree with. "It allows individuals to receive messages electronically from other individuals via e-mail. Moreover, it provides open forums for discussion on a wide variety of topics through discussion lists, bulletin boards, and chat rooms" (Roberts, Wanta and Dzwo 2002, 452). Internet users use the news media as the guideline to important issues that need to be discussed online. Therefore, high media coverage will make not only certain issues appear to be important, but also stimulate enough interest in the topic. Thus, the internet users will feel pressure to bring the topic to the public platform for discussion.

Although the original agenda-setting studies reported a direct relationship between the media agenda and the public agenda, other studies indicate the existence of intervening factors. Besides increasing public concern with certain issues, press coverage can also decrease concerns as a function of the recipient's self-involvement and interest in the issues (Wanta and Hu 1993). Media coverage, then, may have led to more discussions of issues on the internet. This same scenario could be happening in Paraguay. Internet users in Paraguay may use the media to uncover important issues that they will subsequently discuss online. Besides, it is possible for people all around the world to find other people with similar agendas and collaborate with them because of the advances in technology. Therefore, the use of mobile devices and the internet may decrease the influence of agenda-setting on the public. Actual current usage is necessary to determine this.

Electronic Communication Usage in Latin America

Since the 1968 study, published in a 1972 edition of Public Opinion Quarterly, more than 400 studies have been published on the agenda-setting function of the mass media, and the theory continues to be regarded as relevant (McCombs 2005). Although Agenda-setting theory has been studied in developed countries for several years, its power in the developing countries remains an under-explored topic. Aleman (2006) provides a meaningful study by analyzing how the legislative agenda is controlled in Latin American countries. According to PR Newswire (2014), the basic broadband penetration in the Latin American region remains below the world's average level, and the usage of computers is much lower than other developed countries. Accordingly, poverty and low computer use have become two main aspects of hindering social group spending on basic telecom services.

In Latin America, different countries have different levels of broadband development. There are more developed and better broadband services in some southern countries and some of the wealthier cities, while the lower rates usually appear in Central Latin America and the northwest coast. Some countries including Haiti, Paraguay, Nicaragua and Cuba have particularly low penetration. M2 Presswire (2013) has reported the telecom system and current development status in Latin America. Business Wire (2009) has also investigated the internet market and telecom development in Paraguay. They both claimed that Paraguay has the lowest fixed-line telecom system and the internet usage on Latin America. The internet sector in Paraguay is open to the market to some extent. Using ADSL, WiMAX, and cable modem technologies, several companies offer retail broadband in Paraguay, but the current usage of basic broadband is still not optimistic. The speeds are slow, and the prices are high.

Development of Fixed-line and Fixed-broadband in Paraguay

According to M2 Presswire (2010), due to the country's landlocked position, Paraguay's telecommunications have been restricted for several years. Paraguay's interconnection, to a large extent, relies on other neighboring nations' submarine cable networks. Therefore, this situation has driven up the price of fixed-line and broadband services. Besides, as the most important state-owned company, Copaco has retained a monopoly on almost all fixed-line services, such as local telephony and international telephony. Copaco's underinvestment in fixed-line infrastructure made the fixed-line services stand still.

Although Paraguay has the lowest rate of fixed-broadband penetration in South America, the market is growing rapidly, primarily in the major urban cities like Asunción, which is the capital and the largest city of Paraguay. Copaco has controlled the majority of the ADSL market. As Copaco's main competitor in the fixed broadband market, Millicom's Tigo, offers broadband in Asunción and neighboring towns through a hybrid fibre-coaxial network. Tigo also provides broadband through WiMAX and FttP technologies (M2 Presswire 2010).

Development of Cellular Phone in Paraguay

Despite the absence of reliable fixed-line networks restricting the development of mobile market in Paraguay, there still are four main operators who provide mobile services for the public, namely, Millicom's Tigo, Telecom Argentina's Personal, América Móvil's Claro, and Copaco's Vox. They offered about 18 cellular phones in Paraguay for every fixed-line in service, which is the highest proportion in Latin America (M2 Presswire 2010).

According to M2 Presswire (2010) and Business Wire (2009), mobile broadband has become the fastest growing telecom field in Paraguay, and strong growth is expected to continue in this market.

As mentioned before, despite the fact that Agenda-setting theory and internet communication have been studied in developed countries for many years, their power in the developing countries; and especially in Latin America, remains an under-explored topic. Along with all the researches and studies, the condition of the internet and mobile devices usage will be explored by examining the following research questions.

Research Questions

- RQ 1: What age, gender and education levels use electronic devices in Paraguay?
- RQ 2: What types of electronic devices are used commonly in Paraguay?
- RQ 3: Is electronic device usage correlated to economic status?
- RQ 4: Is electronic communication popular in Paraguay to maintain relationships or form new relationships in the real world?
- RQ 5: Have the internet and mobile devices improved individuals' lives in Paraguay?

METHOD

Participants

Since the study was the investigation of the internet and cell phone usage in Paraguay, all thirty six participants included in the study were Paraguayan. Although there were ten Paraguayan students in an American university as key referents, the participant group remained narrow to be representative. As the starting point; therefore, the ten students sent research surveys to their friends, families, domestic classmates and teachers, and other Paraguayan citizens. Each of the ten students contacted at least three individuals as participants, so there were at least thirty additional samples in the study. The total samples were divided into five age groups, namely, 18-25 years old, 25-35 years old, 35-45 years old, 45-55 years old, and 55-65 years old.

Instrumentation and Procedures

Qualitative and quantitative methods were used in the study. The questionnaire consisted of forty questions in English for Paraguayan citizens. Also, there was a translated form in Spanish in order to avoid language misunderstanding. The questionnaire not only investigated the basic information of how people use the internet and cellular phones, but also included questions about why individuals choose certain electronic tools and devices to communicate with others. There were no offensive, discriminatory, biased, or confidential questions that would cause respondents to feel embarrassed or uncomfortable. Besides, based on the voluntary principle, all participants were free to stop or refuse participation at any time. As an online survey application, Survey Monkey provides free, custom-made surveys and correspondence survey programs including sample selection, data analysis, bias elimination, and data representation tools. With the help of Survey Monkey, the goal was to collect at least thirty completed surveys in Paraguay.

After all survey results had been received, the data analysis was performed using Statistical Packages for Social Science (SPSS). As a software package for statistical analysis, SPSS provides survey authoring and deployment, data mining, text analytics, and automated scoring services. In this study, the data was first recorded into SPSS one by one and then analyzed through SPSS Statistics.

Results

The survey data were collected in November, 2014 and April, 2015. In total, 38 respondents participated in the survey. Twenty questionnaires (10 in English and 10 in Spanish) were distributed, 30 links to the Survey Monkey forms were sent to Paraguayan long-term residents. A sample of 38 fully completed forms were returned.

RQ1: What age, gender and education levels use electronic devices in Paraguay?

Survey results show that all respondents use the internet and cellular phone. Twenty-six respondents are 18-25 years old, and this age group constituted the largest percentage of internet and cellular phone usage, with 68.4%. Only two respondents are 35-45 years old, which made up the smallest proportion, with 5.3%. See figure 11.1.

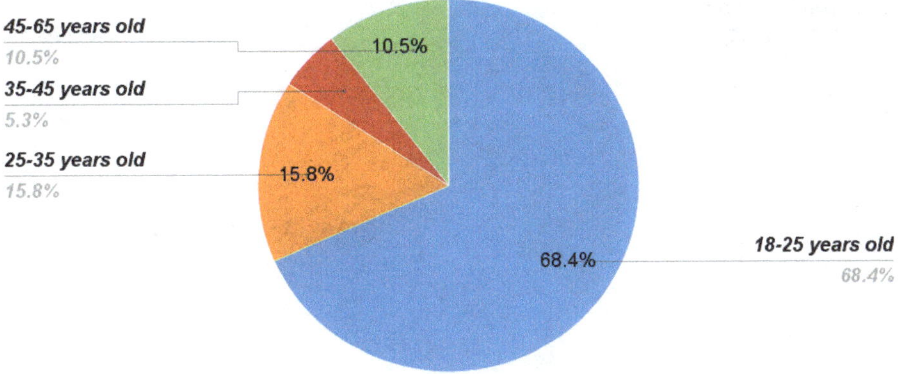

Figure 11.1: Respondent Age Demographics of Paraguay

Among all participants, 26 respondents are female (68.4%) while only 12 respondents are male (31.6%). See figure 11.2.

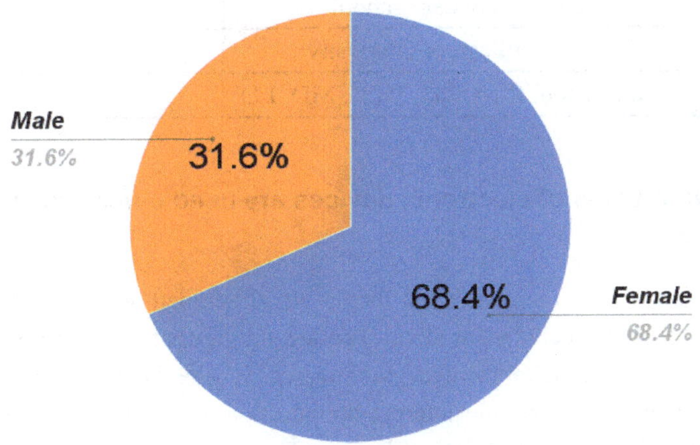

Figure 11.2: Respondent Gender Demographics of Paraguay

Figure 11.3 shows most internet and cellular phone users have completed high or secondary school, which constituted 60.5%. Education level completion was 2.6% for primary school, 31.6% for institute, college or university, and 5.3% for graduate school (MA, MS, Ph.D, MD, LD).

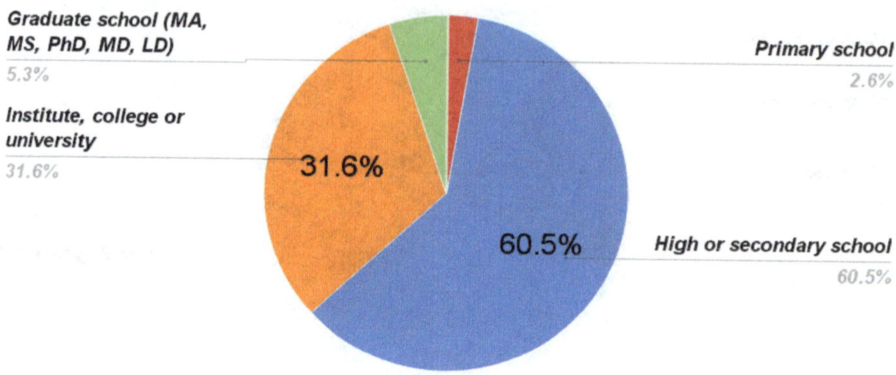

Figure 11.3: Education Level Completion

Legend for Pie Chart

Education Level Completion	Respondents
Primary school	1
High or secondary school	23
Institute, college or university	12
Graduate school (MA, MS, Ph.D, MD, LD)	2

RQ 2: What types of electronic devices are used commonly in Paraguay?

Fourteen electronic communication media were recorded as used at least once. The media included cellular (or mobile) phone, online chat room, radio, television, VCR, satellite dish, digital camera, e-commerce website, banking kiosk or money card, the internet listserv, the internet bulletin board, email account, voicemail, and proxima (or other presentation device). The table 11.1 shows the media use of respondents in Paraguay. Among them, the cellular (or mobile) phone and the email account were used commonly, with 37 respondents (97.4%) and 36 respondents (94.7%), respectively. The least percentage of the electronic communication usage was satellite dish and proxima (or other presentation device), which both constituted 13.2%.

Table 11.1 Types of Electronic Communication Used in Paraguay

Number	Electronic Communication Types	Respondents	Percentage
1	Cellphone, moto, mobile phone	37	97.4%
2	Chatroom	11	28.9%
3	Radio	29	76.3%
4	Television	32	84.2%
5	VCR	6	15.8%
6	Satellite dish	5	13.2%
7	Digital camera	31	81.6%
8	E-commerce website	7	18.4%
9	Banking kiosk or money card	11	28.9%
10	Internet listserv	7	18.4%
11	Internet bulletin board	5	13.2%
12	Email account	36	94.7%
13	Voicemail	26	68.4%
14	Proxima (or other presentation device)	5	13.2%

RQ 3: Is electronic devices usage correlated to the economic status?

No, the electronic communication usage is not directly correlated to the economic status. Thirty-six people responded the question. Among them, 18 Paraguayan electronic devices users have no reliable income (51.4%). These people were students who live in Paraguay or study overseas. See figure 11.4.

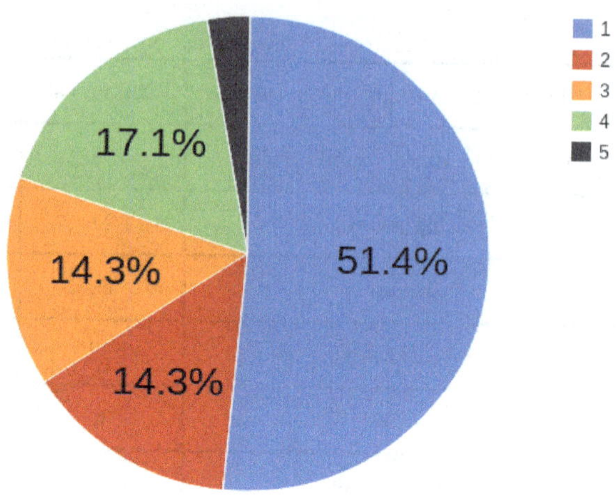

Figure 11.4: The Economic Status of Paraguayan Electronic Devices Users

Legend for Pie Chart

Number	Economic Status	Respondents	Percentage
1	I have no reliable income.	18	51.4%
2	I have a fixed income/pension less than 1,000 USD per month.	5	14.3%
3	I have a job or business providing housing, food, clothing and life necessities.	5	14.3%
4	I have enough to provide for myself and others' needs on a regular basis.	6	17.1%
5	I am wealthy by my country's standards.	1	2.9%

RQ 4: Is electronic communication popular in Paraguay to maintain relationships or form new relationships in the real world?

No, the electronic communication is not popular in Paraguay to maintain relationships or form new relationships in the real world. Thirty-six people responded the question. Among them, 25 respondents prefer to communicate in person (69.4%). See figure 11.5. Most respondents never form new relationships online (63.9%) and never meet new electronic friends face-to-face (61.1%). See table 11.2.

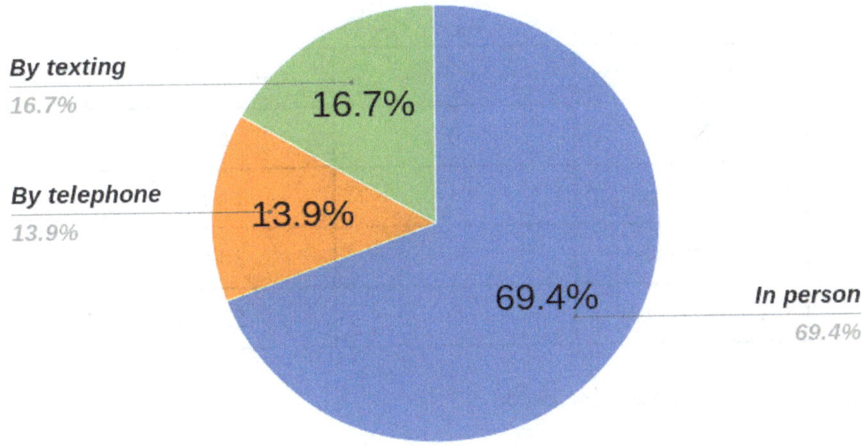

Figure 11.5: Communication Choice

Table 11.2 New Relationships Online

Form New Relationships Online Descriptors	Percentage	Meet New Electronic Friends Face-to-face Descriptors	Percentage
Never	63.9%	Never	61.1%
Hardly ever	30.6%	Hardly ever	25%
Sometimes	5.6%	Sometimes	8.3%
Quite often	0%	Quite often	2.8%
Always	0%	Always	2.8%

RQ 5: Have the internet and mobile devices improved individuals' lives in Paraguay?

When asked whether the internet and mobile devices had improved individuals' lives, 21 respondents chose "tremendously", which made up 58.3% of all participants. The two main reasons of the answer were 1) convenient to contacting with families and friends, and 2) fast to getting information and communicating with others. 11 respondents chose "sometimes" (30.6%), and 4 replied "hardly ever" (11.1%). See table 11.3.

Table 11.3 Paraguayan Good Life Appraisals

Descriptor	Responders	Percentage
Not	0	0%
Hardly ever	4	11.1%
Sometimes	11	30.6%
Tremendously	21	58.3%

The qualitative responses included:

"I can use it [to] connect with my families and friends in my country."

"Because my family is [far] away [from me]. It is easier to use the cellphone when you cannot reach the other person physically."

"Because I can communicate with friends and family. Also, I can use the internet directly whenever I want."

"Because I can find a lot [of] information that I need for work and internet is the best source of communication."

DISCUSSION

Summary

RQ 1: What age, gender and education levels use electronic devices in Paraguay?

All respondents were electronic devices users. Young people in Paraguay were likely to accept using the internet and cellphones. Results of this study showed that 26 respondents are 18-25 years old, and this age group constituted the largest percentage of electronic devices usage, with 68.4%. The gender demographics were 68.4% female and 31.6% male. Most internet and cellular phone users have completed high or secondary school, which constituted 60.5%. Education level completion was 2.6% for primary school, 31.6% for institute, college or university, and 5.3% for graduate school (MA, MS, Ph.D, MD, LD).

RQ 2: What types of electronic devices are used commonly in Paraguay?

Results of this study showed that fourteen electronic communication media were recorded as used at least once. This result means different electronic communication media were used widely in Paraguay. The fourteen electronic communication media included cellular (or mobile) phone, online chat room, radio, television, VCR, satellite dish, digital camera, e-commerce website, banking kiosk or money card, the internet listserv, the internet bulletin board, email account, voicemail, and proxima (or other presentation device). Among them, the cellular (or mobile) phone and the email account were the dominant tools of electronic communication for Paraguayan citizens, with 37 respondents (97.4%) and 36 respondents (94.7%), respectively. The least percentage of the electronic communication usage were satellite dish and proxima (or other presentation device), which both constituted 13.2%. The results indicated that the main reason of using electronic devices in Paraguay was communicating with other people.

RQ 3: Is electronic devices usage correlated to the economic status?

No direct relation was found between the electronic devices usage and the economic status. The results showed 51.4% electronic devices users have no reliable income. These Paraguayan respondents were students who live in their country or study overseas. The results supported the fact that the usage of internet and cellular phone is still common among young students, although the brand broad fees are high.

RQ 4: Is electronic communication popular in Paraguay to maintain relationships or form new relationships in the real world?

The results of the study showed the electronic communication is not popular in Paraguay to maintain relationships or form new relationships in real world. 25 respondents prefer to communicate in person (69.4%). Most respondents never form new relationships online (63.9%) and never meet new electronic friends face-to-face (61.1%). The result means the most important communication way is still the traditional one; namely, face-to-face interpersonal communication.

RQ 5: Have the internet and mobile devices improved individuals' lives in Paraguay?

Most respondents indicated that the internet and mobile devices had tremendously improved their lives (58.3%). Seen from the answers, there were two main reasons of their opinion. First, electronic devices are convenient to contact with families and friends. Second, the internet and cellphones are fast to get information and communicate with others. The responses included connection to friends and family, ease of use, and information availability.

Limitations

There were three main limitations of this study. First, the results of this research lacked the data of rural citizens in Paraguay. All respondents of the survey were urban residents. On account of this, the study cannot describe the difference of electronic communication usage between people who live in cities and those who live in countryside. Second, the participation of men and women was not equal. The gender demographics were 68.4% female and 31.6% male. Female respondents constituted more than two times available male respondents. Males and females may have different reasons when choosing electronic devices. Also, do men and women use different types of electronic communication media? This is still a question that needs to be considered. Third, the results lacked specific answers to the purposes of Paraguayan use of the internet and cell phones. Answers included "Sometime[s], when I use cell phone or internet, I feel more comfortable", "It helps me in everything I need", and simply "convenient". These answers did not supply specific purposes; the respondents talked more about feelings toward using the internet and cell phones, which means the study cannot analyze the main reasons why people choose certain electronic communication media in Paraguay.

Suggestions for Future Research

In Paraguay, the usage of electronic communication media shows a huge difference between accessible urban areas and less accessible rural regions. Electronic survey methodology cannot access those who do not have the technology. Future research may find some reasons related to the difference from various aspects, including the economic factor, the location, the population, the media coverage factor and the public's awareness or knowledge as well.

In this study, many respondents, especially those who study abroad had mentioned that one of the most important reasons for using electronic devices was contact with their families and friends who still lived in Paraguay in a faster, cheaper, and more convenient way.

REFERENCES

Aleman, Eduardo. "Policy Gatekeepers in Latin American Legislatures." *Latin American Politics & Society* 48, no. 3 (2006): 125-55. doi: 10.1353/lap.2006.0028.

Cohen, Bernard. *The Press and Foreign Policy.* New York: Harcourt, 1963.

December, John. "Units of Analysis for Internet Communication." *Journal of Communication* 46, no. 1 (2006): 14-38. doi: 10.1111/j.1083-6101.1996.tb00173.x.

International Religious Freedom Report. "Paraguay (Western Hemisphere Report)." *International Religious Freedom Report*, 2007.

Landolt, Patricia, Luin Goldring, and Judith K. Bernhard. "Agenda Setting and Immigrant Politics: The Case of Latin Americans in Toronto. (Diverse Pathways to Immigrant Political Incorporation: Comparative Canadian and U.S. Perspectives)." *American Behavioral Scientist* 55, no. 9 (2011): 1235-1266. doi: 10.1177/0002764211407841.

Lippmann, Walter. *Public Opinion.* New York: Harcourt, Brace and Company, 1922.

McCombs, Maxwell. "Agenda Setting Function of Mass Media." *Public Relations Review* 3, no. 4 (1977): 89-95. doi: 10.1016/S0363-8111(77)80008-8.

McCombs, Maxwell. *Setting the Agenda: The Mass Media and Public Opinion.* Cambridge: Polity Press, 2004.

McCombs, Maxwell. "A Look at Agenda-setting: Past, Present and Future." *Journalism Studies* 6, no. 4 (2005): 543-57.

McCombs, Maxwell, and Amy M. Reynolds. "News Influence on Our Pictures of the World." In *Media effects: Advances in theory and research*, edited by Bryant Jennings and Zillmann Dolf, 1-18. Mahwah: Lawrence Erlbaum Associates Publishers, 2002.

McCombs, Maxwell, Lance Holbert, Spiro Kiousis, and Wayne Wanta. "The News and Public Opinion: Media Effects on Civic Life." *European Journal of Communication* 28, no. 6 (2013): 728-729. doi: 10.1177/0267323113505802b.

McCombs, Maxwell, Llamas J. Pablo, Lopez-Escobar Esteban, and Rey Federico. "Candidate Images in Spanish Elections: Second-level Agenda-setting

Effects." *Journalism & Mass Communication Quarterly* 73, no. 4 (1997): 703-717.

McCombs, Maxwell, and Marcus Funk. "Shaping The Agenda of Local Daily Newspapers: A Methodology Merging the New Agenda Setting and Community Structure Perspective." *Mass Communication and Society* 14, no. 6 (2011): 905-919. doi: 10.1080/15205436.2011.615447.

Morris, Merrill, and Christine Ogan. "The Internet as Mass Medium." *Journal of Communication* 46, no. 1 (1996): 39-50. doi: 10.1111/j.1083-6101.1996.tb00174.x.

Business wire. "Paraguay Has the Lowest Fixed-Line Teledensity and Internet Penetration in South America." *Business wire*, March 30, 2009.

M2 Presswire. "Latin America Online Payment Methods 2013 - First Half 2013." *M2 Communications*, June 06, 2013.

M2 Presswire. "Paraguay's Telecoms, Mobile and Broadband - The Countrys Telecom Market Has Considerable Potential for Growth." *M2 Communications*, July 13, 2010.

Hiltz, S. Roxanne, and Murray Turoff. *The Network Nation: Human Communication via Computer.* Cambridge: The MIT Press, 1993.

PR Newswire. "Latin America - Broadband and Internet Market." *PR Newswire*, August 04, 2014.

Rauen, Cristiane V. "Technological Change and Public Policy Agenda Setting: The Broadband Regulation for Universalization in Brazil/ *Mudanca Tecnologica e Definicao Da Agenda De Politicas Publicas: Regulacao Para Universalizacao Da Banda Larga No Brasil.*" *Revista De Direito, Estado E Telecomunicacoes* 3, no. 1 (2011): 89-110. http://www.getel.org/GETELSEER/index.php?journal=rdet&page=articl e&op=download&path%5B%5D=41&path%5B%5D=36.

Roberts, Marilyn, Wayne Wanta, and Tzong-Horng (Dustin) Dzwo. "Agenda Setting and Issue Salience Online." *Communication Research* 29, no. 4 (2001): 452. doi: 10.1177/0093650202029004004.

Rogers, Everett M., James W. Dearing, and Dorine Bregman. "The Anatomy of Agenda-Setting Research." *Journal of Communication* 43, no. 2 (1993): 68-84. doi: 10.1111/j.1460-2466.1993.tb01263.x.

Tsebelis, George, and Eduardo Aleman. "Presidential Conditional Agenda Setting in Latin America." *World Politics* 57, no. 3 (2005): 396-420. doi: 10.1353/wp.2006.0005.

Wanta, Wayne, and Yu-Wei Hu. "The Agenda-setting Effects of International News Coverage: An Examination of Differing News Frames." *International Journal of Public Opinion Research* 5, no. 3 (1993): 250-263.

Wanta, Wayne, and Yu-Wei, Hu. "Time-lag Differences in the Agenda-setting Process: An Examination of Five News Media." *International Journal of Public Opinion Research* 6, no. 3 (1994): 225-240.

Zyglidopoulos, Stelios, Pavlos C. Symeou, Philemon Bantimaroudis, and Eleni Kampanellou. "Cultural Agenda Setting: Media Attributes and Public Attention of Greek Museums." *Communication Research* 39, no. 4 (2012): 480-98. doi:10.1177/0093650210395747.

2014 World Population Review. "Paraguay Population 2014". *2014 World Population Review.* Last modified October 19th, 2014. http://worldpopulationreview.com/countries/paraguay-population/.

CHAPTER 14

Cellular Telephone and Internet Usage in the Philippines: A Media Ecology Perspective

XiangXiang Jura Ji

The study of telecommunications and media usage in the developing countries has been drawing the attention of scholars in the area of communication research. This study adopts the perspective of the media ecology theory to look into the current situation of cellular telephone and internet usage in the Philippines in order to present and evaluate its media environment. There are many ways to study the interaction between human beings and their communication technology. Media ecology theory is one of them. In particular, media ecology studies how the communication media affect human perception, understanding, feeling and value; and also how this influences our chance of survival (Postman 1971).

Media Ecology Theory

In this section, I will summarize the Media Ecology theory by introducing its development of the definition, the academic factory, its applications and limits. Media ecology can be traced to McLuhan's interest of promoting a balance in the media ecology (Postman 2000). Afterwards, Postman built it as a formal theory used in communication study. In Postman's definition, he insisted on studying media as the environment in 1971. Furthermore, he explained the environment as a complex message system which imposes on human beings certain ways of thinking, feeling, and behaving. It has three characteristics: Structures we can see, say and act upon; assigned roles and how we play them; and what we are permitted to do and what we are not (Postman 1992).

Ong (1977), who is regarded as one of the twin pillars with Postman in media ecology, enriched this theory in the aspect of ecological concern, which he described as "a new state of consciousness, the ultimate in open-system awareness" (324). In his later years, Ong applied a philosophic concern to media ecology, to explain the relationship between these macro and micro level relationships, concluding that we are living in an age of countless, conspicuous interconnections (Strate 2004).

In the past decades, many communication researchers contributed their definitions of media ecology with different foci. A more active definition of media ecology is from Fuller (2005), who emphasized the connections between actors and processes in media systems at various scales. That not only affects humans in content, but also potentially affects humans who can then take action to influence social systems. As for Strate (2004), media ecology is a perspective of looking at things as an intellectual tradition. It began not as an ideology, but as an approach, an exploration into the uncharted environments of the various media of communication that constitute so large a part of our world (Madianou 2011). As more and more researchers became interested in media ecology study, many academic groups and major departments began to appear in the early 21th century; MEA is one of them.

MEA and the Humanism of Media Ecology

With particular interest in understanding how media ecology affects society, an academic department was set up in 2000 called the Media Ecology Association (MEA). They hold convention annually with significant topics in the media ecology field. In 2014, the topic was Confronting Technopoly: Creativity & the Creative Industries in Global Perspective. MEA also has its academic journal, *Explorations in Media Ecology*, which publishes essays, commentary, and critical examinations relevant to media ecology as a field of study and practice.

In the first convention of the MEA, Postman (2000) asked four crucial questions which I think reflected the intrinsic value of the study of media ecology: To what extent does a medium contribute to (1) the uses and development of rational thought, (2) the development of democratic processes, (3) greater access to meaningful information, and (4) enhancement or diminishment of our moral sense? (13-15). Undeniably, Postman had a deep consciousness of humanity in media ecology. However, other media ecology researchers have presented different perspectives. For example, McLuhan concentrated on media itself without moral evaluation, while his student, Walter Ong, emphasized the philosophy of the human person and relationships. Strate (2006) notes that media ecology is a multidisciplinary perspective with many ideas contributing to its rich intellectual landscape. I think openness is one of the most prominent characteristics of media ecology; and thus, it requires media ecology scholars to work dialectically, using contrasts to understand media (Strate 2011, 1-16).

Application of Media Ecology: Valcanis as an Example

Fuller (2005) regarded ecology as the most expressive language to describe the massive and dynamic interrelation of processes and objects, beings and things, patterns and matter. Because of the utility and openness of this term, media ecology has been applied in communication study for various topics: New media, technoculture, global culture, nations, young generations, and literature study. Hereon, I will summarize a research article which applied this theory as an example, to explore how it can be applied concerning the topics of new media and global culture.

Media Ecology theory became known to more people because of its conception of analyzing dynamic processes rather than an object (Fuller 2005). It also encouraged researchers to rethink technological determinism. Tom Valcanis (2011) is one of them. His article began with a typical media ecology question: To what extent has new media created a common globalized media environment and culture? Valcanis assumed that there exists a globalized culture shaped by new media (35). Based on this, his article used media ecology as the analytical framework to develop three sections: a new media ecology; no space in any time; and it is everyone's turn, all the time. In the first section, Valcanis defined the conception of media ecology and introduced its application scope by citing famous media ecologists' opinions, such as Marshall McLuhan and Lance Strate (35). Also, Valcanis tried to prove that media ecology is an interdisciplinary area, and has a close bond with culture and globalization (36).

Valcanis came to essential elements of media: time and space. Compared with the old time media, information can be transmitted in cyberspace, in which human beings arrange themselves into networks through computers and smartphones (37). Nowadays, smartphones deal directly with the ecological interrelationships among the virtual space of the internet (Pedersen 2008). According to Valcanis, these new media technologies have transformed the global culture into a participatory culture (2011, 38-39).

In his treatise, Valcanis (2011) explained how the cultures participate and interact when they grow in new media, which I also think is a meaningful topic to study in media ecology. Interestingly, Strate (2003) compared the computer to the clock, a self-operating machine that has no physical product, yet creates economically effective links between people and information. The conceptual cyberspace also gives rise to cybertime, which McLuhan regarded as a characteristic of electronic cultures (Strate 2003). While presently, I think timeline on Facebook might be a typical example of it. Valcanis drew the conclusion that multimedia not only forms new platforms for communication, but also shapes a new way of organizing and regulating ideas. As I can see, this is the core proposition of media ecology: a medium is a technology within which a culture grows; that is to say, it gives form to a culture's politics, social organization, and habitual ways of thinking (Postman 1971).

Shortcomings of Media Ecology Theory

Media ecology opened a new perspective to study communication phenomena; however, it also has its own limit. For example, Chris Wood (2011), the author of "Media Missing the Message," pointed out that the new media ecology delivers more morsels, but less nourishment. He claims that media ecology is a nifty metaphor, a pure confection of human imagination (2011). It seems to be true, but I do not agree with his statement because of the valuable perception it had brought into the communication study field. However, I appreciate his advice that new academic study of media ecology should pay attention to biological ecology's tendency, and learn to preserve the essential functions of an increasingly cluttered microcosm. Media ecologists still have a long way to go, and I think the microcosm might be a potential area to study.

COMMUNICATION STUDY IN THE PHILIPPINES

As a typical developing country with different nationalities, the Philippines are characterized by its multicultural and intensely migrant society, which also combines Eastern and Western cultures. According to the United States Census Bureau, immigrants from the Philippines made up the second largest group after Mexico that sought family reunification (2009). Although its democracy and economy are under development, the Philippines are considered a middle power. The percentage of internet users was approximately 30% (International Telecommunication Union 2013); justifying communication study of the Philippines in the fields of gender, politics, agriculture, and education, particularly focusing on female communication and democratic progress.

Females in Interpersonal Communication Study

Samarajiva, Zainudeen, Iqbal, and Ratnadiwakara (2008) studied five developing countries in Asia. By conducting cross-country comparisons, they found that the gender divide in access to telephones, which is significant in Pakistan and India, is not obvious in the Philippines. After five years, this claim will be tested in my research. The articles I mention focus on the study of females' cell phone usage, and how it influences Philippine society.

As ten percent of the population in the Philippines is working abroad, researchers show a great interest in the role telecommunication plays in family separation. Madianou (2011) uses qualitative research methods, to compare the cellphone experiences of UK-based Filipina migrants and their children left in the Philippines, to draw the conclusion that the phone enables mothers to reconstruct their role as parents, while their children are ambivalent about its consequences.

Cabanes and Acedera (2012) also look into how the mobile phone matters in the relations of left-behind fathers and migrant mothers. They interviewed ten pairs of fathers and children from mother-away families, to find the mobile phone tends to enlarge the difficulties of dealing with the changed realities Filipino families face, as opposed to their traditional expectations.

Democracy Progress: Youth and Blogs

Since new media technologies have gained critical mass in the Philippines, the political engagement activists also have changed. David (2013) studies how the engagement has been redefined by the young, and how information and communication technologies contribute to their political life. By conducting focus group discussions and in-depth interviews, he found that the opinions online are valued as a political activity, and act as a consequential action by gathering support and attention of the traditional media and politicians.

While young activists use ICTs and social networks to amplify their voices, some Filipinos also use political blogging. Mirandilla-Santos (2011) conducted a study using a list of 30 Filipino political bloggers and 64 of their readers. Most of the bloggers and readers were male, college educated, high-income, veteran internet users. They found that blogs support democracy by allowing expression and participation despite the opposition in Philippine politics.

Cellphone Education: A Creative Way

In 2007, Libreroa, Ramosb, Rangac, Triñonab and Lambert studied the potential of cell phone and short message service (SMS) techniques for education in the Philippines. In 2013, an agency (DDB DM9JaymeSyfu) planned programs to solve the issue of the decreasing use of textbooks by condensing content into SIM cards. While many developed countries' students use Kindle or iPad to do their reading or assignments, students in the Philippines can also learn by using cellphones as the means: Their bags became 50% lighter and attendance has increased to 95%. I think it is a creative idea to apply telecommunication technology in education this way. It shows that even low-end mobile technology can lead to high-end problem solving. Communication study in the Philippines also encompasses commercial, health, and agricultural fields. Most of these studies related the influence of cellular usage to traditional media such as newspapers, television, and radio.

Two Perspectives

The purpose of a technology and its usage are not always coordinate with each other, and cellphone purpose and usage are like this in the Philippines. For one thing, the mobile was invented for talking, but in the Philippines, it is used for writing (Pertierra 2013). Texting is one of the most common ways for Filipinos to get in touch with each other. It is mainly used in the private sphere, but that will not continue indefinitely. Pertierra illustrates a case study about how text strangers became text mates. Even the government now encourages people to text public agencies (24). This phenomenon has much to do with the relative high level of cellphone usage. According to Pertierra, about 85% of Filipinos have easy access to cellphones, and there were 65 million cellphone subscribers out of the population of 90 million by 2013. Nearly all, 99% of the population, had been covered by the mobile cellular network (*Philippine Daily Inquirer* 2010).

In terms of utility, a cellphone is more than a mobile. It is a multimedia platform which is highly interactive. From paper media to television, it connects almost all existing electronic services. Some researchers notice this character of new media. Mirca Madianou and Daniel Miller (2013), for example, compared people from the transnational families of the Philippines and Caribbean. Studying how they use different new media as a communicative environment rather than discrete technologies, they assumed that the primary concern of shifting an individual medium is based on the social, emotional, and moral consequences of choosing among the different media.

Raul Pertierra (2013) looks into the influence cellphones and the internet on Philippine society. He pointed out that new media can not only enable Filipinos to communicate with each other easily, but also promote their ego identity. Moreover, he found that cellphones and the internet play the role of transforming traditional media, helping them penetrate and influence increasing aspects of everyday life. He illustrated these influences with three typical case studies; however, he drew the conclusion that beside the positive ability new media promise, new technology also gives rise to new divisions and inequalities. The main factor of inequality is Filipinos' ability to access telecommunication tools. In order to certify these influences, the following research questions have emerged.

- RQ 1: How many Filipinos will respond to electronic survey questions?
- RQ 2: What age groups are more likely to use cellular telephones?
- RQ 3: Are males more likely to use electronic communication than females in the Philippines?
- RQ 4: Is education level reflected in cellular telephone usage?
- RQ 5: Are time span, frequency, gender and age related to cellular telephone usage?
- RQ 6: What correlations will appear for problematic electronic usage, relationships, and kinds of media?

METHOD

From October 24 to November 14, 2013, surveys were collected. The respondents were residents who have lived in the Philippines for more than three years. All participants were aged 18 to 65, and were able to complete the survey questions in English. Confidentiality was maintained for their recorded responses by taking no personal information such as names, addresses, places of contact, email or telephone numbers.

The respondents were expected to answer the posed research questions on Survey Monkey links placed on Facebook, QQ, and by email. Age qualified respondents shared online, or were sent the link of the online survey. The snowball method reached additional participants. No remuneration was provided as an incentive to those surveyed. Participants received self-knowledge when answering and thinking about their usage of cellular telephone and internet communication.

There were four sections asking 40 questions in total to measure different aspects of the cellular telephone and internet usage in the Philippines by citizens or sojourners. The non-parametric measure, Spearman's rho, was used to correlate age and internet usage indicators; and Pearson's Correlation Coefficient was used to correlate (1) cellular and internet chat usage, (2) the five internet usage indicators, (3) education level and the internet usage indicators, (4) positive cellular comments and cellular usage, (5) positive internet comments and email usage, and (6) age and cellular usage.

The survey was not perfectly accomplished because participants could discontinue at any time, skip questions, and disregard the survey request. All the collected data were entered and analyzed in an SPSS database.

RESULTS

Forty-seven people were invited to complete the survey, with most of them from Facebook. They were members of several groups, such as churches and Chinese students studying in the Philippines. People seldom replied when receiving the survey link the first time when we were strangers. Thus, researcher respondent communication was necessary to build an indirect friendship with respondents before conducting the survey. The snowball method worked well, due to familiar friends' recommendations, to increase the response rate to 83%. However, it was hard to find the proper responder to start the snowball method. One problem the results show was a decrease in response as the four survey segments were accomplished. See table 12.1.

Table 12.1 Effective Rates of all Respondents (n=28)

Sections	part one	part two	part three	part four
Responses	28	21	21	16
Rate	100%	75%	75%	57%

A balance in basic demographics, such as gender, age, and education was sought. However, the online approach missed responders aged 45 to 65. This group had only 1 person (see figure 12.1); and also failed to balance the education level (see table 12.5). A large number, 88.2% of the respondents, came from urban areas. Some data were missing (see table 12.2). These elements provide a direct restriction to the research analyses. However, the gender dimension was well balanced (see table 12.4); therefore, development of the analysis in the aspect of gender was attempted. See tables 12.2-12.5 and figure 12.1.

Table 12.2 Filipino Respondent Age Groups (n=28)

		Frequency	Percent	Valid Percent
	18-25	10	35.7	37.0
	23-35	13	46.4	48.1
Valid	35-45	3	10.7	11.1
	45-65	1	3.6	3.7
	Total	27	96.4	100.0
Missing	System	1	3.6	
Total		28	100.0	

Table 12.3 Valid and Missing Responses (n=28)

		education	gender	age
N	Valid	27	26	27
	Missing	1	2	1

Table 12.4 Filipino Respondent Rate by Gender (n=28)

		Frequency	Percent	Valid Percent	Cumulative Percent
Valid	male	12	42.9	46.2	46.2
	female	14	50.0	53.8	100.0
	Total	26	92.9	100.0	
Missing	System	2	7.1		
Total		28	100.0		

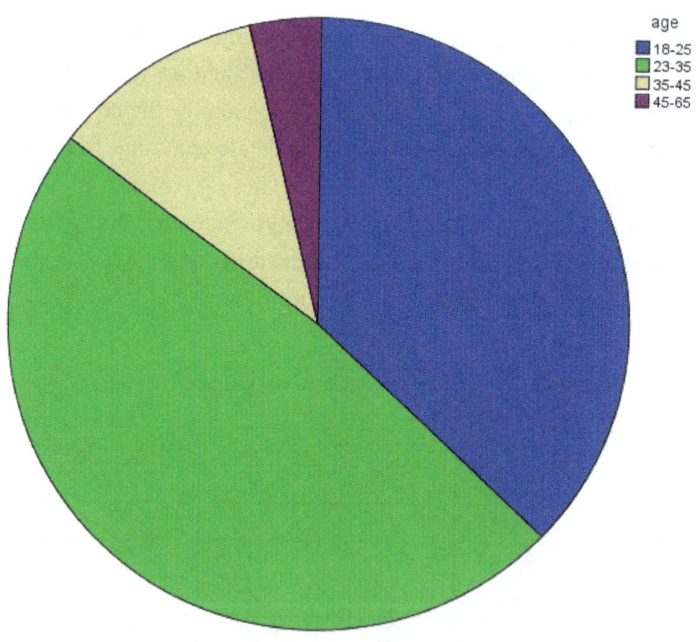

Figure 12.1: Filipinos Respondent Age Demographics (n=28)

The largest age grouping was 25-35 years old and second largest was 18-25; accounting for 82.1% of the total sample. As age increased from 35 years old, electronic survey respondents decreased.

Table 12.5 Filipino Respondent Education Level (n=28)

		Frequency	Percent	Valid Percent	Cumulative Percent
Valid	high/secondary	4	14.3	14.8	14.8
	college/university	20	71.4	74.1	88.9
	graduate-PhD	3	10.7	11.1	100.0
	Total	27	96.4	100.0	
Missing	System	1	3.6		
Total		28	100.0		

Cellphone Usage by Gender

Female cellphone usage is one focus of interpersonal communication study in the Philippines. In this study, the data shows that females use cellphones earlier than males (see figure 12.2). However, females and males have equal access to telephones in general, which supports the view of Samarajiva and Others (2008) seven years previous to this study. The data also reflect the significant characteristics of telecommunication usage in different genders. Nearly all sampled had used cellphones five years or more. Less than one year of usage (3%), and 2-3 years of usage (3%) were reported.

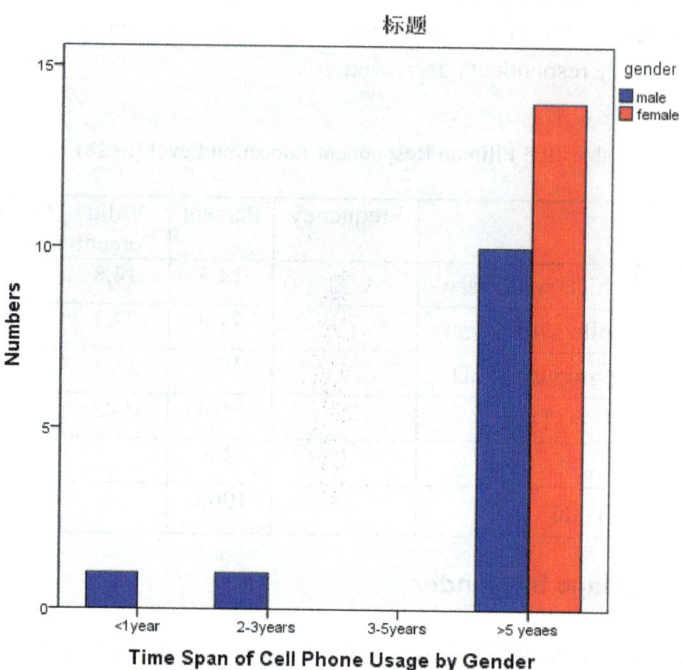

Figure 12.2: Time Span of Cell Phone Usage by Gender (n=28)

The bar diagrams came from the qualitative data of two questions (question 3 and question 4 in part four of the survey). It is clear for all age categories that females usually make more phone calls, and also receive more or equal amounts of phone calls than males. The contrast to males is marked in females aged from 18 to 25 years. Young men reported themselves as more passive in terms of cellphone usage. As age increases to 25-45 age categories, Filipino males tend to use cellular telephones for receiving calls the same as females. See Figures 12.3-12.4.

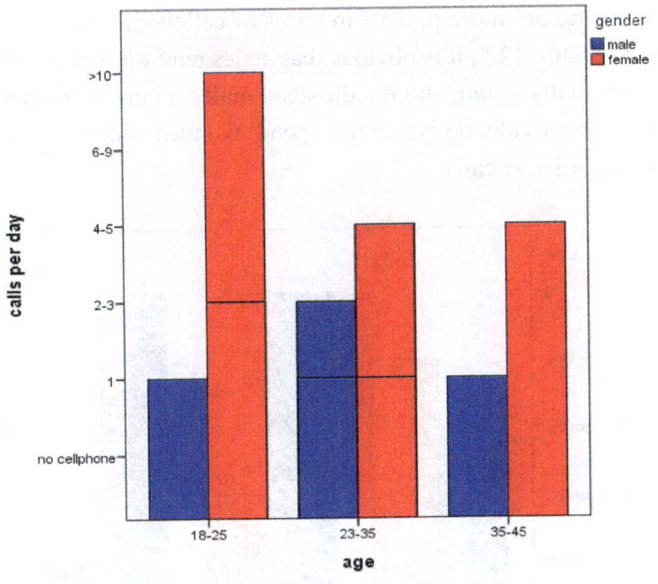

Figure 12.3: Frequency of Making Calls per Day by Gender and Age (n=28)

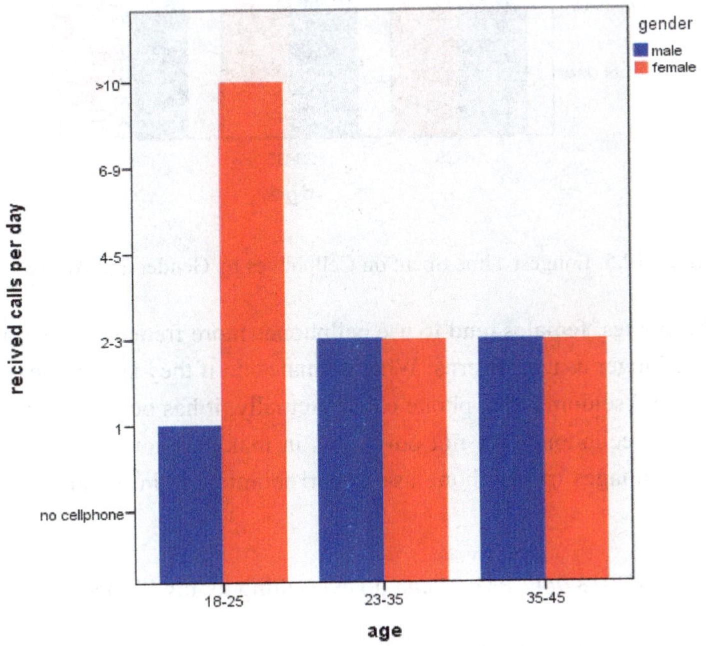

Figure 12.4: Frequency of Receiving Calls per Day by Gender and Age (n=28)

Is it true that males are more passive in terms of cellphone usage? Not really. By explanation of Figure 12.5, it is obvious that males tend to spend a longer time on cellphones, especially young and middle-aged males. Females, though they make and receive frequent calls daily, do not spend as much time as males do on the individual or cumulative calls.

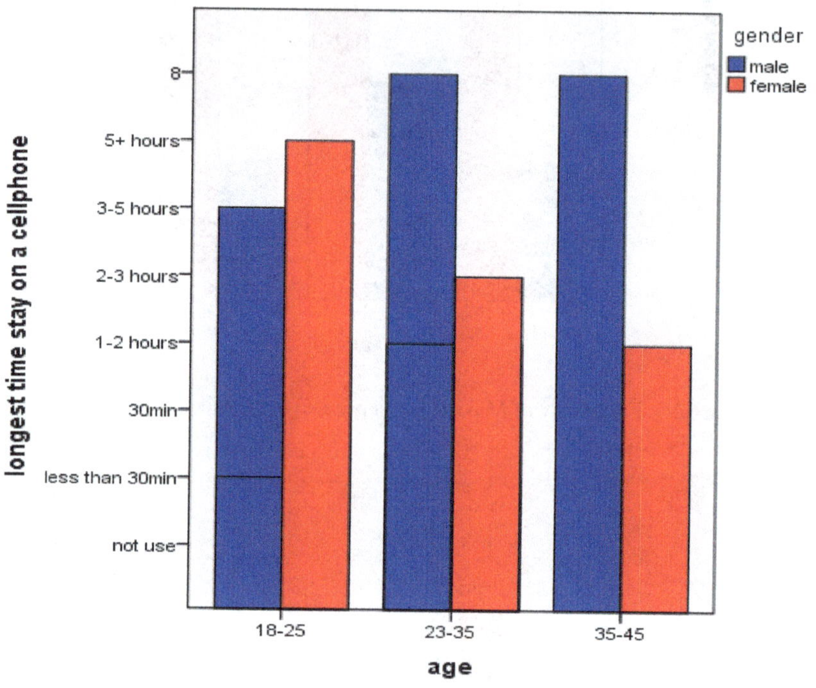

Figure 12.5: Longest Time Spent on Cellphones by Gender and Age (n=28)

In the Philippines, females tend to use cellphones more frequently for calls; while males have longer usage patterns. What do males do if they spend a lot of time on cellphones and seldom make phone calls? Actually, it has become common sense to people that cellphones are not only tools in making phone calls. Respondents gave the advantages of cellphone use to corroborate this in voluntary qualitative answers:

10/24/2013 9:42 PM: "I can do everything by my iPhone."

11/11/2013 11:56 PM: "Using a cellphone helps me (1) communicate with people (2) enables me to connect with the internet, and (3) transact business anytime and anywhere."

With access to the internet, cellphones can deal with ecological interrelationships in virtual space (Valcanis 2011).

Internet Usage Correlation Effect Analyses

Postman (2000, 1992) points out that the environment of messaging systems can affect people's feelings and behaviors. According to the current survey, 58% of the respondents acknowledged that they had lost sleep because of electronic usage (cellphone and internet). An overlapping 81% of them admitted that electronic communication is sometimes easier than face-to-face communication. The analysis divided the impact into two divisions: the positive and the negative correlations of internet usage in various aspects of life, to see how these Filipino's lives are effected by the media they use. See table 12.6.

Table 12.6 Internet Usage: Positive Correlations (n=17~27)

		kinds of media usage	more interesting relationship	more freedom in relationship	use internet hour per week
kinds of media usage	Pearson Correlation	1	.600**	.608**	.506**
	N	27	18	18	26
more interesting relationship	Pearson Correlation	.600**	1	.827**	-0.086
	N	18	19	19	17
more freedom in relationship	Pearson Correlation	.608**	.827**	1	0.168
	N	18	19	19	17
use internet hour per week	Pearson Correlation	.506**	-0.086	0.168	1
	N	26	17	17	26

** Correlation is significant at the 0.01 level (2-tailed).

Table 12.6 shows that positive feelings about personal relationships and electronic communication have a high correlation with the degree of one's media diversity usage (>.6), but are not necessarily correlated with the time one spends on the internet (<.2). Surprisingly, "interesting relationship because of electronic communication" had a slight negative correlation with one's time spent on the internet (-.08). This outcome indicated that spending too much time on the internet does not guarantee that one would have an interesting relationship with others. On its reverse side, spending too much time on the internet may lessen the "interesting relationship," be it a relationship in reality or in a virtual world.

As for the negative correlations, this study measured the dependency of three elements according to the responses to the survey: "stay on internet longer than intend to," "stay on cellphone longer than intend to", and "lose sleep because of electronic usage." Table 12.7 indicates that people who stay on the internet longer than intended were more likely to stay on cellphones longer (.749). The respondents' losing sleep because of electronic usage is related to their overtime using habit, especially on the internet (.702).

Table 12.7 Internet Usage: Negative Correlations (n=21)

		stay on internet longer than intended	stay on cellphone longer than intended	lose sleep because of electronic usage	use internet hours per week
stay on internet longer than intended	Pearson Correlation	1	.749**	.702**	-0.002
	N	21	21	21	20
stay on cellphone longer than intended	Pearson Correlation	.749**	1	.642**	-0.05
	N	21	21	21	20
lose sleep because of electronic usage	Pearson Correlation	.702**	.642**	1	0.212
	N	21	21	21	20
use internet hours per week	Pearson Correlation	-0.002	-0.05	0.212	1
	N	20	20	20	26

** Correlation is significant at the 0.01 level (2-tailed).

The average negative Pearson correlation of .698 is higher than the average positive correlation (.678), which means that one negative element/habit was more likely to induce the others. However, people frequently accessing the internet do not necessarily stay on the internet or cellphones longer than they intend. The data (<0) shows that they may even have better control of their time spent on electronic communication tools. This may be due to the respondents' occupation. For example, one of the respondents introduced himself as an IT engineer. He uses the internet because his work requires it, but he seldom stays on the internet outside of work. This is a special example and according to these respondents, we can hardly see any significant correlation between overtime usage and education, gender, age, or economic state.

DISCUSSION

These results indicate that the influential relationship between electronic media usage and the Philippine users are mutual and dynamically balanced. The gender factor acting on cellphone usage is clearly represented by the in-depth sample of qualitative and quantitative data, and the functions of cellphones have also given rise to user dependence. The internet usage brings about both positive and negative effects upon Filipino's lives. The outcome suggests that it may not follow the preconceptions of empiricism, which makes it worthy of further study.

However, I personally doubt the result that the access to cellphones and the internet had no significant relationship to respondents' basic classifications, such as education, age and economic state. The reason for the failure to test the effects of these elements may be due to the insufficient consideration of sampling.

Besides overcoming this deficiency, further studies are suggested to find more traits concerning media usage in the Philippines, to figure the reasons that these traits exist; and hopefully, to form a media ecology of regular pattern that fits developing countries. Those in Asia are especially critical for our expanding network of commerce, social services, and education.

REFERENCES

Cabanes, Jason Vincent A., and K.A.F. Acedera. 2012. "Of Mobile Phones and Mother-Fathers: Calls, Texts, and Conjugal Power Relations in Mother-away Filipino Families." *New Media & Society* 14: 916-30. Accessed October 29, 2013. doi: 10.1177/1461444811435397

Castles, Stephen, and Mark J. Miller. (2009). "Migration in the Asia-Pacific Region". *Migration Information Source.* Accessed October 28. www.migrationinformation.org/feature/display.cfm?ID=733

Cruz, Deirdre De La. 2009. "Coincidence and Consequence: Marianism and the Mass Media in the Global Philippines." *Cultural Anthropology* 24: 445-48.doi: 10.1111/j.1548-1360.2009.01037.x

David, Clarissa C. 2013. "ICTs in Political Engagement among Youth in the Philippines." *International Communication Gazette* 75: 322-37. Accessed October 29, 2013. doi: 10.1177/1748048512472948

Fuller, Steve. *The Intellectual*. Cambridge: Icon Books, 2005.

Kincaid, Lawrence D. 2000. "Mass Media, Ideation, and Behavior: A Longitudinal Analysis of Contraceptive Change in the Philippines." *Communication Research* 27: 723-63. doi: 10.1177/009365000027006003

Librero, Felix, Angelo Juan Ramos, Adelina I. Ranga, Jerome Triñona, and David Lambert. 2007. "Uses of the Cell Phone for Education in the Philippines and Mongolia." *Distance Education* 28: 231-44. Accessed November 4, 2013. doi: 10.1080/01587910701439266

Madianou, Mirca. 2011. "Mobile Phone Parenting: Reconfiguring Relationships between Filipina Migrant Mothers and Their Left-behind Children." *New Media & Society* 13: 457-70. Accessed October 29, 2013. doi: 10.1177/1461444810393903

Madianou, Mirca, and Daniel Miller. 2013. "Polymedia: Towards a New Theory of Digital Media in Interpersonal Communication." *International Journal of Cultural Studies* 16: 169-87. doi: 10.1177/1367877912452486

Ong, Walter. *Interfaces of the Word*. Ithaca, New York: Cornell University Press, 1977.

Otremba, Jolene. 2012. "Social Media Report: Philippines the Social Capital." *Campaign Asia-Pacific*. Business Source Premier, EBSCOhost (78950336). Accessed October 7, 2013.

Pathak, Shareen. 2013. "Why TXTBKS Won At Cannes." *Advertising Age* 84:22-22. Accessed November 4, 2013. http://search.proquest.com/docview/1426819077?accountid=27424

Postman, Neil. 2000. "The Humanism of Media Ecology." *Proceedings of the Media Ecology Association* 1: 10-16. Accessed October 6, 2013.

Postman, Neil. *Technopoly: the Surrender of Culture to Technology*. New York: Alfred A. Knopf, 1992.

Postman, Neil, and Charles Weingartner. *The Soft Revolution: A Student Handbook for Turning Schools Around*. Delacorte Press, 1971.

Pertierra, Raul. 2013. "We Reveal Ourselves to Ourselves: The New Communication Media in the Philippines." *Social Science Diliman* 9: 20-40. Accessed October 23, 201 http://journals.upd.edu.ph/index.php/socialsciencediliman/article/viewAr ticle/3738

Mirandilla-Santos, Mary Grace. 2011. "A-List Filipino Political Bloggers and Their Blog Readers." *Media Asia* 38: 3-13.

Strate, Lance. 2003. "The Cell Phone as Environment." *Explorations in Media Ecology* 2:1, 19-24.

Strate, Lance. July 2011. Further Thoughts on Online Education. *Blog Time Passing.* http://lancestrate.blogspot.com/2011/07/further-thoughts-on-online-education.html

Strate, Lance. *Echoes and Reflections: On Media Ecology as a Field of Study.* New Jersey: Hampton, 2006.

Strate, Lance. 2004. "A Media Ecology Review." *Communication Research Trends* 23: 3-48.
Accessed September 21, 2013. http://cscc.scu.edu/trends/v23/v23_2.pdf

Scolari, Carlos A. 2012. "Media Ecology: Exploring the Metaphor to Expand the Theory." *Communication Theory* 22: 204-25. Accessed October 6, 2013.
doi: 10.1111/j.1468-2885.2012.01404.x

Samarajiva, Rohan, Zainab Zainudeen, Tahani Iqbal, and Dimuthu Ratnadiwakara.
2008. "Who's Got the Phone? The Gendered Use of Telephones at the Bottom of the Pyramid." *New Media & Society* 12: 549-66.

Valcanis, Thomas. "An I-phone in Every Hand: Media Ecology, Communication Structures, and the Global Village." *ETC: A Review of General Semantics* 68, no. 1(2011): 33.

Wood, Chris. "Media Missing the Message." *Alternatives Journal* (2011). http://www.readperiodicals.com/201105/2338253151.html

Cellular Telephone and Internet Usage in Russia

Julia Daiche

This chapter will deal with cellular telephone and internet usage in Russia. The focus of attention is directed to the most frequently used types of communication media, the time spent by survey participants on cellular telephones and the internet, as well as the period of time these devices were used. Furthermore, part of the research is survey participant demographics, the usage of electronic devices such as cellphones and the internet, and reported advantages of using cellphones and the internet. Quantitative and qualitative data was collected and documented by percentage descriptors of respondents. Qualitative data were derived from individual comments provided in the questionnaires.

Russia is the world's largest state in terms of area, but with a population of 143 million it is considered to be one of the world's less densely populated states (Internet World Stats 2013). This fact makes Russia a country of great diversity and controversy. Moscow and St. Petersburg are Russia's main agglomerations in terms of rapid economic growth, industrial and educational development, population density, and income per capita. There are many rural and isolated areas, which are slower in their development than Moscow and St. Petersburg.

Economic development, democracy and free communication in the Russian culture affect mobile usage which has been rapidly and widely adopted in the country (Salmi and Sharafutdinova 2012, 385). Factors like low income, unregulated markets and low demand on mobile technology are influencing mobile usage. Russia is described as a transformation country using the GSM standard since 1992 (Ponder and Markova 2002, 173). Education of the customer is regarded as the key to successful promotion of mobile internet (Skvortsova 2005, 2). The mobile device has become more and more important in the everyday life of many Russians and has dramatically changed their lives. It has emerged as an information portal and source of mass communication. One way to explain the adoption of such a device is the Diffusion of Innovations theory.

Diffusion of Innovations Theory

The Diffusion of Innovations theory was first introduced by Everett Rogers, a professor of rural sociology, in the 1960s. Since its introduction the Diffusion of Innovations theory has been used and tested in thousands of studies over the last five decades (Robinson 2009). The theory has not only been used to explain the development of various products and ideas, but also to explain its ability to influence individuals, groups, and societies (Farr and Ames 2008, 376).

Rogers describes diffusion as a kind of social change. He defines diffusion as "the process by which an innovation is communicated through certain channels over time among the members of a social system" (Rogers 1995, 5). It is a special type of communication—all the messages are about new ideas. The newness of ideas gives diffusion its special character (Rogers 1995, 6). Rogers (1995) describes four crucial elements involved in the diffusion of new ideas: innovations, channels, time and social systems. An innovation is a new idea, practice, or object (Rogers 1995, 11). Something is regarded as an innovation when individuals perceive it as new. Individuals' perception of the newness determines their reaction to it. Farr and Ames (2008) emphasize that innovations diffuse more quickly when individuals of a social system perceive them to be better and more beneficial than current ideas or practices (377).

One important process in Rogers' Diffusion of Innovations Model is the innovation-decision process. It is a process through which a person passes from first knowledge of an innovation to developing attitudes towards it. The four stages included in the innovation-decision process are the knowledge stage, the persuasion stage, the implementation stage, and the confirmation stage (Rogers 1995, 36).

The rate of adoption is determined by the characteristics of an innovation as perceived by the members of a society. Rogers (1995) describes the attributes of an innovation as: relative advantage, compatibility, complexity, trialability, and observability. "*Innovativeness* is the degree to which an individual or other unit of adoption is relatively earlier in adopting new ideas than the other members of a system" (22). He outlines five different categories of people on the basis of their involvement in change and innovativeness: innovators, early adopters, early majority, late majority, and laggards (37 f). The critical question that arises is how does the theory integrate individuals who will never adopt a particular innovation e.g. Facebook, iPhone, Kindle, etc.? The theory does not seem to include those individuals and does not explain their behavior. Concerning this target group, the Diffusion of Innovation theory appears to be incomplete.

Another important characteristic of the theory and specific way in which the time dimension is involved is the rate of adoption. Rogers (1995) describes it as the relative speed with which members of a social system adopt an innovation. When a large number of people adopts and use a new idea, the idea rises on a cumulative frequency over time. The distribution results in an S-shaped curve, which is a 'trademark' of the theory. At the beginning only a few individuals (innovators) adopt an innovation. As more and more individuals (early adopters and early majority) adopt the innovation, the curve begins to climb. As fewer and fewer individuals (laggards) remain who have not yet adopted the innovation, the trajectory of adoption levels off and begins to decline. The S-shaped curve reaches its asymptote, and the diffusion process is completed (22 f). See figure 13.1.

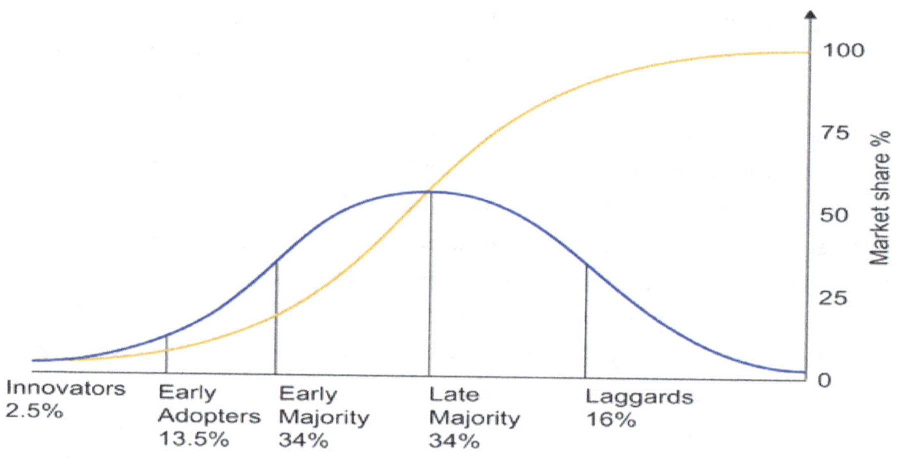

Figure 13.1: Standard Rogers S-curve for Diffusion of Innovation
(Sharma, Sharma, Hemant, and Singh 2011)

Hogg, Lomicky, and Hossain (2008) used Rogers' theory as a framework to analyze the mass media's role in the spread of information. In their study the innovation was a blog or blogging. Focus of analysis was how mass media have portrayed and introduced the innovation of blogging to the public. The total number of articles reporting about blogging resembles an S-shaped curve, which is represented in Rogers' Diffusion theory. This curve denotes the typical rate of adoption of a new innovation. The study shows that media serve an important function of providing information and knowledge about blogs as an innovation. Without the mass media this innovation would not have reached such a large audience and would not have spread so rapidly. Another point of critique arises in

Rogers' S-curve. His S-curve is static and does not react to dynamic processes or innovative technological changes. Especially technological gadgets are constantly improved and newly developed. It might happen that the S-curve will never be completed because new innovations will replace old ones, which will never be adopted by the majority.

There are two other important factors concerning the importance of opinion leaders in the innovation-decision process. The first factor is the importance of opinion leaders in peer-to-peer networks while the second factor is their role of 'reinvention'. They are regarded as being responsible for implementing change and refining the innovation to meet the needs of risk-averse society members (Li and Edwards 2013, 394). Opinion leaders usually disseminate ideas through the social system. By identifying and using their influence one can accelerate the adoption of an innovation (Lomas and Others 1991).

These theoretical constructs are confirmed in Barker's (2004) research study. He applied the Diffusion of Innovation theory to demonstrate its application with different populations, sectors and regions. The three countries to which the theory was applied were Haiti, Nepal, and Mali. In these examples Barker demonstrated that known and trusted opinion leaders in a community have an enormous influence when it comes to the community's acceptance or rejection of new innovations. Once opinion leaders are convinced of the benefits of a particular innovation, they are able to influence the entire community in favor of that innovation. The conclusion is that when the source of information is highly credible, the Diffusion of Innovations theory will be most effective.

Li and Edwards' (2013) study examined the change in English Language Education (ELE) in Western China after the impact of a UK-based professional development program on the Chinese curriculum. The Diffusion of Innovations theory was used as a conceptual model to examine how innovative approaches to English as a Second Language (ESL) pedagogy are adopted by teachers. By applying the Diffusion Model in particular stages of the adoption process, the authors showed that the success of an innovation depends on an 'opinion leader' who introduces the innovation. The critical question that arises here is what would happen in a worst-case scenario when opinion leaders are against a particular innovation? Will the same S-curve be formed inverted and run negatively? Will market shares drop? Can this phenomenon be related to image damage of a particular brand and their products, losing credibility, and trust at the business market? What happens with a popular innovation when the opinion leader will lose his or her popularity? Will that penalize the innovations' success?

Nguyen's (2012) study was about online news diffusion. A forecasting model of online news adaptation is used to support this argument that, even if the internet becomes easily accessible to everybody, a 'digital delay' will not occur. Online news adopters show social and economic differences from non-adopters displayed in differences in education, personal income, gender, age, combined income, and living area. Later adopters of online technologies will be left further and further behind, as they oftentimes tend to be less socio-economically and psychologically successful. Earlier adopters will gradually establish an extensive market demanding more complex skills, which will become a guideline for technology producers. Later adopters who missed the early stages of online communication will be further left behind and abandoned. Nguyen's argument is another weak point of the Diffusion Model especially in online technologies or technologies in general. The gap between earlier adopters and later adopters grows bigger and bigger. The dynamic character of technologies makes it difficult to keep up with, and socio-economically successful individuals are in an advantage to outpace those who are socio-economically less successful.

Westlund (2008) examined the diffusion and adoption of the mobile phone. The empirical focus was on mobile news services in Sweden. The function of mobile phones has changed since its early adaptation. Nowadays, the mobile phone is no longer a simple telephone but integrates communication and multimedia functions. Westlund (2008) claims that diffusion research should focus on mobile phone usage, not possession. The study shows that people's patterns of news consumption differ between men and women, in age, and by educational level. The hypothesis in the Diffusion theory is confirmed that early adopters have specific characteristics. They are likely male, 15-49 years old, are well educated and have a great interest in technology. Here again, a weak point of the Diffusion Model can be recognized. It is important to focus on peoples' usage of diffusion and not only on their possession.

Russia

According to Vershinskaya (2002) one of the major factors in a country's readiness to become an information society is its access to a modern telecommunication infrastructure. Fixed and cellular telephone communications are main indicators for a country's information infrastructure (139). Russia's country area covers around 12.5% of the Earth's territory (Vershinskaya 2002, 140), but only 1.987% of the world's population live there (*World Population Review* 2013). Russia's population density, which is an important indicator for information and communication technologies (ITC) dissemination, is rather low

(8 persons per sq km) compared to other countries (Internet World Stats 2013). Furthermore, the population density differs from region to region, and varies between rural and urban areas (Vershinskaya 2002, 140).

Rice and Katz (2003) claim that the poorest segment of a population is often the earliest adopter of mobile phones in developing countries (602). They argue that "the prices of the ICTs do not serve as a substantial barrier, a finding of surprising import when considering the subsidy policies of many Western governments relative to the internet" (602). Various indicators e.g. macro-economic indicators, social indicators, and even individual life stories reveal the importance of mobile phone technology. Different national surveys (Vershinskaya 2003), that have analyzed the mobile phone explosion in post-communist Russia and the micro-social capital analysis of Ling and Hadden (2003), who analyzed the use of mobile phones in helping teens (and sub-teens) to do their homework and manage their household moral economies, demonstrate cellphones' importance (Rice and Katz 2003, 602).

Vershinskaya (2003) showed how social and economic changes in the Russian society have resulted in an ICT revolution in the mid-1990's. Russia was perceived as lagging in the sphere of information technology from the 1960's to the 1980's. In the late 1980's and early 1990's Gorbachev's restructuring, democratization, and openness opened the door for ICTs. The internet and the mobile phones disseminated (Campbell 2007, 347). The Soviet Union's development of telecommunications suffered from grave under-investment because the main preference in the domestic industry was given to heavy industry and official policy, which promoted public access to telephones (Regli 1997, Rantanen 2001). As a consequence, in the first half of the 1980's, the Soviet Union had around 28 million phones, while the USA had 170 million with about the same population size (Ganley 1996, Rantanen 2001, Lonkila 2008).

Mobile communication is still perceived to be a new ICT in the world. While in Europe its rapid penetration started at the beginning of the 1990's (Vershinskaya 2002, 139), in Russia the mobile communication boom started only in the second half of 1999-2000. Prior to that time, less than 1% of Russia's population used mobile phones (141). Before 1999, cellphones were considered largely as status symbols in Russia, which only few wealthy Russians could afford (Lonkila 2008, 277). Cellphone users were mostly male, elite and 38-45 years old (Vershinskaya 2002, 141). The elite character of mobile communication was reflected by high charges and high prices of cellphones (Lonkila 2008, 277). According to Lonkila (2008) the Russian domestic telecommunication market began to develop rapidly after financial crisis in 1998. When tariffs declined and the cellphone lost its symbolic value, it became more affordable for the larger part of the Russian population (277). The number of cellular connection network customers grew rapidly during this time.

Representative surveys carried out by the Levada Centre (2007) reveal that, in 2001 and 2003, 46% of Russians had a landline telephone at home, while in 2007 there were 57%. According to Lonkila (2008) this small increase suggests that the cellphone was becoming increasingly popular among the population in comparison to the landline (276).

While in the past the first cellphone users were elite, male and 38-45 years old, nowadays the main trend in mobile communication development is more universal. The most active users of mobile phones, which make up the majority of users, are younger people. Since the mobile phone became less expensive, it is affordable for the middle class and a necessity (Vershinskaya 2002, 149). Today, the cellphone has become an essential part of Russians' daily lives. Nobody would want to leave home or workplace without their cell phones (Lonkila 2008, 281). Lebedeva's (2011) survey has shown that a major portion of the students in her study (74%) own a computer and a cell phone. "Gender differences between the respondents in regard to ICT use are not really significant. ... The region of residence also has very little influence on ICT use" (Lebedeva 2011, 7). See figures 13.2 and 13.3.

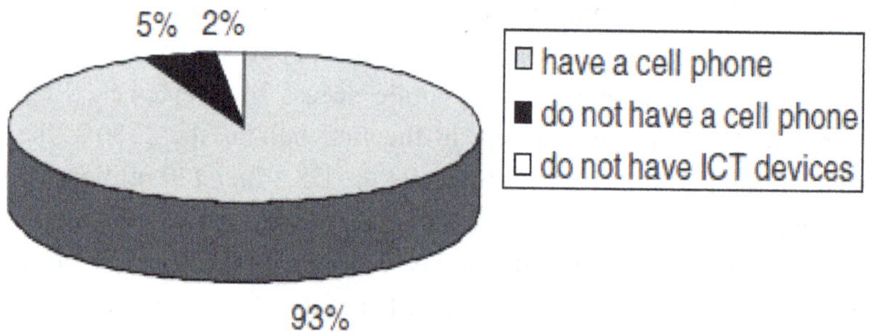

Figure 13.2: Possesion *of Cellphones*
(Lebedeva 2011, 7)

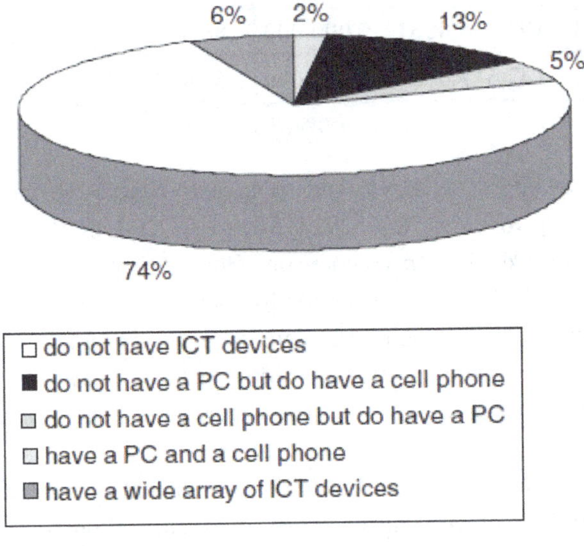

Figure 13.3: Distribution into Groups by Level of ICT Use
(Lebedeva 2011, 8)

At the beginning of 2000, the number of cellphone subscribers in Russia (indicated by the number of SIM-cards) was 1.3 million. Only three years later it was 17.8 million and 140.3 at the end of 2006. In 2001, 2% of the Russian population had a cellphone. In 2003, 9% and 61% in 2007 (Levada Centre 2007, cited in Lonkila 2008, 277). According to Internet World Stats (2013) in December 2011 the number of Internet users was 59,700,000 i.e. a 42.8% penetration according to the International Telecommunication Union (ITU). In a national comparison Russia is on the 6th rank in the top 20 countries with the highest number of internet users. See table 13.1.

Table 13.1 The Top Countries with the Highest Number of Internet Users – June 30, 2012
(Internet World Stats 2013)

#	Country or Region	Population, 2012 Est	Internet Users Year 2000	Internet Users Latest Data	Penetration (% Population)	Users % World
1	China	1,343,239,923	22,500,000	538,000,000	40.1 %	22.4 %
2	USA	313,847,465	95,354,000	245,203,319	78.1 %	10.2 %
3	India	1,205,073,612	5,000,000	137,000,000	11.4 %	5.7 %

4	Japan	127,368,088	47,080,000	101,228,736	79.5 %	4.2 %
5	Brazil	193,946,886	5,000,000	88,494,756	45.6 %	3.7 %
6	**Russia**	**142,517,670**	**3,100,000**	**67,982,547**	**47.7 %**	**2.8 %**

(Internet World Stats 2013)

The trend of owning a cellphone and using it to access the internet grows in Russia. According to Leontew's (2013) webpage, 51% of Russians own more than one cellphone device. In comparison: Only 17% of Americans own more than one cellphone. The main factor for Russian consumers to buy a cellphone depends largely on its appearance/design. In a survey conducted by J'son & Partner Consulting in October 2012, 43% of smartphone users and 57% of tablet PC users are planning to spend more time on the web in the next year than they do now.

Even when the internet is becoming increasingly important to Russian communication and sociality in the urban areas, the cellphone still remains an important link "between individuals and their household, sustaining social networks—just as it does in less developed countries" (Pelckmans 2009, Morris 2013, 189). Especially in Russia, cellphones fulfill a number of technological and social functions beyond making calls and texting (Strathern 2002, Morris 2013). Nowadays, cellphones are multifunctional devices. A survey conducted by the Levada Centre (2012) in 2011 shows that mobile phones (including Smartphones) are even used as electronic media to read electronic books (331). In 2010, 30% of survey respondents provided information that they make use of the internet on their phones, while 44% do not use it at all (337).

Based on the information provided by the literature review of Russia, five research questions about its cellular telephone and internet usage arose. This study answers the following research questions:

- RQ1: What are respondents' age, gender and educational demographics who use electronic devices?
- RQ2: What type of communication media is more frequently used by the respondents?
- RQ3: What advantages are reported by respondents of using the internet and cellphones?
- RQ4: What amount of time was the longest respondents have stayed on a cellphone and on the internet?
- RQ5: For how long have respondents been using cellphones and the internet?

METHOD

For the research 42 completed surveys were collected on the usage of cellular telephones and internet communication in Russia. Respondents of the electronic survey must have been either residents of Russia, nationals or sojourners who have spent at least three months in the country. Thus, most respondents of the study are international, although it was not necessary that they were current residents of Russia or took the survey there. Participants must have been between 18-65 years old to be able to participate in the study. Age, gender and ethnicity were not screening factors in participants' selection and no other screening criteria were used. It was desirable that participants were equally divided between male and female, and urban and rural residents. It was hoped participants would also vary in their age range and educational background.

Surveys were distributed between 24 October and 18 November, 2013. Social media was used to post recruitment information for the survey. Networks such as ICQ, Skype, Facebook and E-mail were used by posting the survey and the Recruitment Information Sheet (RIS). As the Principal Investigator (PI), I depended on the snowball method of dispersal once key Russian subjects, nationals and sojourners were contacted. The snowball method is especially quick and easy to apply due to the nature of social media.

The survey contained 40 questions. Participants needed approximately 15-20 minutes to answer all questions. The web survey tool Survey Monkey was used to manage responses, which were split in four parts. Each of the four provided Survey Monkey links containing 10 questions.

Survey participants were informed about their voluntary participation and the right to discontinue participation at any time or skip any question they wanted. The survey results were placed in aggregate form by Survey Monkey, and analyzed by the primary investigator to retain anonymity and confidentiality. No identification of respondents was possible or desired. Responses were not linked to any name as names or identifiers were never used.

Participants have not directly benefited from this study and were not awarded any compensation for participating. It is hoped that they have benefited from the opportunity to reflect upon their own behavior. Participants received self-knowledge when answering and thinking about their usage of cellular telephone and internet communication. Benefits for the discipline were obtained from the knowledge gained by the study and may benefit other colleges, universities, and future students.

Each of the respondents had an explanation of respondent rights, and each was given the option of voluntary survey completion. There was no written documentation of consent. Consent was assumed when participants took the survey. No debriefing was mandated.

RESULTS

RQ1: What are respondents' age, gender and educational demographics who use electronic devices?

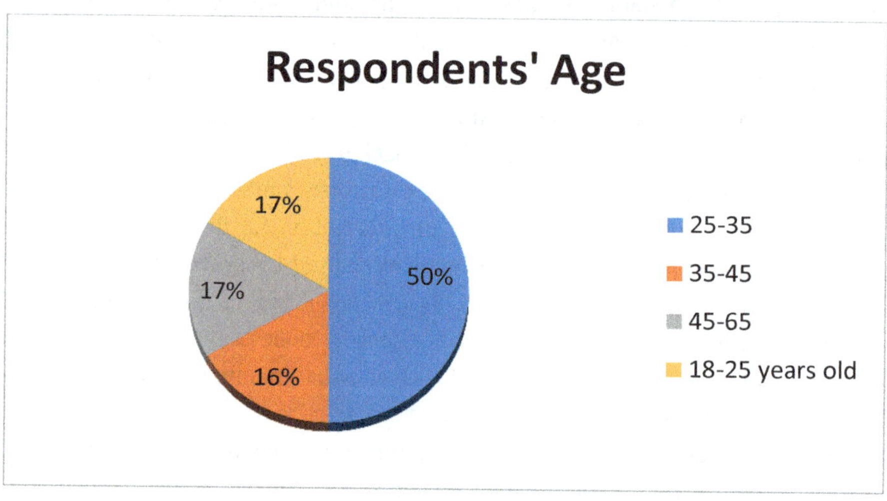

Figure 13.4: Respondents' Age Demographics (n=42)

Respondents' age demographics are presented in Figure 13.4. The majority (50%) of the respondents is between 25-35 years old. The first of the two second largest groups of respondents (17%) is between 18-25 and the second group (17%) is between 45-65 years old. The smallest group of respondents (16%) is between 35-45 years old.

Table 13.2 Respondents' *Gender* (n=41)

Answer Choices	Responses
male	43.90%
female	56.10%

Respondents' gender is represented in Table 13.2. The majority (56.10%) of respondents is female, while 43.90% is male.

Table 13.3 Respondents' Educational Backgrounds (n=42)

Answer Choices	Responses
Primary school	0%
High or secondary school	11.90%
Institute, college or university	54.76%
Graduate school (MA, MS, PhD, MD, LD)	33.33%

Respondents' educational background is represented in Table 13.3. The majority of the respondents (54.76%) have finished an institute, college or university. A large number (33.33%) have graduated from a graduate school. While none of the respondents have finished only the primary school, 11.90% graduated from a high or secondary school.

RQ2: What type of communication media is more frequently used by the respondents?

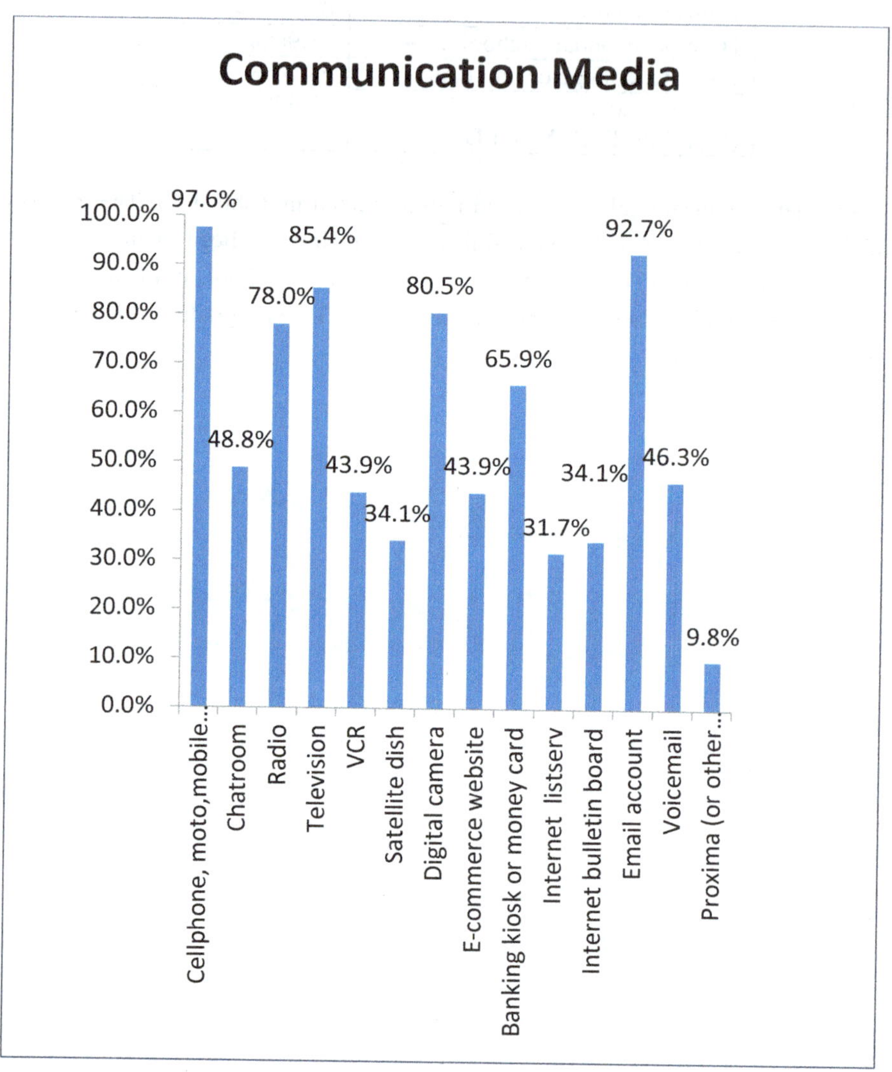

Figure 13.5: Types of Communication Media Used by Respondents (n = 41)

Different types of communication media which were used by the respondents at least once is represented in figure 13.5. The cellphone (97.6%), the email account (92.7%), the television (85.4%) and the digital camera (80.5%) were four communication media used at least once by the majority of respondents.

RQ3: What advantages are reported by respondents of using the internet and cellphones?

Table 13.3 The Most Positive Advantages of Using Cellphones as Reported by Respondents in Their Chronological Order

1) It is an easy way to communicate and share information (e.g. videos, pictures, music, etc.).
2) It bridges distance.
3) It is useful in cases of emergency.
4) It is convenient, as it is not bound to a place.

Table 13.3 represents the four most frequently positive evaluated advantages of using cellphones by the respondents. Some of the respondents' qualitative responses to the question "I like to use the cellphone because …. Its advantages are …." are listed:

"You can talk to people you cannot meet in person because of the distance between you (e.g. when living in different countries)."
"It is easy to use."
"To call anybody from everywhere you like."
"In critical situations it is always on my side."
"I can stay in contact with all my friends all around the world."
"It is convenient to use in case of an emergency."
"You can share information e.g. by sending SMS and share pictures, music, etc."

Table 13.4 The Most Frequent Advantages of Using the Internet as Reported by Respondents in Their Chronological Order

1) Fast and easy access to information.
2) It is a cheap communication tool (e.g. with people living in different countries).
3) Synchronous and asynchronous types of communication are possible.
4) It is fun.

Table 13.4 represents the four most frequently positive evaluated advantages of using the internet by the respondents. Some of the respondents' qualitative responses can be found below to the question "I like to use the Internet because …. Its advantages are …."

"It enables you find very quickly almost every type of information you
    need."
"It is a cheap opportunity to communicate with friends abroad."
"It is easy to use. I can send a message if my opponent is offline."
"I can get a lot of information in a timely manner."
"I can get information on whatever I need. It is fun and easy
    communication."

RQ4: What amount of time was the longest respondents have stayed on a cellphone and on the internet?

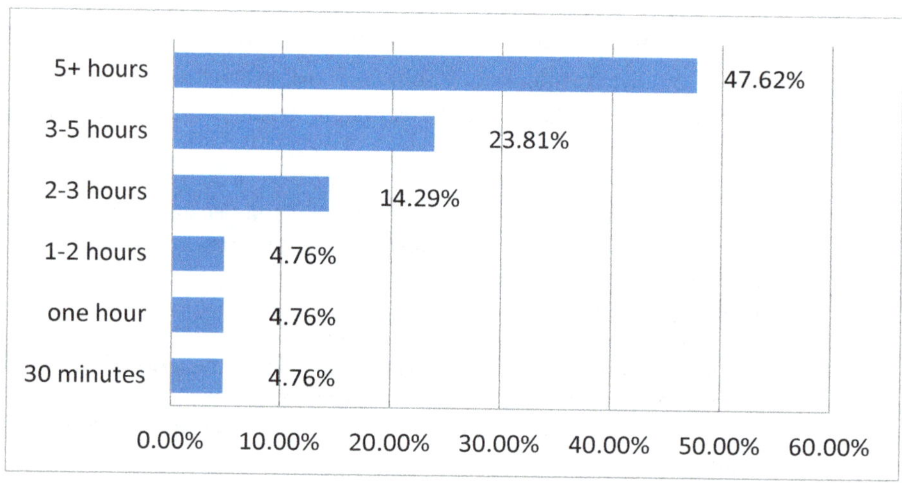

Figure 13.6: Amount of Time Respondents Have Stayed the Longest on the Internet (n=42)

Figure 13.6 shows the amount of time respondents have stayed the longest on the internet. The majority (47.62%) has stayed five or more hours the longest on the internet. The second largest group (23.81%) stayed 3-5 hours online. The third largest group (14.29%) stayed 2-3 hours the longest on the internet.

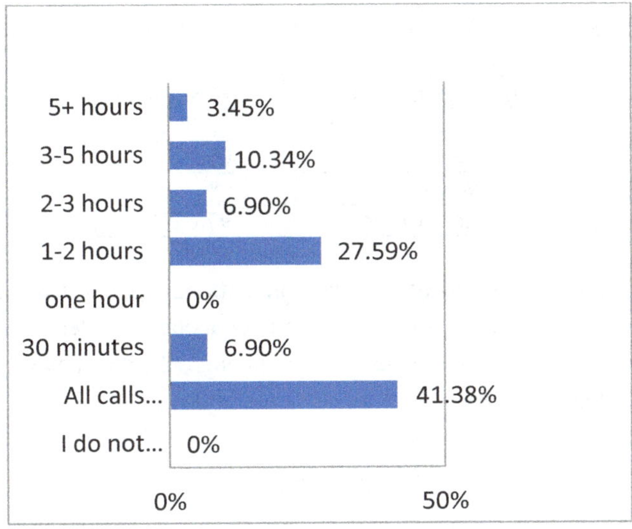

Figure 13.7 Time Stayed on a Cellphone (n= 28)

Figure 13.7 shows the amount of time respondents have stayed the longest on the cellphone. The majority (41.38%) makes all calls less than 30 minutes. The second largest group (27.59%) stayed 1-2 hours the longest on a cellphone. The third largest group (10.34%) stayed 3-5 hours the longest on a cellphone.

RQ5: For how long have respondents been using cellphones and the internet?

Table 13.5 Period of Time Respondents Have Been Using a Cellphone (n=42)

Answer Choices	Responses
Less than one year	4.76%
2-3 years	0%
3-5 years	0%
5 years or more	95.24%

Table 13.5 shows the period of time respondents have been using a cellphone. The majority (95.24%) of respondents have been using cellphones for 5 years or more. Only a small percentage of respondents (4.76%) have been using a cellphone less than one year.

Table 13.6 Period of Time Respondents Have Been Using the Internet (n=42)

Answer Choices	Responses
Less than one year	2.38%
2-3 years	0%
3-5 years	0%
5 years or more	97.62%

Table 13.6 shows the period of time respondents have been using the internet. The majority (97.62%) of respondents have been using the internet for five years or more. Only a small percentage of respondents (2.38%) have been using a cellphone less than one year.

DISCUSSION

RQ1: *What are respondents' age, gender and educational demographics who use electronic devices?*

In the sample, half of the respondents, 50% indicated to be between 25-35 years old. Sixteen percent of the respondents were 35-45 years old, and the two groups of the 45-65 and 18-25 years old respondents both represent 17% of the total sample. The majority of users of electronic devices are young people between 25-35 years old. This finding correlates with Vershinskaya's (2002) claim that the majority of cellphone users are young people and not male, elite and between 38-45 years old as was the case before 1999 as described by Lonkila (2008). However, as Vershinskaya (2002) does not tell what particular age group she defines as being young people, it remains unclear whether a correlation in respondents' age exists.

Another difference between cellphone users before 1999 and the respondents of this survey is that the majority of respondents who use electronic devices in this sample are female. Females represent 56.10% of the sample, while males represent 43.90%. The higher number of female respondents in the sample does not simultaneously mean that females use more electronic devices than males.

The majority of respondents in the sample have high educational background. Most of the respondents, 54.76%, have finished an institute, college or university. Graduate school (MA, MS, PhD, MD, LD) was completed by 33.33%, while only 11.90% finished high or secondary school and none of the respondents finished only the primary school.

These results indicate that the sample is uniform in regard to respondents' educational background as they do not vary greatly but the majority has a higher degree. Due to the lack of a greater variety in respondents' educational background the sample is one-sided and biased. Conclusions about respondents' usage of electronic devices can only been drawn for this particular group but not for less educated people.

RQ2: What type of communication media is more frequently used by the respondents?

Cellphones, email accounts and the television were the three most frequently used communication media. The cellphone was used by 97.6%, the email account was used by 92.7% and the television was used by 85.5% of all survey respondents. Other media used by more than 50% of the respondents were the digital camera with 80.5% and the radio with 78%.

Respondents' usage of communication media provides us with information about Russian people media preferences. It is important to bear in mind that Figure 8.5 was based on the respondents' instruction "I have used (indicate all that apply) the following communication media at least once" this means that Figure 8.5 can, but does not necessarily have to represent respondents' most frequent usage of various communication media. The data were not established on respondents information of how frequently they use some of the communication media, but if they have used some of the communication media at least once in their life. Although it is unlikely, it might still be possible that even when respondents have used one of the communication media at least once, they do not use it on the daily basis.

RQ3: What advantages are reported by respondents of using the internet and cellphones?

The most frequent stated advantage of cellphones as stated by survey respondents is that cellphones offer an easy way to communicate and share information. Cellphone users can easily share videos, pictures, music and other types of electronic messages between each other. Today's technology offers many possibilities to distribute data. Cellphones are easy to use and do not require technical knowledge or preliminaries.

Cellphones' second advantage is that they bridge distance between interlocutors. As stated by one respondent "You can talk to people you cannot meet in person because of the distance between you (e.g. when living in different countries)." This advantage is offered by landline telephones too and not only by cellphones.

The third and fourth advantages of cellphones are related with each other. Respondents claim that cellphones are useful in cases of emergency and are convenient to use because they are not bound to a particular place like landline telephones for example. Cellphones are always at respondents' side and are convenient to use when in need for using them.

Respondents' most frequently reported advantage of using the internet is that it presents a fast and easy opportunity to access information. As stated by one respondent "It enables you to find very quickly almost every type of information you need." People do not have to spend much time and look information up in books but can easily access the internet and just type in any catch-phrase they are looking for and search engines like Google will provide you with all the information the World Wide Web has to offer.

Similar to the cellphone the internets' second advantage is that it is a cheap communication tool. While several years ago people had to pay much money to make long distance calls, today there are programs like Skype, which allows people to communicate via internet for free. This is especially advantageous when people live in different countries and long distance calls are expensive. Furthermore, emails and other various social networks or communication platforms allow people to communicate in a written form for free.

Another advantage is that the internet does not only offer the chance to communicate in synchronous but also in asynchronous ways, which makes it very convenient to use. The receiver of a written message does not have to be online at the same time as the writer of the message to be able to get the information. The receiver can read and answer the message at any convenient time.

The last advantage of the internet as stated by respondents is that it is fun to use. Of course each participant has another definition of fun and especially the perception of something being fun. The impression gained here is that respondents use the internet for recreation. It is a tool for fun and stress relief.

RQ4: What amount of time was the longest respondents have stayed on a cellphone and on the internet?

The longest amount of time the majority of respondents (47.62%) have stayed on the internet was five hours or more. Here it would be interesting to know for what purpose respondents have used the internet for such a long time. Was their internet usage work related or was it for private purposes and recreation? Furthermore, by choosing the option of "5+ hours is the longest I have stayed on the internet" it remains unclear how much time respondents have actually stayed on the internet. The range might vary dramatically between respondents. Some might have stayed on the internet only six hours while other might have stayed online for a much longer period of time. Thus, an option to report respondents individual internet use if it is over 5+ hours would have been helpful to compare respondents in their internet usage.

About 23.81% of all respondents have stayed 2-3 hours on the internet the longest. Only a minority of respondents (4.76% for each description) have stayed 1-2 hours, only one hour or 30 minutes on the internet the longest. In total, 85.72% of all respondents have spent more than 2 hours the longest on the internet, which raises the question whether this amount of time is their usual internet consumption or is rather an exception.

Different to respondents' amount of time using the internet, the longest amount of time the majority (41.38%) of respondents' cellphone calls are less than 30 minutes. The second largest number of respondents (27.59%) has stayed 1-2 hour on the phone. Three to five hours on the cellphone was the longest 10.43% have stayed on the cellphone. It is interesting that none of the respondents has indicated not to use a cellphone at all. In comparison to respondents' internet usage it is striking that only 3.45% have stayed 5 hours or more on a cellphone. The usage of the two electronic devices is opposed to each other. While the majority of respondents spend 5 or more hours on the internet, the majority of respondents use a cellphone primarily for short calls under 30 minutes.

RQ5: For how long have respondents been using cellphones and the internet?

The majority of respondents (95.24%) indicated that they have been using a cellphone for 5 years or more. Only 4.76% of the respondents use a cellphone less than one year. None of the respondents uses a cellphone for 2-3 or 3-5 years. These results are in line with Lonkila's study (2008) that the cellphone became more affordable for the larger part of the Russian population after the financial crisis in 1998. It would have been interesting to compare respondents' cellphone usage with their landline telephone usage. That way one could find out whether Lonkila (2008) was correct to interpret Levada Centre's (2007) findings that landline telephones at home had only a small increase. Lonkila claims that this suggests that cellphones are becoming increasingly popular among the Russian population in comparison to landline telephones. In Russia the majority of people have started to use cellphones before 2008. Thus, the majority of the respondents indicated that they have been using cellphones longer than 5 years. The 4.76% of respondents who indicated that they are using the cellphone less than one year are laggards. The small percentage of laggards shows Russia's developed cellphone communication.

In comparison to cellphone usage, 97.62% of the respondents have been using internet for 5 years or more. Only 2.38% reported to use the cellphone less than one year. None of the respondents have been using the cellphone for 2-3 or 3-5 years. Similar to the usage of cellphones, the majority of respondents are using the internet for a longer period of time than 5 years. Here, a more categorization going more back in time would be beneficial to find out how long exactly respondents are using the internet. They might have given qualitative responses. It is nevertheless likely that a slightly bigger majority have used cellphones more than other online tools on the internet (e.g. email accounts, chatrooms, etc.). One hypothesis to explain this phenomenon might be that the cellphone is primarily a hardware device which can be used to use other communication media like the email account for example. As the electronic communication media offers various ways to communicate and can be accessed either through a computer or a cellphone, it is not surprising that the cellphone is listed first, especially when the PC was not part of the response option. Different respondents prefer different ways of communication media, but it seems that the majority is using one tool, the cellphone, to access those media. Even when the PC was not part of the response option about respondents' types of communication media used at least once, it is nevertheless important to mention that cellphones are in most cases much more affordable than computers. Thus, cellphones might be used and owned by a larger number of people than it is the case with computers.

Summary

The information gathered for this survey describes the usage of electronic devices by a particular sample. Thus, all the demographic factors collected apply only to the sample rather than being representative for an entire population. The majority of respondents (50%) were between 25-35 years old. Sixteen percent of the respondents were 35-45 years old, and the two groups of the 45-65 and 18-25 years old respondents represented 17% of the total sample each. The majority of electronic devices users are young people between 25-35 years which correlates with Vershinskaya's (2002) claim that the majority of cellphone users are young people.

The majority of the sample was female 56.10%, while 43.90% were male. The higher number of female respondents in the sample does not simultaneously mean that females use more electronic devices than males.

The majority of respondents in the sample have high educational background. Most of the respondents (54.76%) have finished an institute, college or university. Graduate school was completed by 33.33%, while only 11.90% finished high or secondary school and none of the respondents finished only the primary school. The sample did not vary in regard to respondents' educational background. Thus, conclusions about respondents' usage of electronic devices can only been drawn for higher educated respondents but not for a less educated group.

Cellphones, email accounts and the television were the three most frequently used communication media. The cellphone was used by 97.6%, the email account was used by 92.7% and the television was used by 85.5% of all survey respondents at least once.

The most frequent stated advantage of cellphones as stated by survey respondents is that cellphones offer an easy way to communicate and share information. Cellphones' second advantage is that they bridge distance between interlocutors. Their third and fourth advantages of cellphones are related with each other. Respondents claim that cellphones are useful in cases of emergency and are convenient to use because they are not bound to a particular place like landline telephones. Cellphones are always at respondents' side and are convenient to use when in need.

Respondents' most frequently reported advantage of using the internet is that it presents a fast and easy opportunity to access information. Similar to the cellphone the internets' second advantage is that it is a cheap communication tool. Another advantage is that the internet does not only offer the chance to communicate in synchronous but also in asynchronous ways, which makes it very convenient to use. The last advantage of the internet as stated by respondents is that it is fun to use.

The longest amount of time the majority of respondents (47.62%) stayed on the internet was five hours or more. About 23.81% of all respondents stayed 2-3 hours on the internet the longest. Only a minority of respondents (4.76% for each description) stayed 1-2 hours, only one hour or 30 minutes on the internet the longest. In total, 85.72% of all respondents spent more than 2 hours the longest on the internet.

Contrary to respondents' amount of time using the internet, the longest amount of time the majority (41.38%) of respondents made cellphone calls was less than 30 minutes. The second largest number of respondents (27.59%) has stayed 1-2 hours on the phone. Three to five hours on the cellphone was the longest 10.43% have stayed on the cellphone. It is interesting that none of the respondents has indicated not to use a cellphone at all. In comparison to respondents' internet usage it is striking that only 3.45% stayed five hours or more on a cellphone.

The majority of respondents (95.24%) indicated that they have been using a cellphone for five years or more. Only 4.76% of the respondents use a cellphone less than one year. None of the respondents use a cellphone for 2-3 or 3-5 years.

In comparison to cellphone usage, 97.62% of the respondents have been using internet for five years or more. Only 2.38% reported to use the cellphone less than one year. None of the respondents have been using the cellphone for 2-3 or 3-5 years.

Application of the Theory

It is difficult to apply the Diffusion of Innovations theory to the survey about cellular telephone and internet usage in Russia as the survey's questions were not conducted to be used for this particular theory. Nevertheless, it was tried to direct the research questions towards some characteristics of the theory.

The Diffusion of Innovations can only partly be used to explain the development of cellular telephone and internet usage in Russia because nowadays cellphones and the internet are not new innovations anymore. The crucial elements involved in the diffusion of new ideas: innovations, channels, time and social systems (Rogers 1995) cannot be found in the survey as it rather focuses on respondents' current electronic devices usage than their developments. According to the definition of an innovation, something is regarded as an innovation when individuals perceive it to be new, better and more beneficial than current ideas or practices (Farr an Ames 2008, 377). Although, cellphones and the internet are not regarded to be new innovations anymore, respondents of the survey evaluate cellphones and the internet positively. Respondents describe both electronic devices with various advantages.

Although, the survey does not draw any comparisons with older communication media (e.g. landline telephone) respondents' descriptions indicate that the internet and the cellphone are both perceived to be good and beneficial (in comparison to other devices). Cellphones popularity is depicted in Figure 8.5 very well. It is the most frequently used communication media. Although the internet is not listed individually as a communication medium, 92.7% of all respondents use email accounts, which can only be retrieved on the internet. Respondents' frequent usage of email accounts reflects its ascribed high value and necessity to use.

No innovation-decision process can be distinguished in the survey. None of the questions ask respondents for the processes their passed from first knowledge of an innovation to developing attitudes towards it. Thus, the four stages included in the innovation-decision process cannot be applied.

Respondents' rate of adoption cannot be evaluated as all respondents have adopted the internet and the cellphone by now. It is not possible to distinguish between innovators, early adopters, early majority, late majority and laggards in this survey as the time spam is too short to analyze an already well-established innovation like the internet and the cellphone. Today, the majority of respondents use the internet and the cellphone. Only a small percentage (internet 2.38% and cellphone 4.76%) have been using both electronic devices less than one year. Thus, this minority of users might be described as laggards, because they have started to use the internet and the cellphone much later than most respondents of the survey.

A time dimension, on which the Diffusion of Innovation theory is based on, is not part of the survey (or at least not to the desired degree). The longest span of years asked by the questionnaire refers to respondents' usage of the internet and cellphones. The answer possibilities for the questions "I have used the internet for …" and "I have used the cellphone for…" are: less than one year, 2-3 years, 3-5 years, and 5 years or more. The two electronic devices have been used for more than 5 years by the majority of respondents. More answer choices, going back in time, are needed to be able to refer back to the point in time where respondents started to use the internet and cellphones. The missing time dimension is also a reason why Rogers' rate of adoption (better known as the S-shaped curve), which is described as the relative speed by which members of a social system adopt an innovation, cannot be analyzed, as well.

In his study about online news diffusion Nguyen (2012) claims that even if the internet becomes easily accessible to everyone, a 'digital delay' will not occur. The results of Figure 8.5 disprove Nguyen's argument. As presented in Figure 8.5, not every survey respondent has used all types of communication media at least once. A chatroom was used only by 48.8% of the respondents, e-commerce website by 43.9%, internet bulletin boards by 34.1%, etc. The lower respondent rates imply that not all respondents are either familiar with all the presented communication media or they did not have the chance to use it at least once. This would imply that contrary to Nguyen's argument a 'digital delay' did occur. Those respondents who are not using those communication media used by others are delayed.

Nguyen claims that online news adopters show demographic differences in terms of age, education, personal income, etc. from non-adopters. The Diffusion of Innovation theory claims that early adopters have specific characteristics. Westlund's (2008) study about the diffusion and adoption of the mobile phone shows that people's pattern of news consumption differs in its users' demographics. But as the sample size of this the present study is too small to make generalizations and no demographic information about late adopters is available this hypothesis cannot be supported by this study.

Limitations of the Study

A major limitation of this research project was the difficulty to access the desired target group. As a result, the survey might be biased as it consists of a convenience sample and only a small sample size of 42 respondents could be collected. The number of cases (n) for all questionnaires varies from 42 for the first questionnaire and 28 for the last one. Although, all respondents met the established criteria for participation in this survey and were either residents of Russia, nationals or sojourners who have spent at least three months in the country, this set of criteria had one main disadvantage. The majority of the respondents participating in the survey did not live in Russia anymore, but have lived in other countries for a long period of time. As a consequence, even when respondents still remember and even practice the Russian way of life and its habits of using electronic devices, they are still influenced by the culture of their current country of residence. Furthermore, respondents were similar in many demographics. For example, the majority had a similar educational background. A large proportion (33.33%) of the respondents attended a graduate school.

Another limitation of the survey is that age classification was not clearly distinguished. Respondents might have trouble in deciding what age category they might belong if they were 25, 35, or 45 years old as those were listed twice (e.g. 18-25, 25-35, 35-45, 45-65). This inobservance might have influenced the results regarding respondents' age.

As the 40 questions of the survey were spread over four Survey Monkey links it increased the chance that respondents only fill out the first questionnaire and skip the others. While 42 respondents filled out the first questionnaire with the first 10 questions, only 30 respondents completed the second questionnaire, 29 the third one and only 28 the fourth questionnaire. By splitting the questionnaire in four parts the connection between the individual survey parts was lost. It is not clear which respondent filled out that part of the survey (demographics and answers were mixed).

The question "I have used (indicate all that apply) the following communication media at least once" should be changed to "I frequently use (indicate all that apply) the following communication media" or this question should be added to the questionnaire as it would provide more insight in respondents communication media use. Only because a respondent has used one of the communication media at least once does not mean that they are constantly using it.

Another limitation of the research project was that the survey and its questions were not developed for the chosen theory, Diffusion of Innovations. Thus, it is difficult to adopt the theory to the survey and evaluate its results.

Suggestions for Future Research

For future research it is recommended to use 1) a random sample and 2) a larger sample size. The respondents should be more diverse and differ more in their age, educational background, living areas, etc. It is desirable not to split up the survey in different parts but to conduct a continuous questionnaire. That will decrease the probability that participants might skip questionnaires because they lose interest or get tired of completing them. Another advantage not to split the survey in several parts is that respondents' demographic information and their answers can be correlated and various relationships might be found.

Statistical tests to find various relationships between genders and other variables (e.g. internet and cellphone abuse) could provide other demographic related correlations. In the question about communication media: adding telephone and Skype as options to compare cellphone usage as communication devices would be beneficial.

REFERENCES

Barker, Kriss. 2004. "Diffusion of Innovations: A World Tour:" *Journal of Health Communication* 9: 131-137. doi: 10.1080/10810730490271584.

Campbell, Scott W. 2007. "A Cross-Cultural Comparison of Perceptions and Uses of Mobile Telephony." *New Media & Society* 9(2): 343-363. doi: 10.1177/1461444807075016.

Farr, Celeste A., and Natalie Ames. 2008. "Using Diffusion of Innovation Theory to Encourage the Development of a Children's Health Collaborative: A Formative Evaluation." *Journal of Health Communication* 13: 375-388.

Hogg, Nanette, Carol S. Lomicky, and Syed A. Hossain. 2008. "Blogs in the Media Conversation: A Content Analysis of the Knowledge Stage in the Diffusion of an Innovation." *Web Journal of Mass Communication Research* 12: 1-16.

Internet World Stats. Usage and Population Statistics. last modified September 23, 2013, accessed October 29, 2013, http://www.internetworldstats.com.

J'son & Partners Consulting. "Обзор российского рынка мобильного интернет-доступа. Использование мобильного интернета на смартфонах и планшетных ПК." [Overview of the Russian Market of Mobile Internet Access. The Use of the Mobile Internet on Smartphones and Tablet PCs]. last modified 2013, accessed October 22, 2013, http://www.json.ru/poleznye_materialy/free_market_watches/analytics/o bzor_rossijskogo_rynka_mobilnogo_internet-dostupa_ispolzovanie_mobilnogo_interneta_na_smartfonah_i_planshetn yh_pk/.

Levada Centre. "Modern Means of Communication" accessed 15 November, 2007, www.levada.ru/press/2007080602.html.

Levada Center. "Russian Public Opinion." last modified 2012, accessed October 29, 2013, http://en.d7154.agava.net/sites/en.d7154.agava.net/files/Levada2011Eng.pdf.

Li, Daguo, and Viv Edwards. 2013. "The Impact of Overseas Training on Curriculum Innovation and Change in English Language Education in Western China." *Language Teaching Research* 17(4): 390-408. doi: 10.1177/1362168813494124.

Ling, Rich and Leslie Haddon. 2003. "Mobile Telephony, Mobility, and the Coordination of Everyday Life." In James E. Katz (Ed.) *Machines that Become Us: The Social Context of Personal Communication Technology* (pp. 245–265). New Brunswick: Transaction.

Lomas, Jonathan, Murray Enkin, Geoffrey M. Anderson, Walter J. Hannah, Eugene Vavda, and Joel Singer. (1991). "Opinion Leaders vs. Audit and Feedback to Implement Practice Guidelines." *Journal of the American Medical Association* 265: 2202–2207. doi:10.1001/jama.1991.03460170056033.

Lonkila, Markku, and Boris Gladarev. 2008. "Social Networks and Cellphone Use in Russia: Local Consequences of Global Communication Technology." *New Media & Society* 10(2): 273–293.

Morris, Jeremy. 2013. "Actually Existing Internet Use in the Russian Margins: Net Utopianism in the Shadow of the 'Silent Majorities'" *Regional Studies of Russia, Eastern Europe, and Central Asia* 2(2): 181-200. doi:10.1353/reg.2013.0014.

Nguyen, An. 2012. "The Digital Divine Versus the 'Digital Delay': Implications from a Forecasting Model of Online News Adoption and Use." *International Journal of Media & Cultural Politics* 8 (2&3): 251-268. doi: 10.1386/macp.8.2-3.251_1.

Pelckmans, Lotte. 2009 "Phoning Anthropologists: The Mobile Phone's Reshaping of Anthropological Research," In Mirjam Bruijn, Francis B. Nyamnjoh, and Inge Brinkman, *Mobile Phones,* 28.

Ponder, Jaroslav K., and Markova, Ekaterina N. 2002. "Bridging the Eastern European Digital Divide: Significance of Mobile Telecommunications in Poland and Russia." *Communications & Strategies*, 45(1): 171-200.

Rantanen, Terhi. 2001. "The Old and the New Communications Technology and Globalization in Russia." *New Media & Society* 3(1):85-105.

Regli, Brian J. W. 1997. Wireless: Strategically Liberalizing the Telecommunications Market. Mahwah, NJ: Erlbaum.

Rice, Ronald E., and James E. Katz. 2003. "Comparing Internet and Mobile Phone Usage: Digital Divides of Usage, Adoption, and Dropouts." *Telecommunications Policy* 27: 597-623.

Robinson, Les. 2009. "A summary of Diffusion of Innovations." Last modified Jan 2009, accessed October 17, 2013. http://www.enablingchange.com.au/Summary_Diffusion_Theory.pdf.

Rogers, Everett M. 1995. *Diffusion of Innovations.* NY: The Free Press.

Salmi, Asta, and Sharafutdinova, Elmira. 2012. "Culture and Design in Emerging Markets: The Case of Mobile Phones in Russia." *Journal of Business and Marketing,* 23(6): 384-394.

Sharma, Dhirendra, Hemant Sharma, and Vikram Singh. (2011) "Growth and Diffusion of E-journal Usage in Universities of the Western Himalayan Region of India." *Program: Electronic Library and Information Systems* 45(2): 149–172. doi: 10.1108/00330331111129705.

Skvortsova, Svetlana. 2005. "Challenges and Opportunities Associated with Making Internet Mobile in Russia" accessed December 6, 2013. http://www.itu.int/osg/spu/youngminds/2005/YoungMinds05_Skvortsova.pdf.

Strathern, Marily. 2002. "Abstraction and Decontextualization: An Anthropological Comment," In Stephen Woolgar (Ed.) *Virtual Society? Technology, Cyberbole, Reality*. Oxford: Oxford University Press: 303.

Vershinskaya, Olga. 2002. "Mobile Communication. Use of Mobile Phones as a Social Phenomenon – The Russian Experience." *Revista de Estudios de Juventud* 57(2): 139-149. http://www.itu.int/osg/spu/ni/ubiquitous/Papers/Youth_and_mobile_2002.pdf#page=137.

Vershinskaya, Olga. 2003. "Information and Communication Technology in Russian Families." In James E. Katz (Ed.), *Machines that Become Us; The Social Context of Personal Communication Technology* (pp. 105-116). New Brunswick: Transaction.

Westlund, Oscar. 2008. "From Mobile Phone to Mobile Device: News Consumption on the Go." *Canadian Journal of Communication* 33(3): 443-463.

World Population Review 2013, last modified 2013, accessed October 29, 2013, http://worldpopulationreview.com/russia-population-2013/

CHAPTER 16

Electronic Device Usage and the Internet in Turkey

Baturalp Hamit Gursoys

The Republic of Turkey is a parliamentary republic country located in both Asia and Europe. The majority of the land is in Anatolia and Thrace, and it is spread over a small portion of the Balkans. Bulgaria is to the northwest, Greece the west, Georgia in the northeast, Armenia in the eastern side; and Iran, Azerbaijan, Iraq and Syria are southeast neighbors of Turkey. In the south lies the Mediterranean. The Aegean Sea is to the west and the Black Sea is on the north. The Sea of Marmara connects Europe and Asia by the Bosphorus Bridge, and the Dardanelles with Anatolia to Thrace separate Asia from Europe. Turkey has been an important, centralized place for Europe and Asia. It has a geopolitical power necessary for four areas of the world.

With the landmark decisions on January 24, 2001 in Turkey, it has begun the transition to a market economy. In 1983, the military came to power after the political parties began structural changes in the economy. This period was followed closely by developments in the world, and Turkey has made rapid strides in adapting to technological changes. Development in the telecommunications sector has been the most visible in Turkey in the last 15 years, since 2000.

This research focuses on the electronic device usage and internet usage in Turkey based on (1) economic, social and political effects of internet usage, (2) internet abuse among the young and its relations to internet usage patterns and demographics, (3) mobile phone usability in Turkey, and (4) ICT usage in Turkey. The classic mathematical theory by Claude Shannon and Warner Weaver is explored in this study, one of the important ways of understanding communication. The data collected is analyzed using the quantitative method of data analysis. At least 65 questionnaires will be distributed to Turkish citizens by using Survey Monkey.

Literature Review

In 1995, the EU-Turkey Customs Union started the broad liberalization of trade in Turkey with tariffs which were opened. It has become the country's foreign trade policy. This constitutes one of the most important cornerstones of development for Turkey. In 2014, exports compared to the previous year increased by 4% to $ 157.6 billion. Turkish brands consist of Beko and Vestel, one of Europe's largest consumer electronics manufacturing companies. These two important brands have been developing a significant amount of investments in the industry by finding new technologies.

Another important part of the Turkish economy is banking, construction, home appliances, textiles, oil refining, petrochemical products, food, mining, iron, and steel and the machine industry.

Turkey's population, strategic location, and economy have developed mobile phones, and made a significant leap in internet services in Europe. Fluctuations in the sector over the global recession cause only temporary economic damage, more crucial to the economic potential of the telecommunications sector is long-term growth. The structure of the telecommunications services market shows that operators in the country are have long-term planning of activities. They target markets that direct an international isle.

Theory

Claude Shannon and Warner Weaver developed a mathematical theory of communication while they were working in the lab of Bell Telephone Company. These two important researchers were actually doing work in fields such as mathematics and electronics, but their work in World War 2 and afterwards, gave direction to mass communication studies. Shannon and Weaver's work is considered classic theory. Harold Dwight Lasswell raised his communication model as an empirical sciences understanding of positivism in this model. It contains mathematical formulas. Lasswell was very impressed by the empirical sciences, and Shannon had just finished developing his theories related to cryptography. Therefore, they were well aware that human communication is a mix of randomness and statistical dependencies. Letters in messages were obviously dependent on previous letters, to some extent.

In 1949, Shannon published the groundbreaking paper, "A Mathematical Theory of Communication." Shannon used Markov models as the basis for how people can think about communication. Moreover, he started with a toy example. Imagine you encounter a bunch of text written in an alphabet of 'A', 'B' and 'C'.

Perhaps you know nothing about this language. However, you can notice A's seem to clump together, while B's and C's do not. Then, he shows that you could design a machine to generate similar looking text, using a 'Markov chain'. Shannon touches on how this method was found in the past. Moreover, his examples show what kinds of experiments have been used to certify a mathematical communication method (Richard and Hajek 1948).

According to the Weaver-Shannon model of the mathematical theory of communication process - there are three types of noise that prevent proper communication. The model, used in communication, is a very technical process of meaning and content. This structure in the mass media, although the vanguard, could not prevent the development of new theory. However, the importance of developing this theory is necessary in the electronics field, especially as a very important position in the computer and software systems. It is necessary to understand the basics of a system of communication science. Warren Weaver (1949) explains the capacity of a communication channel is to be described in terms of its capability to broadcast what is created out of the source of given information. In the next chapters, Weaver explains coding, noise and messages by using mathematical formulas.

Communication in education is defined as a network of relationships needed to create the targeted behavior change. Noise affects the teacher-student interaction in a negative way. It is necessary to determine the cause in order to solve the noise problem. Factors that make the difference between reaching the goal include the transmission path from the information sources called noise. Noise is important in the communication process model. The communication process depends on complete and accurate processing of the noise factor. Noise is generally defined as re-occurring or artificial sounds that annoy people. Technical noise is defined as the overlap of the anarchic sound waves. Noise makes havoc on the human body. It can cause visual and psychological damage (Flensburg 2009).

In 1948, an American mathematician named Claude Shannon wrote an article he published in the *Bell Electrical Journal*. Later on, Shannon collaborated with an electronic engineer named Warren Weaver to apply his theory to communication. That article mentions that the mathematical theory or model of communication was effectuated by entropies and communication, coding, redundancy, thermodynamics, digitalization, human communicators and misconceptions. The information sent from a source using a particular signal by a carrier tool, indicates that destination where the source is a decision-making position. That is the real person or entity initiating the communication. An information source that people want will send a message that it wants to target. These sources may also be formal organizations as well as a person. Contacting the beginning of information implies two ways or interactive communication. Roughly, information means if it has one meaning at the source, the content of the subject and original meaning is communicated (Krippendorff 2009).

Exchange and development of media communication has been in conjunction with the development of the technology revolution. In addition, telegram, telex, telephone, radio and cinema effected a major change to the basic elements of communication, and mass communication such as television. This demonstrates matter in abandonment, and new habits that are acquired. Computer usage is a displaced case with the latest technology products, such as the internet and smart phones. Basic concepts that occur in the context of the media are a current controversy, and consequently definitions change to accommodate multiple carriers, senders and receivers.

Communication is not perceived today as in the 1940s or 1950s with Shannon and Weaver. Today people are facing more individualistic, normative interactions with a communication network and structure. Therefore, the first publications printed; and then the radio, the cinema and activities based on the preferences of television dominated. Scientific studies formed in accordance with the individual preferences of people. An effective medium for today's communication in this context is the internet or social networking.

One of the basic elements of communication; the mathematical theory of communication by Shannon and Weaver, is accepted as a communication model. It includes source, record, transfer and receiver. The main purpose of this communication includes the key messages and feedback that is recycled. The immediate responses to messages transmitted through this recycling provide the basic elements of the communication, and emphasize the importance of the impact of the message based on the mass movement of thought. We sort the basis of today's mass media, television replaced by the internet and newspapers that are observed. They could be read electronically as an extension of the internet. Because of this change, the theory advocated in the 1950s changed many opinions and thoughts; it became evaluated in different layers. The Shannon and Weaver communication definition also mentions three levels of communication problems, which are technical problems, semantic problems and effectiveness problems.

Economic, Social and Political Effects of Internet Usage in Turkey

People in all developing countries are affecting the economic and transformational character of the world by shaping social and political life. The internet is a symbol of the 21st century. In this study, the use of internet economics, social and political activities are in accordance with development aims to demonstrate the effects of today's interconnected world. Primarily, the internet economy is addressed in the transition from a traditional economic structure. The internet community changes cultural dimensions, and makes vulnerable individuals after their radical transformation (Deviren and Yildiz 2014). This research examines life in the world of internet use, and its role in technology after political tensions subside.

Development in information technology can place traditional economic activities parallel with the regulations of one of the most affected areas—the economy. The economy has brought long-running debate in Turkey, until the concept of a new digital economy became a common denominator. Economic understanding concludes that everyone is a compromise. The impact of information technology and the new economy has given birth to the internet, rather than people's muscle strength. It represents a world where brainpower is most important.

Currently in Turkey, computer and internet usage rates of individuals in the 16-74 age groups were 53.5% generally. However, this ratio was 62.7% in men and 44.3% for women. Computer and internet usage rates were 49.9% in 2013, which was highest in the age groups between 16-24 for computer and internet usage. Computer and internet use is higher among men in all age groups in Turkey (Nursen Deviren and Yildiz 2014).

The vast majority of internet users in Turkey are going online daily. The highest rate of the users is in the 16-24 age range (84%), followed by 25-34 years (77%), then 35-44 years (62%), and 45-54 years of age (41%). The daily rate for internet users over 55 years of age has increased to 62% (Gencer and Koc 2012).

Households without access to the internet at home (42.8%) said they do not need to use the internet. The second factor was specified as the cost of the internet connection (31.9 %). The proportion of households with broadband internet access was 57.2%. Accordingly, 37.9% of households have ADSL, cable TV or cable internet through the infrastructure. Access to the internet via fixed broadband connection, such as fiber, was 37%. Access to the internet was also via mobile broadband connection. The share of narrowband connection stayed at 6% (Gencer and Koc 2012). In the 16-74 age groups, internet in the first quarter of 2014 was 79.1% at home. This access was with relatives, friends (38.7%), with shopping (30.2%), and airport (23.3%), and internet cafes (14.3%). Turkish people are also using internet outside of the home and workplace (58%). Individuals in the first three months of 2014 were using mobile phones to connect to the internet wirelessly, or using a smart phone; and 28.5% used a laptop, net book, or tablet used as a portable computer (Gencer and Koc 2012).

Mobile Usage in Turkey

Another feature most commonly used in Turkey is simultaneous messaging. Those that use messaging comprise 31% of the world. One out of every two internet users in Turkey is using the internet for simultaneously messaging (Aryana and Clemmensen 2013).

On the other hand, the internet is used for learning about a desired product, making bank transactions, and on-line buying. The rate of use for Europe and the world is much higher than in Turkey. People purchased and obtained information about products, (56% of the world, while 70% in Europe), whereas this ratio was 27% in Turkey. Likewise, 39% of the world ratios of bank transactions were made through the internet, while 56% were detected in Europe and 12% in Turkey. One of every two internet users in Europe stated that they use the internet for shopping; this ratio remained at 7% in Turkey because websites are not trustworthy enough for shopping (Aryana and Clemmensen 2013).

According to their survey, almost all over the world and Europe, with 92% of households in Turkey, people have at least one mobile phone. Nearly 100% of mobile phone users in Turkey call and use cell phones to search something; 77% percent of the sample respond to chat, 42% listen to music and radio, 33% make time schedules and plans, and 26% prefer mobile phones for simultaneous messaging. In Turkey, 63% of mobile phone users say, "It is important to me no matter where people are accessible". In Turkey, 76% pay attention to monitoring current news. Especially teenagers have the latest models of smart phones to carry which creates a pleasant impression on others for them, while 7% were identified as needing a mobile phone (Aryana and Clemmensen 2013).

Transformation efforts, the information society in Turkey, and world developments parallel the beginning of the 2000s, especially in Turkey, The European Association started as a result of perspective. In 2003, information was gathered nationwide for the individual transformation of society. The efforts of "e-Transformation Turkey Project" has been implemented, a byproduct of this effort. The State Planning Organization's emergence as a strategy was released in 28 July 2006. That strategy was realized with the document called Information Society Strategy 2006-2010 (Demirli 2013).

Important topics in the information society were transformation efforts, e-Government applications, and how to disseminate them. Since the end of the 2000s and the beginning of e-Government in Turkey, applications and uses have been under taken. The first momentum in the period show investments are only focused on the physical technological infrastructure and the community in general use of ICTs (Demirli 2013). Because of not enough emphasis, the development of competencies could not be maintained at the desired level. Strategy was improved in recent years because of updates, and coordination of e-government efforts on November 1, 2011. The Ministry of Information Society Development and Department Transportation on the Presidency, Shipping and Communications were transferred to the new ministry. Moreover, a Science and Technology board meeting was held on January 16, 2013 for a new e-Government portal. The satisfaction of the citizens started a new era for Turkey.

Sixty percent of Twitter users in Turkey are men. Twitter users under the age of 34 were 64% of those surveyed, and half had completed university education. Twitter is active in Turkey with social media users, and the world of fashion also has a mass following of those who are dealing with other cultures (Gunuc, Misirli and Odabasi 2013).

According to a survey conducted in March 2013; during 30 days of following the behavior of 1,000 Twitter users, 76 % of people who use Twitter every day interact several times a day. Females (56%) use Twitter as an agenda for fashion brands or products, while most of the men preferred news and soccer. User ratings show that 60% of countries that have internet access are using Twitter. The survey reveals that 59% of Twitter users tweet while watching television (Gunuc, Misirli and Odabasi 2013).

According to the report, 61% of mobile users on Twitter follow a brand and company, dealing with quite a wide range of brands. Sports teams and news / media organizations followed two categories: channels, and political parties. Nearly half (46%) of the brand followers subscribe to promotions and are turning their Twitter account to place items to be aware of (45%) with retweets or favorites. On Twitter, users always remember to check promoted tweets seen by 87% of Twitter users (Gunuc, Misirli and Odabasi 2013).

Transportation–Communication Capital and Economic Growth in Turkey

Turkey had improvement in the telecommunications sector from 1983 until 1993. In this time of technological development in the world, rapid integration has been achieved in the telecommunications sector. Turkey, within the European Union countries sector, was slower when compared to EU countries. In some indicators that capture the EU member states, it has emerged late (Eruygur, Muhtesem and Merter 2013).

The Turkish telecommunications sector and the European Union telecommunications sector turned quite backward compared to the industry in terms of quality of service. The privatization initiatives are current and constant time, and after the tenth year, Turk Telekom privatized sufficient investment services. Ensuring competition in the telecommunication industry, in some areas of privatization, reinforces the expectation of rapid development in other areas for the future. Finally, the telecommunications authority in the sector with fulfilling its regulatory function, attempts to eliminate the competition even though there are prohibitive regulations (Karademirlidag 2010).

In the last twenty to thirty years of telecommunications services in many developed and developing countries, liberalization and economic growth contribute significantly to the competition. Innovative services and a competitive telecommunications infrastructure have paved the way towards development, competitiveness and increased employment. Accelerated growth ultimate benefits the consumer. The following research questions further this study.

- RQ1: What age, gender and education demographics use mobile phones in Turkey?
- RQ2: What types of electronic communication are used in Turkey?
- RQ 3: How long have Turkish people used mobile phones for internet access?
- RQ 4: What are the main reasons for using the internet in Turkey?
- RQ 5: Has internet usage become a problem for Turkish users?

METHOD

First, survey participants were Turkish students who study for bachelor or master's degrees in the United States, beside participants who live in Turkey. Eight Turkish students were found in a Midwestern US University to assist in filling the survey to the best of their capability regarding electronic device and internet usage. These participants helped in distributing the questionnaires to their close friends, siblings, or relatives who live in Turkey. Therefore, people that helped to fill this survey were from large cities such as Istanbul, Ankara, Izmir and Bursa in Turkey.

The numbers collected were analyzed via the quantitative method of information analysis, using frequency percentage. Sixty copies each of English and Turkish typed questionnaires were administered to participants from Turkey and filled copies were collected instantly. Furthermore, explanations were specified to those who were not clear as to how to do it. For pre-testing this study, the researcher filled a copy of the survey, to make sure understanding and clarification of individual objects. The survey contains simple and straightforward questions, easy enough for the participants to understand. Questions were considered in line with previous research, in order to complete sufficient information on the electronic device and internet usage in Turkey. Participants who felt uncomfortable filling the survey could discontinue, and were under no requirement to complete it. Survey Monkey was also used in the collection of information from respondents from Turkey to target at least 50 replies. Once data using distributed questionnaires and Survey Monkey was collected, an analysis by SPSS was used, a program for statistical analysis.

RESULTS

RQ 1: What age, gender, and education demographic use mobile phones in Turkey?

Most of the participants for this survey were males (64.71%) and females (35.29%). Age groups of 25-35 (45.45%) and 18-25 (38.18%) were the highest surveyed. In the perspective of education levels of Turkish survey participants, the highest group reported college or university degree (49.09%) followed by high school degree (36.36%). See figures. 16.1-16.3.

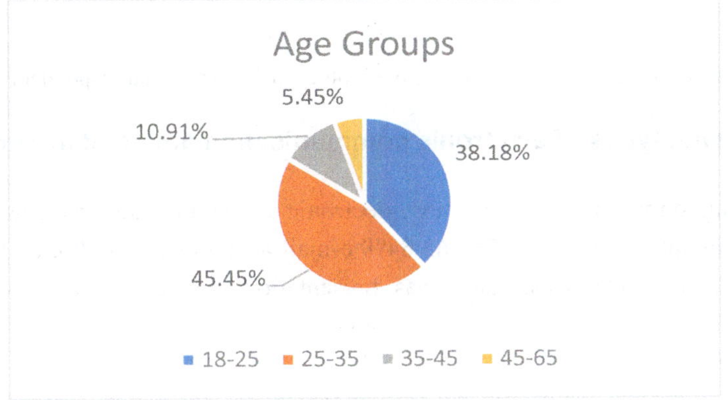

Figure 16.1: *Age Demographics of Turkey* (Total respondents 55).

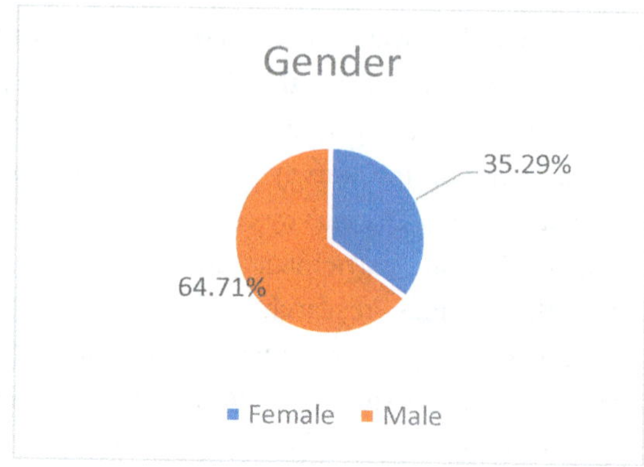

Figure 16.2: *Gender Demographics of Turkey* (Total respondents 51).

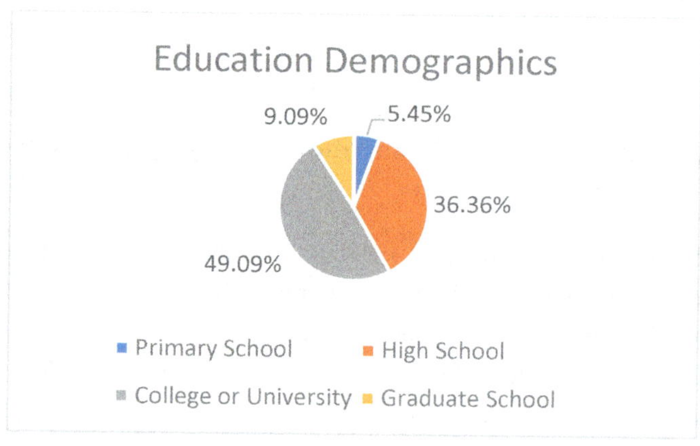

Figure 16.3: Education Level of Demographics in Turkey (Total respondents 55).

RQ 2: What types of electronic communication are used in Turkey?

According to the survey results, most electronic communications rated as follows: 74.55% cellphone, 69.09% TV, 63.64% e-mail accounts and radio, and 54.55% digital camera. The least rated was presentation devices, which is Proxima. (7.27%)

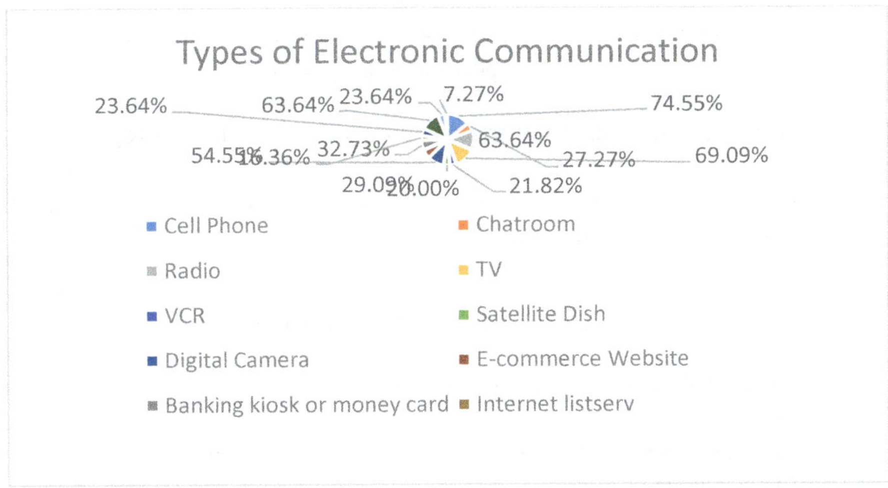

Figure 16.4: Types of Electronic Communication Used in Turkey (Total respondents 55).

RQ 3: How long have Turkish people used mobile phones for internet access?

All of the respondents replied to this survey question by entering at least "less than one hour", so all survey users currently use internet on their mobile phones (100%).

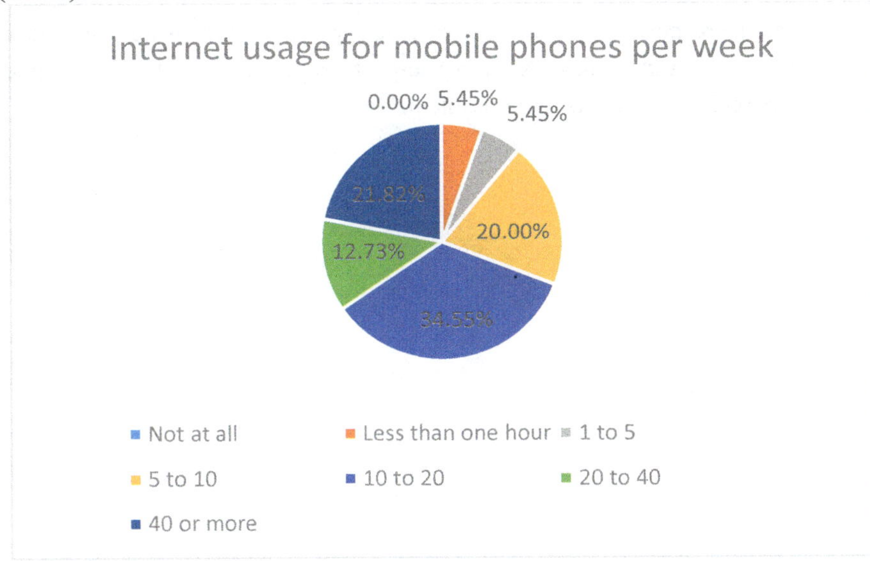

Figure 16.5: Internet Usage of Turkish People for Mobile Phones per Week (Total respondents 55).

RQ 4: What are the main reasons for using internet in Turkey?

Since the question was not created in Likert scale format, all participants answered uniquely as seen in Table 16.1.

Table 16.1 Using Internet as Advantage (Total respondents 9).

Positive Reasons for Using Internet
"Fast"
"It is possible to get information instantly about anything."
"I can find everything"
"Easiest way to attainability"
"World Wide"
"Unlimited information"
"Compulsory"
"Google"
"That is the new social life"

RQ 5: Has internet usage become a problem for Turkish users?

As expected, respondents avoided extreme answers, and were undecided. Therefore, most of the answers of participants reported that they sometimes (44.44%) perceive internet as a life problem.

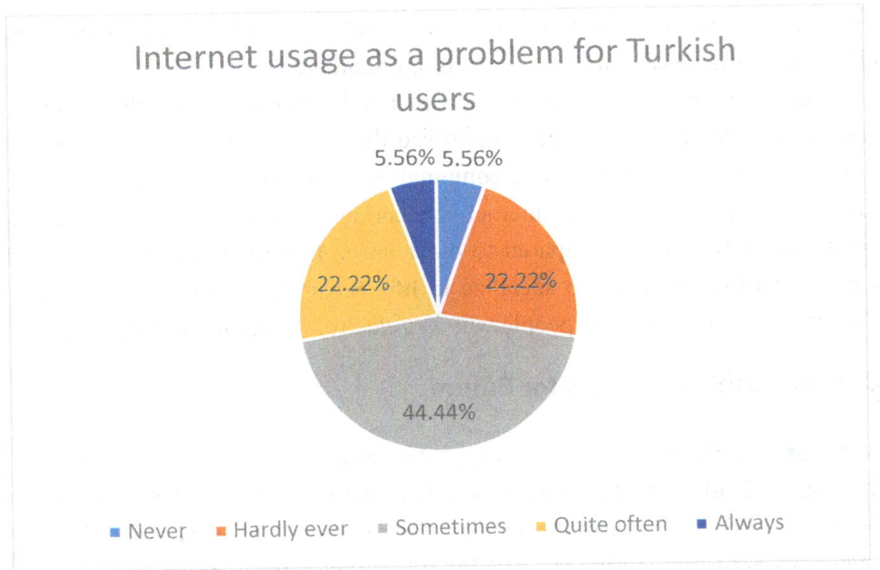

Figure 16.6: Internet Usage Perceived as a Problem by Turkish Users
(Total respondents 36)

DISCUSSION

The survey results showed that the most of users of electronic communication are within the ages of 18-25 years with 38.18% and 25-35 age group with 45.45%. There were not enough responses from other age groups which were 35-45 and 45-65. This survey results verify the rate of young generation users who have adopted technology at a high frequency. Gender demographics were 64.71% males, and 35.29% female respondents. Educational demographics of the use of electronic communication in Turkey showed that, 49.09% of the survey respondents have college experience or university degrees, 36.36% have high school diplomas, 9.09% have completed graduate school, and 5.45% completed only primary school.

Fourteen electronic devices were listed for participants to indicate which of them are most widely used in Turkey. The devices included cellular (mobile), radio, online chat room, VCR, television, digital camera, satellite dish, banking kiosk or money card, e-commerce website, the internet bulletin board, the internet listserv, voicemail, e-mail account and presentation device (proxima). Five forms of electronic communication gained the highest percentages in Turkey: cell phones, televisions, e-mail accounts, digital cameras and radios. VCR, internet listerv, internet bulletin board, and presentation devices had the lowest response.

This survey indicates that Turkey users prefer communicating with people electronically according to world technology standards.

More than 90% of the respondents feel and articles reported that electronic communication has tremendously improved their lives in Turkey; however, 10% of the respondents feel electronic communication is not really important in their life. Results also showed that internet is becoming very important day by day for young generation users compared to older users, to keep in touch with friends, family, and the world; or to socialize in life. Smartphone and internet usage in Turkey can be addiction for people who use electronic communication tools.

Conclusion/Suggestions for Future

Although significant progress towards the introduction of competition in the telecommunications sector, Turkey is still in the early stages of the competition and needs to do a lot of work to improve the regulatory framework. Furthermore, development of competition can provide substantial gains in prosperity.

The Turkish telecommunication sector has revealed that the European Union, compared with the telecommunications sector in terms of quality of service has increased rather slowly. After the tenth year of privatization initiatives, Turk Telekom has privatized sufficient investment services, and can be considered a cause of slow development. Ensuring competition in the sector; in some areas in the realization of privatization, in other areas to reinforce the expectation of rapid development in the future. Finally, the Telecommunications authority in the sector is fulfilling its regulatory function; it attempts to eliminate prohibitive regulations.

REFERENCES

Aryana, Bijan, and Torkil Clemmensen. "Mobile Usability: Experiences from Iran and Turkey." *International Journal of Human-Computer Interaction* 29 (2013): 220-42.

Blahut, Richard E., and Bruce, Hajek. *The Mathematical Theory of Communication: Claude E. Shannon and Warren Weaver.* University of Illinois Press Urbana and Chicago: 378-501, 1948.

Demirli, Cihad. "ICT Usage of Pre-service Teachers: Cultural Comparison for Turkey and Bosnia and Herzegovina." *Educational Sciences: Theory & Practice* 13 no: 2, (2013): 1095-1105.

Deviren, Nursen V., and Yildiz, Onur. "Economic, Social and Political Effects of Internet Usage." *Electronic Journal of Social Sciences* 13 no: 51, (2014): 52-76.

Eruygur, Aysegul, Muhtesem Kaynak, and Merter Mert. "Transportation–Communication Capital and Economic Growth: A VECM Analysis for Turkey." *European Planning Studies* 20, no. 2 (2012): 341-63.

Flensburg, Per. "An Enhanced Communication Model". *The International Journal of Digital Accounting Research."*. 9(2009): 31-43.

Gencer, Suzan. Lema., and Koc, Mustafa. "Internet Abuse among Teenagers and Its Relations to Internet Usage Patterns and Demographics." *Educational Technology & Society* 15 no: 2, (2012): 25–36.

Gunuc, Selim, Ozge Misirli, and Ferhan H. Odabasi. "Primary School Children's Communication Experiences with Twitter: A Case Study from Turkey." *Cyberpsychology, Behavior, and Social Networking* 16 no: 6 (2013).

Krippendorff, Klaus. " Mathematical Theory of Communication". In S. W. Littlejohn and K. A. Foss (Eds.), *Encyclopedia of Communication Theory*, (2009): 614-618. Los Angeles, CA: Sage. Retrieved from http://repository.upenn.edu/asc_papers/169.

Shannon, Claude E." A Mathematical Theory of Communication." *The Bell System Technical Journal* 27, (1948): 379-423, 623-656.

Suher, İdil Karademirlidag. "Turkish Corporate Communication Executive's Attitudes and Opinions about Employee Communication." *Journal of Yasar University* 5, 18 (2010): 3062-3080.

Weaver, Warren. *Recent Contributions to the Mathematical Theory of Communication.* Harvard University, September 1949.

CHAPTER 17

Ugandan Cellular and Internet Usage: Changing Health Care and Education Needs

Connie Eigenmann and Trisha Capansky

Those who are new to communication devices either draw back or plummet forward in acceptance, often in innovative and unusual ways. Ugandan usage trends in handheld devices depend upon energy-saving strategies to combat limited battery life and electricity barriers (Wang, Mutafungwa, Puvvala, and Manner 2012). Even rural households are willing to sacrifice travel and store-bought food to use cell phones, and female entrepreneurs have risen in ownership (May and Diga 2007). The Uganda Communications Act of 1997 allowed mobile usage to outnumber fixed line subscription by ½ million in September 2002, and by three million in March 2007. Several reasons for this include (1) the migrational nature of Ugandans, (2) low start-up costs, (3) the trend of liberalization, (4) privatization, and (5) increased competition in the telecommunications sector. Uganda's Poverty Eradication Action Plan with development goals of universal education and health care accelerated the usage of mobile and internet for its 33,641,000 citizens even further in 2012. This necessitates the study of Uganda in its use of electronic communication as a representative of developing nations.

Of particular interest is the informal, non-remunerative, shared access to technology where some privileges of ownership are retained in Ugandan rural areas (Burrell 2010). Women can be restricted in cellular use, and non-owners in life-threatening situations can be enabled by radios, televisions and mobile telephones. Observations of phone access, sharing, and use in rural Uganda offer evidence that issues of equality in access are complex: Cellular usage provided ease in trading, farming, fishing, taxi driving and carpentry for 19-77 year olds (Burrell 2010). Ugandans may become multifaceted in using information and communication technologies (ICTs) as: (1) purchaser, (2) owner, (3) possessor, (4) operator, or (5) user of the phone. The telephone can be used as a life line for calling in favors and support among the poorest segment of society (Horst and

Miller 2006). Therefore, age, gender, and income barriers are disappearing as developing countries adopt electronic communication. The first sector affected is agricultural.

Agricultural Usage

By integrating the internet and Short Message Service (SMS) technology to relay market information, the mobile telephone has been modified into a market information tool. Info-trade has collected, processed and disseminated agricultural market information to brokers and suppliers in the industry since 1999 when Uganda became the first country on the African continent where the number of mobile subscribers passed the number of fixed-line users. The ratio is now more than 18:1.

The info-trade base in Wandegeya, Kampala, disseminates information on the most recurrent pests and diseases affecting over 46 commodities like cassava, potatoes, vegetables and bananas to individual farmers in 21 districts. Mobile texting has revamped the agricultural industry especially among dairy farmers in the West. FIT Uganda Ltd. recently won the overall Uganda Communications Commission (UCC) award 2012, in recognition of their efforts to reach farmers through information and communication technologies (Nalubega 2012). FIT reached over 50,000 subscribers who received the text messages, 250,000 web to phone subscribers, and 30,000 radio listeners who included farmers, traders, and wholesalers in the agricultural sector.

The village phone-model pioneered by Grameen Bank in Bangladesh provides small businesses with a portable communication set. With loans, entrepreneurs can buy a mobile phone, a car battery to charge it, and a booster antenna that uses a base station within 15 miles. The handset has software that tracks revenues from every call. The loan providers order the equipment and transport it to those who cannot afford to travel long distances. Uganda imported the model in 2003 by brokering the relationship between mobile operator MTN Uganda and Ugandan microfinance institutions (Anderson 2007). Muto and Yamano (2009) studied 856 Ugandan households in 94 communities to posit that market participation of farmers in remote areas who produce perishable crops is increased by cellular telephone usage. These personal and ethnic networks in rural Uganda impacted mobile usage and job search techniques from 2003 to 2005 (Muto 2012).

Ferris, Engoru and Kaganzi (2008) discovered 94 percent of farmers interviewed owned a radio and 25 percent of farmers owned mobile phones. Up to 52 percent of farmers indicated that receiving Market Information Services

(MIS) had a positive impact on their business, and 39 percent stated it impacted decision making and stabilized incomes. Daniel and Reynolds (2012) highlight voice application to combat low literacy rates in the Kabale district delivery of agricultural updates. This replaces governmental extension service centers that are understaffed and distanced from farmers. Another sector affected by increased electronic communication is Ugandans requiring medical intervention.

Medical Usage

Transmitting radiographs over the telephone has been possible since 1929, so use of the internet has been slow in adoption (Johnson, Goel, Birtwistle, and Hirst 1998). However, developing countries have limited land lines for telephony. Johnson et al. (1998) published the first case study using the world wide web in an emergency to transmit medical images requesting advice of a senior doctor (3). This opened the possibility of wireless telephony for medical intervention directly to those most needing it by crossing age, gender, and income barriers.

Ybarra, Biringi, Prescott, and Bull (2012) studied use of the internet for reduction of risk behaviors associated with HIV. They tested CyberSenga, a prevention program for secondary school students. Findings of limited skills, available bandwidth, need for interaction, and lack of electricity hindered internet-based delivery. Swendeman and Rotheram-Borus (2010) asserted that information and communication technology such as the internet and mobile phones can deliver behavioral components for STD/HIV prevention and care to more people at less cost. Mobile phone usage is demonstrating an 'inverse digital divide' with nearly half of African–Americans, Latinos and youth accessing the internet and e-mail on their mobile phones compared with just over a quarter of whites in the United States (Contreras 2009). Direct messaging, speed of delivery, and confidentiality are key supports for mobile usage in both developing and developed countries use of medical alerts.

Mitchell, Bull, Kiwanuka, and Ybarra (2011) direct their attention to cellular usage among adolescents for HIV prevention program delivery at low cost. Easy, simple, culturally relevant messages can be successfully delivered to large audiences in Uganda. They review 14 programs using mobile phones to promote smoking cessation, physical activity, diabetes and asthma self-management, hypertension medication, and cholesterol lowering. Text messaging and mobile usage was found to be connected to mother's education level in 1503 participants aged 12 to over 18. Although Uganda has 43 living languages, English is used for education and is the official language. Ybarra, Kiwanuka, Emenyonu, and Bangsberg (2006) researched resource-limited settings for internet-based health interventions to discover an increase in maternal education produced an increase in weekly adolescent (ages 12-18) internet use in searching, emailing, chatting and gaming. The internet was found to provide inexpensive, widely accessible, and anonymous yet individually tailored information.

Internet usage characteristics were similar among males and females, except for a male preference for gaming, and a female preference for emailing. Mbarara's (rapidly growing city 180 miles south of Kampala, population of 70,000) internet usage was somewhat limited by access while being on the leading edge of electronic adoption. Siedner, Haberer, Bwana, Ware and Bangsberg (2012) collaborated in structured interviews of 50 patients served by a Mbarara clinic. They found cellular telephone messaging was highly acceptable as a cost-effective method of receiving abnormal laboratory results. From 2000 to 2008, cell phone connectivity increased 5 to 70%. In 2011, 38 subscribers per 100 inhabitants were reported. The study participants were 56% female with a median age of 38 years. At least 80% reported multiple cellular phones in their households, and 48% reported owning multiple Subscriber Identity Module (SIM) cards.

Luscombe (2012) describes the 131 Ugandan hospitals as inadequate to provide healthcare to 36 million people for diseases such as malaria. Supply lines to clinics can be facilitated by cellular phones costing about $7 which Ugandans share and use. Public fear in epidemics can be alleviated, and toll-free hotlines can be anonymous. Drug supplies can be quickly re-directed to areas suffering most. Mothers can be alerted to free vaccinations, misdiagnosing can be rapidly corrected, and the government can become more responsive to citizens through the Ministry of Health. Educational usage also allows the crossing of age, gender, and income barriers in electronic communication.

In 2000, the Ebola virus that is now prominent in Liberia and Sierra Leone spread through Uganda, causing the worst outbreak in history until 2014. Just as Liberia and Sierra Leone today are lacking in the means to communicate treatment and prevention practices to mass audiences, until recent years Uganda also failed to have in place similar communication systems. Today, Tom Frieden, director of the Centers for Disease Control and Prevention, calls Uganda a model for low- and middle-income countries in terms of better detection, response and prevention (McKay et al. 2014). In countries currently battling an Ebola outbreak the World Health Organization recommends that health officials use radio to promote education about Ebola, including symptoms, causes, treatment, and geographical areas where outbreaks exist because radio is the single reliable information source in many of these countries. Ugandans today are initiating electronic and digital programs to contain Ebola outbreaks. Using ICTs, the Uganda Virus Research Institute participates in a joint project with the U.S. Centers for Disease Control and Prevention to create awareness of a hemorrhagic fever program and laboratory for diagnosing blood samples at the onset of illness. Neighboring countries already battling Ebola outbreaks are beginning to adopt Uganda's approach for conveying information and awareness. For instance, Liberia mobile network providers send daily text messages about Ebola to their customers.

In 2009, the CDC established a laboratory in Uganda to test for viral hemorrhagic fevers such as Ebola and Marburg. Test results are typically known within 24 hours. Prior to the Ugandan-based laboratory, international testing facilities based primarily in the United States received blood samples from Uganda, taking several weeks to diagnose infectious diseases to weeks.

To combat recent Ebola outbreaks in Africa using mass communication techniques where possible, IntraHealth International and UNICEF are connecting with mHero, an SMS-based mobile app originally developed in Uganda to combat malaria outbreaks. The app sends knowledge-sharing messages to health workers on issues related to emerging cases, training materials, testing, and coordination techniques. The mHero app is also compatible with DHIS2, a data management, mapping, and logistics system used by several governments, and by health officials for reasons such as tracking pregnant mothers living in rural areas.

Educational Usage

Uganda's commitment to addressing education through ICT development began in the 1990s when the country expanded its ICT usage to address poverty and educational concerns. Uganda faces similar issues addressed by other developing countries: poorly developed ICT infrastructure, high bandwidth costs, and an unreliable supply of electricity. Particularly in rural areas are students and educators limited in access to wireless network capability and mobile phone availability in their school systems. Funding for development in rural schools are a part of overall budgets for several governmental departments such as the Uganda Communication Commission and the Rural Communication Development Fund to create resources for incorporating multimodal communication platforms into pedagogy.

The Uganda Digital Education Resource Bank project provides an online repository for Uganda's secondary schools for training in web application and design, and links that connect users to quality-of-life resources such as the ASK program (Access, Services, Knowledge), aimed at providing information to youth about sexual and reproductive health. The Nabugabo Updeal Joint Venture project, also a link, promotes community cleanliness through tips on establishing garbage collection and sanitation services. The link Deforestation: African Resources in Decline educates students about environmental issues related specifically to Africa.

Schoolnet Uganda, an organization that conducts technical training for school IT coordinators, teachers, and students, trains users to access material on the Internet in a time-saving, cost efficient manner. Although many schools have Internet access, the high cost of Internet in Uganda limits the amount of time students and teachers can access it, which subsequently prohibits computer-skill development. Additionally, much of the material available through popular search engines lacks peer review. Underserved districts in rural areas such as Bugiri, Bushenyi, Kiboga, Ntungamo, and Kumi now have established ICT resource centers. Challenges that these centers frequently experience are increases in demand for training combined with a limited number of computers, and a dependable source for electricity since theft of electric power cables and cutting down electric poles are common practices among vandals.

To correct for low performance rates in science subjects from female students, the project Inspiring Science Education for Girls Using Information Communication Technology offers gender-specific instruction related to science and technology, ICT camps, and science fairs. The Ministry of Education and Sports, one of several partners for the project, helps fund ICT training for teachers and offers development money for constructing local centers. "A lack of, or inadequate, instructional materials like books, equipment, and chemicals can be addressed through the use of electronic books, virtual science labs, simulations, and video clips" (Kakinda 2007). Several schools participate in the project including Gayaza High School and Aggrey Memorial (Wakiso), Bweranyangi Girls School (Busenyi), and Muntuyera High School (Ntungamo).

More positive attitudes towards using media (audio, video, gaming, internet, email and social interaction) were found in students who owned computers and had previous computer experience (Ndawula, Ngobi, Namugenyi and Nakawuki 2012, 154). Uganda developed its initial ICT national policy in 2003 highlighting lifelong education for all citizens. Katahoire, Baguma, and Etta (2001) describe collaboration and delivery of Ugandan primary and secondary school curriculum by ICTs then in operation. Rapid growth in demand for secondary education and limited resources mandated Ugandan development of computer skills for students. The effort began in 1996, and by 1998, 10 schools, 55 teachers and their administrators were trained. By 2001, learners over age 18 (n=42) comprised of 52.4% at "using a computer well" and 47.6% at "using a computer very well." Users aged 13-17 (n=77) reported 100% "well or very well" usage skills after training and acclimation. Forty-nine per cent of this youthful age cohort reported using a computer 1-2 hours daily. Gender disparity was shown for males using computers (91.5%) and females (76.1%) in a sample of 71 respondents from each gender.

Johns Hopkins School of Medicine and affiliates have advocated the use of information and communication technology for health education in low-resource settings like Uganda (Chang et al. 2012). Extensive mobile phone usage in a program in Uganda in 2009 (98%) made internet access possible for research, learning and communication. Economic development flourished.

Economic Usage

In 2006, the Ugandan government formed a Ministry of Information and Communication Technology to provide strategic opportunity and coordination for e-government development, and in 2009, an Act of Parliament established the National Information Technology Authority also to implement e-government information and services. Besides policy-related implementation and infrastructure establishment, Robert Waiswa and Constant Okello-Obura (2014, 2-3) list several laws established to encourage and secure investments in Uganda's ICT development. Laws include cyber regulations pertaining to electronic transactions and signatures; the Computer Misuse Act of 2011; and a law that bans abusive language and tactics targeted toward other users. Waiswa and Okello-Obura report that while Uganda's regulatory approach for establishing ICT development and e-governance is strong, the country is challenged by its dependence on donor-funded ICT initiatives (10). "Such initiatives are associated with sustainability shocks once the period of donor support expires, rendering continuity impossible." Conversely, mobile phone e-governance initiatives tend to sustain success because providers are not dependent upon external funding. In 2014, Uganda ranked 107th out of 138 countries surveyed (World Economic Forum, "Network Readiness Index" 2014), two rankings above 2013. The NRI provides information on a country's ICT readiness and usage for investors and stakeholders to determine global potential for economic competitiveness.

The mobile internet is becoming an aid to exiting poverty. Minges (2007) estimated 10% usage of mobile telephones for the world population, and that by 2017, mobile subscriptions would supersede fixed telephone subscriptions for the world. There are more than thirty countries—both developed and developing—

where this transition has already taken place. Uganda had done so in 1999 when subscriptions rose from 26,000 mobile/57,000 fixed subscriptions to 87,000 mobile/59,000 fixed. The penetration for all Ugandan telephone usage was 67% by January 1, 2000, making an affordable, accessible, computer-capable device available to most. The industry is operating third generation (3G) mobile networks with a global standard (IMT-2000) and broadband internet access. Twelve or more national and regional standards were in place around the world which affected compatibility and pricing. Currently, bandwidth minimum speed of 144 kbit/s and 2 Mbit/s operates under low mobility environments, and installment is rapid in unwired places.

Another factor that makes mobile internet progressive for developing countries is the potential of m-commerce, the ability to buy goods and services using a mobile phone. Many developing nations have a majority who do not have credit cards or qualify for them due to poverty. This limits internet purchasing, but mobile phones can be used as electronic account managers with internet purchases deducted from mobile bills or pre-paid balances (SIM cards).

The Republic of Uganda is an agricultural country, and its southern region includes a substantial portion of Lake Victoria. Over 85% of its citizens live in rural areas; therefore, Uganda's Gross Domestic Product (GDP) per capita is less than $300, making it a Least Developed Country. It had one of the lowest levels of telephone penetration in the world (POTS, fixed or landlines) before cellular development. Government initiatives of privatization and foreign investment have allowed its telecom sector to become one of the most liberal in Africa. Uganda licensed a private GSM mobile operator, CelTel, in May 1995; and introduced a second operator, MTN Uganda, in October 1998, to triple overall telephone density between 1995 and 1999 (.21 telephone subscribers per 100 to .67). The MTN full service license allowed it to offer all telecommunication services, fixed and mobile telephony, which is rapidly installed. Prepaid cards allowed MTN to become the largest network operator in Uganda over CelTel and UTL. In July 1999, Uganda became the first African country where there were more mobile than fixed telephone customers (Mobile Internet for Developing Countries 2007).

Akpan-Obong et al. (2009) date Ugandan internet usage to April 1993 at Makarere University. They identify internet usage in the north in a case study of internally displaced person camps which gave conflict victims a voice. Uganda was one of the first sub-Saharan countries to have full internet connectivity. The Uganda Communication Act of 1997 culminated in systematic policy on ICTs as fundamental tools for socioeconomic development in 2005. The commission's major mandate was to integrate universal access in ICT delivery throughout the country.

Akpan-Obong et al. (2009) find the internet to be a coping mechanism for conflict resolution, and power to achieve socioeconomic goals. They examine many electronic means of communication: the telephone, cellular phone, and computer-based information systems. At the time of their study, Ugandans lived primarily in rural southern regions. Total population for the entire country was approximately 32 million, and agricultural exports were about 80% of its economy. These included labor intensive cotton, tea, and tobacco. At that time (2009) Kampala and its university led Ugandan cities at 53% internet usage in the areas of medicine, education, and business. Subsidies, solar powered batteries and technology advances have contributed to the current Pearl of Africa's internet development.

John (2012) stated that between 2000 and 2010, internet usage in Uganda increased 7,900 percent. As the first African country to witness higher mobile phone penetration than landline availability, usage grew from 0.2 percent in 1995 to 23 percent in 2008. See tables 17.1 and 17.2 for Miniwatts (2014) and United Nations Department of Economic and Social Affairs data (2014).

Table 17.1 Ugandan Internet Statistics

	Population	Users	% Penetration	% of Africa	Facebook
2013	35,918,915	5,818,864 (Dec)	16.2 %	2.4 %	
2012	33,640,833	4,376,627 (Jun)	13.0%	2.6%	562,240
2011	34,612,250	4,178,085 (Dec)	12.1%	3.0%	

Miniwatts (2014)

Africa Internet Statistics were updated for December 31, 2013 and African Facebook subscribers were updated for December 31, 2012 (Miniwatts Marketing Group 2014). Uganda is listed within the top 10 African countries using the internet. African penetration is at 21.3%, and the rest of the world is at 42.3% (Miniwatts 2014 estimated users). Previously, UN data revealed marked growth in internet usage and population growth. The International Telecommunication Union (ITU) is a United Nations agency based in Geneva, Switzerland. Its membership includes 193 states since its 1865 establishment.

Table 17.2 Uganda Internet Usage and Population Statistics (ITU)

	Population	Users	Penetration of Population
2013	36,824,000	5,965,488	16.2%
2010	33,398,682	3,200,000	12.5%
2008	31,367,972	2,000,000	7.9%
2006	28,574,909	500,000	2.53%
2000	24,400,000	40,000	0.16%

United Nations Department of Economic and Social Affairs (2014)

Selected years were shown to highlight the greatest surges in internet usage. By 2013, Ugandans were using cellular telephones to access the internet effectively, and the communication media were interchangeable.

Cellular Usage

Cellular telephony in Uganda's telecommunications industry began in 1995, with growth surges following in 1998 and 2001. Uganda was one of the first countries in sub-Saharan Africa to gain full internet connectivity. Both fixed-line operators, Uganda Telecom and MTN Uganda offered a range of data services including ISDN, ADSL and local and international leased lines. Several Internet Service Providers (ISPs) offered wireless broadband access. Recent research indicates the impact mobile telephone use has on economic development of households in Uganda (Blauw and Franses 2011). There is strong support that mobile phone usage positively impacts economic development. Mutambi, Byaruhanga, Kariko-Buhwezi, Trojer, and Okidi-Lating (2011) support transferring best practices for Ugandan innovation and sustainable growth. They cite the 2010 National Development Plan (NDP) to transform Uganda into prosperity by 2040. Mutambi and Others (2011) sampled 300 businesses and 200 institutions by survey and stakeholder workshops to discover innovative firms and performed research and development.

Even remote rural citizens can use secure banking on cellular telephones to transcend the practice of receiving government salaries in cash (MapSwitch Uganda 2010). This brings about improved health services, opens educational options and promotes gender equality in Uganda. Equity, universal access, transparency and mutual benefit are possible with an electronic, biometric identification system on cellular telephones costing very little. Elections, education fee payments, mobile banking, web-based tools and money transfers are made available to the entire population through Point of Sale devices and hand-held telephones. This solves one of the main causes of poverty: limited access to banking services. Electronic delivery of salaries can reduce corruption, and electronic delivery of health advice can promote a better standard of life for Ugandans. Educational delivery and equal gender access to open resources can increase social capital and reduce corruption.

Social Capital and Crime

Cybercrime continues to hinder e-government usage and popularity within the Ugandan population. Online services provided by businesses, government, and banks are frequently targeted by hackers that is considered to be a group of Ugandan university students who call themselves *Tusobola Net,* meaning "we handle the Internet" (Waiswa and Okello-Obura 2014).

Katungi, Edmeades and Smale (2006) report men as having better access to social capital in rural Uganda. Males were seen to have reduced cost of information, reduced uncertainty about information reliability, and increased willingness to share information about banana production (n=377 households). Buskins and Webb (2009) posit that empowerment through ICTs occurs in African countries; and Uganda in particular, as supported by home country research teams. Email, listservs, chat rooms, blogs, wikis, intranet, conferencing and presentation devices were examined. Empowerment and agency were aided by wind-up radio and mobile telephony, age, and marital status. Energy and ICT poverty were found to be genderized to a degree. Isolated rural women were sometimes required to climb trees for optimal mobile phone connections; however, the internet access created refuge spaces for women in education and business. The minimal infrastructure of mobile networking allowed choice, social justice, and self-actualization.

Madanda (2010) who posted on the report of the Makerere University seminar to end violence against women, presents ICTs as fueling or mitigating domestic gender disparities and vice. Madanda, Noglobe, and Amuriat (2009) outlined key issues on violence against women (VAW) and ICT usage in Uganda, especially in educational institutions. Spousal control by monitoring and SMS stalking is growing from a 2006 demographic and household survey showing 60% of Ugandan women have experienced physical violence since 15 years of age. Although domestic violence can be triggered by mobile usage, mobile accessibility can decrease or diffuse domestic violence. The current problem is male control of mobile telephones (Madanda et al. 2009, 8).

An added problem in curbing VAW is that most mobile subscribers use prepaid SIM cards that do not record caller information. Email interaction is a proven way to combat VAW (Madanda et al. 2009, 14). Twenty-one stakeholders in the technology and VAW interplay are noted in the report. Research areas of priority include usage demographics and self-awareness of electronic usage.

Research Questions and Hypotheses

Based on trends of technology usage, information access, and patterns of mass communication in the developing country of Uganda, the mobile device has emerged as more of an information portal and source of mass communication. Consequently, the following demographic, descriptive and social impact hypotheses and research questions about cellular telephone and internet use in Uganda have emerged:

- RQ 1: What age, gender and education demographics use electronic devices in Uganda?
- RQ 2: What types of electronic communication are used in Uganda?
- RQ 3: How does electronic usage in Uganda compare to that in Oman, China, Thailand, Pakistan, India, and South Sudan?
- RQ 4a: How long have Ugandans used cellular telephones?
- RQ 4b: How long have Ugandans used the internet?
- RQ 5: What are the main reasons for using the internet in Uganda?
- RQ 6: Has internet usage become a problem for Ugandan users?

- H1: Time spent living or studying outside Uganda will be positively correlated with electronic communication usage.
- H2: Age will be positively correlated to cellular and internet usage, and number of media choices.

Quantitative data will be documented by percentage descriptors of respondents, and qualitative data will include individual comments. Pearson's Correlation Coefficient will be used to correlate significance of (1) outside influences and electronic communication usage, and (2) internet and cell usage, age, and media choices. Problems with using the internet in Uganda will be determined by open response complaints, documentation of time extensions, lack of sleep, inability to curtail use, negative work comments, and self-perceived freedom in using electronic communication. All research questions and areas of data collection are to be compared with previous and subsequent data compellations of developing countries in the *Global Studies Series*.

METHOD

It was hoped to collect at least 30-50 completed surveys. Age, gender and ethnicity were not screening factors. The respondents had to be Ugandan residents, nationals or sojourners to be included, and able to complete the 38 survey questions, preferably in English. Participants also had to be adults 18-65 years of age.

Surveys were planned to be distributed July 8-12, 2013, in the urban and rural areas surrounding Kampala, and SE Bulgiri, but this was not feasible despite an extensive trip to Uganda and neighboring countries. Several Ugandan nationals from Berea College in Kentucky and the Kuchanga Foundation of Minnesota were used to distribute surveys, translate respondent rights and questions, and collect data from Ugandan nationals and sojourners accessible from the US.

Participants received self-knowledge when answering and thinking about their usage of cellular telephone and internet communication. No remuneration was provided as an incentive to those surveyed. The disciplines of communication studies, development and sociology; and perhaps social work, will receive current documentation of electronic use and social implications for an emerging, developing economy. Diffusion of Innovation theory was reinforced, and hopefully advanced. A publication for the study results, and presentation at the International Technology, Knowledge and Society conference is planned. It is good for those completing the survey to realize everyone is necessary for the responsible use of electronic communication and society's development.

There were no foreseeable risks to participants who could discontinue at any time, skip questions, or disregard the survey request. No questions were offensive, embarrassing, confidential or invasive. The data collection process and survey were Minimal Risk: The probability and magnitude of harm or discomfort anticipated in the research were not greater in and of themselves than those ordinarily encountered in daily life or during the performance of routine physical or psychological examinations or tests.

Each of the respondents was given an explanation of their respondent rights, and the option of voluntary survey completion. The survey was easily understood, and possible answers could be supplied in English, or a more comfortable language if desired. Respondents required minimal writing and reading skills, or translators were available to read the questions aloud and record the respondent's answers on the survey. It was possible Bantu and Samia languages were preferred by respondents. The entire process required about 15-20 minutes.

No follow up is possible as all participants are anonymous and not linked to answers. Connie Eigenmann (PI) drafted the original survey of 38 items in 2002. It was IRB approved for Oman respondents in paper May 2002, and Brazilian respondents in paper and an electronic version (surveymonkey.com) July 2012. It was approved for paper or electronic versions, in any language for Ugandan participants in November 2012.

Consent was assumed if participants actually completed the surveys. Non English-reading subjects will simply ignore or discontinue their electronic compliance, and all survey questions and requests were supplied by translators in the requested language if necessary. No debriefing was mandated. The research does not involve protected health information, and there is no conflict of interest allowed for the PI or Ugandan nationals aiding in survey collection. Compliance means respondents have assented to take the survey after understanding the procedures and minimal risk. No names or distinguishing data were assigned to the surveys.

RESULTS

RQ1: What age, gender and education demographics use electronic devices in Uganda?

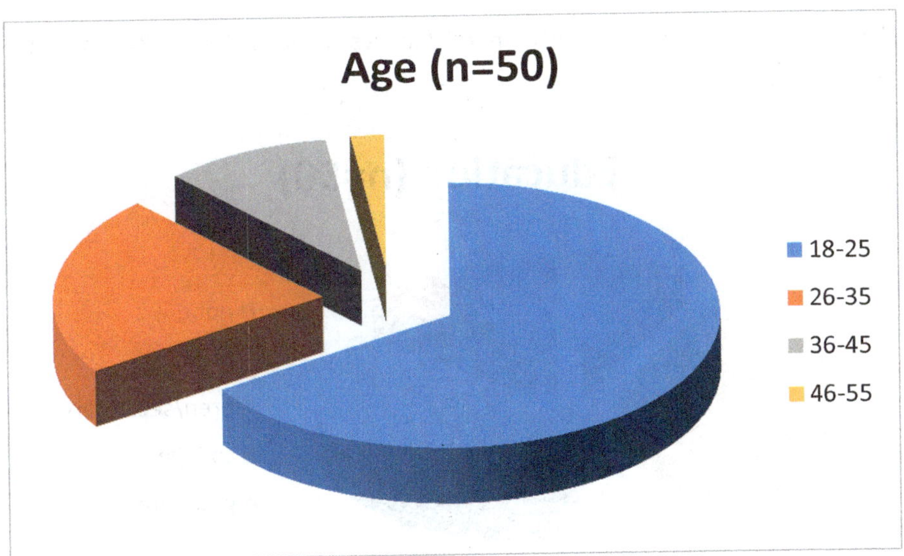

Figure 17.1: Ugandan Age Demographics

Participant age was predominantly within the 18-25 year cohort with 66%. The remaining age cohorts of 26-35 (22%), 36-45 (10%) and 46-55 (2%) made up the total sample. No respondents were available for analysis in the 56-65 age cohort.

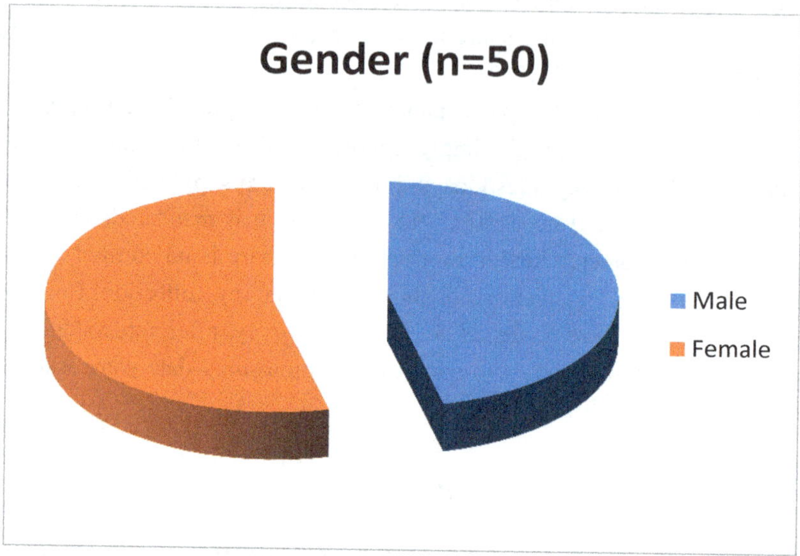

Figure 17.2: Ugandan Gender Demographics

The sample contained slightly more females than males; 54% and 46% respectively.

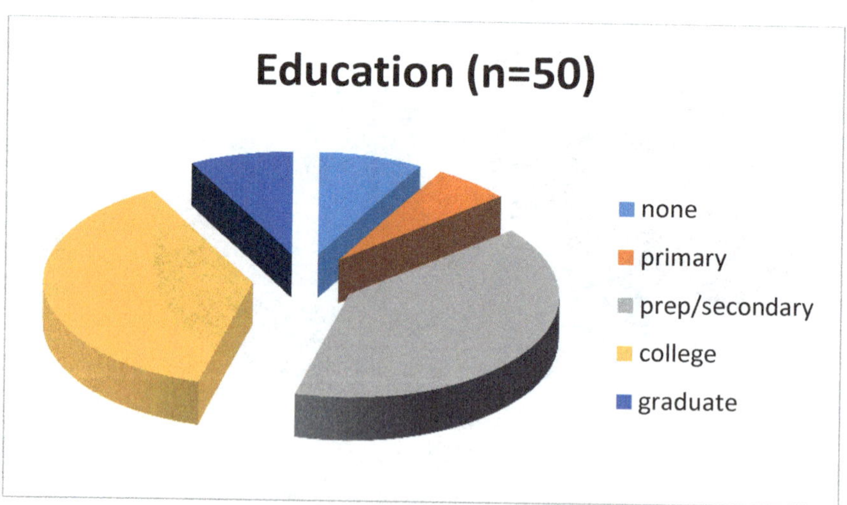

Figure 17.3: Ugandan Education Demographics

The largest categories of education completed were for prep or secondary school (40%), and institute, college or university (38%). No school completion, and graduate school (MA, MS, PhD, MD, LD) completion were noted for 8% of respondents each. Six per cent reported completing primary school only.

RQ2: What types of electronic communication are used in Uganda?

Table 17.3 Number of Media Devices Used by Ugandans (n=50)

Media Devices Used	Percentage of Respondents
None	4
1	22
2	14
3	8
4	10
5	4
6	6
7	8
8	12
9	6
10	6

Ten electronic devices were listed so that respondents could check all that they had used at least once. Devices included:

1. Cellular telephone
2. Chat room
3. Radio
4. Television
5. VCR
6. Satellite dish
7. Digital camera
8. Internet
9. Email account
10. Voicemail

Four percent had used none, and the greatest percentages were for one or two devices used. These devices were most frequently reported as the radio (80%) and cellular telephone (72%); television (52%) and internet (48%) were reported next in frequency of use. See table 17.4.

Table 17.4 Ugandan Electronic Usage by Device and Percentage of Users (n=50)

Device	Number of Users	% of Users
Cellular telephone	36	72
Chat	13	26
Radio	40	80
TV	26	52
VCR	6	12
Dish	11	22
Digital camera	18	36
Electronic web	24	48
Email	19	38
Voice mail	14	28

RQ3: How does electronic usage in Uganda compare to that in Oman, China, Thailand, Pakistan, India, and South Sudan?

Table 17.5 Cellular Telephone Usage Comparisons (n=420)

Country	Usage Period of Greatest Frequency	# of Respondents	% of Respondents
OMAN	2-3 years	60	28
CHINA	2-3 years	79	37
THAILAND	5 or more years	143	57
PAKISTAN	Less than 1 year	92	30
INDIA	5 years or more	12	86
SOUTH SUDAN	5 years or more	17	46
UGANDA	5 years or more	17	34

Uganda has one of the longest usage periods of frequency for five years or more for 34% of respondents. India, South Sudan, and Thailand were comparative for the longer period of five years or more, but their percentages were much higher at India 86%, South Sudan 46%, and Thailand 57%. Pakistan and Oman were the only developing countries that evidenced lower percentages of respondents in the cellular telephone periods of greatest frequency.

Table 17.6 Internet Usage Comparisons (n=447)

Country	Usage Period of Greatest Frequency	# of Respondents	% of Respondents
OMAN	2-3 years	60	28
CHINA	2-3 years	80	37
THAILAND	5 or more years	153	61
PAKISTAN	Not used	101	33
INDIA	5 years or more	13	93
SOUTH SUDAN	5 years or more	17	46
UGANDA	5 years or more	23	46

Internet usage in Uganda is quite similar. Uganda's longest usage period of frequency was again for five years or more, but 46% were recorded as in this period. Uganda also compares to India, South Sudan, and Thailand at 93%, 46% and 61% respectively. Pakistan, China and Oman evidenced lower percentages of respondents in the internet periods of greatest frequency.

Table 17.7 Media Usage Comparisons of Highest Frequency Chosen (n=484)

Country	# of Items Used	# of Respondents	% of Respondents
OMAN	6-9	131	61
CHINA	5-8	69	40
THAILAND	5-7	103	41
PAKISTAN	3-6	132	43
INDIA	6-9	8	57
SOUTH SUDAN	1-2	23	62
UGANDA	1-2	18	36

Ugandans' media usage compares to South Sudan for number of electronic devices and modes of delivery. Both countries had a high percentage of respondents who have used one to two devices at least once. Uganda's cohort is smaller at 36% than South Sudan's at 62%. No other cohort is smaller by percentage than Uganda. The largest cohort percentage of developing countries studied is for Oman (61%) and South Sudan (62%). Six to nine items used were recorded for Oman and India in their highest frequency cohorts.

RQ 4a: How long have Ugandans used cellular telephones?

Table 17.8 Cellular Telephone Use in Uganda (n=49)

Survey Wording for length of cellular use: How long have you used a cellular telephone?

		Frequency	Percent	Valid Percent	Cumulative Percent
Valid	0- no use	8	16.0	16.3	16.3
	Less than 1 year	2	4.0	4.1	20.4
	2-3 years	12	24.0	24.5	44.9
	3-5 years	9	18.0	18.4	63.3
	5 years or more	18	36.0	36.7	100.0
	Total	49	98.0	100.0	
Missing	System	1	2.0		
Total		50	100.0		

Approximately 16% of the sample had not ever used a cellular telephone, and 4.1% had used one less than one year. Nearly one fourth of the sample had used a cellular telephone between two and three years. The highest percentage (36%) had used a cellular telephone for five years or more in Uganda.

RQ4b: How long have Ugandans used the internet?

Table 17.9 Internet Use in Uganda (n=50)

Survey Wording for length of internet use: How long have you used the internet?

	Frequency	Percent	Valid Percent	Cumulative Percent
Valid 0- no use	5	10.0	10.0	10.0
Less than 1 year	4	8.0	8.0	18.0
2-3 years	12	24.0	24.0	42.0
3-5 years	6	12.0	12.0	54.0
5 years or more	23	46.0	46.0	100.0
Total	50	100.0	100.0	

Ten percent of those surveyed reported no internet usage, and 8% reported less than a year of usage. Twenty-four percent reported usage between two and three years. The highest percentage (46%) had used the internet for five years or more in Uganda.

RQ 5: What are the main reasons for using the internet in Uganda?

There was a 90% response rate for this open-ended question. Main reasons were communication, research, entertainment and utilitarian tasks. Over 50% mentioned the top reason: Staying in touch with friends and family. Respondents catalogued Facebook, chatting, email and Skype as the means.

"It is easier than contacting people face2face."

"It is great for guidance and counseling with friends."

"I can communicate with people all over the world."

Research, reading and writing (news and homework) were mentioned by 49% of the respondents.

"I can search for good jobs and courses all over the world."

"Google searching is interesting and useful for schoolwork, especially hard questions."

Third in importance was entertainment (movies, YouTube videos, music, games, photos) as cited by 31%. Shopping and looking at web sites were also mentioned (12%). Least mentioned were work and utilitarian tasks (discussion, storage, record keeping) due to less expense and necessity (2%). The common response was, "It is quick and saves time to pay bills."

RQ 6: Has internet usage become a problem for Ugandan users?

Table 17.10 Problematic Usage in Uganda: Chat Duration (n=50)

Survey Wording for chat duration response: I often stay on the internet chat rooms longer than I intend to.

		Frequency	Percent	Valid Percent	Cumulative Percent
Valid	0-never	6	12.0	12.5	12.5
	hardly ever	5	10.0	10.4	22.9
	sometimes	10	20.0	20.8	43.8
	quite often	8	16.0	16.7	60.4
	always	19	38.0	39.6	100.0
	Total	48	96.0	100.0	
Missing	System	2	4.0		
Total		50	100.0		

Internet usage for chatting duration was never a problem for 12% of Ugandan respondents. Another 10% reported that it was hardly ever a problem. The highest percentage (38%) reported that chatting was always a problem in internet usage. Collapsed responses for sometimes, quite often, and always (noticeable problematic chat usage) totaled 74%, nearly three quarters of the sampled responses.

Table 17.11 Problematic Usage in Uganda: Hours Spent on the Internet (n=50)

Survey Wording for hours spent on the internet response: I use the internet
... hours per week.

		Frequency	Percent	Valid Percent	Cumulative Percent
Valid	Not used	5	10.0	10.0	10.0
	Less than one hour	8	16.0	16.0	26.0
	1-5	15	30.0	30.0	56.0
	5-10	6	12.0	12.0	68.0
	10-20	6	12.0	12.0	80.0
	20-40	5	10.0	10.0	90.0
	40 or more	5	10.0	10.0	100.0
	Total	50	100.0	100.0	

Internet usage was not used by 10% of Ugandan respondents. Another 16% reported less than one hour usage per week. The highest percentage (30%) reported that they used the internet from one to five hours per week. Collapsed responses for 10 or more hours of usage per week totaled 32%, nearly one third of the sampled responses.

Self-reported problems encountered when using the internet were indicated by six descriptors listed on the survey. The first item, Neg Net, allowed for open response of negative reasons why the internet was not liked as a communication venue. See figure 17.4.

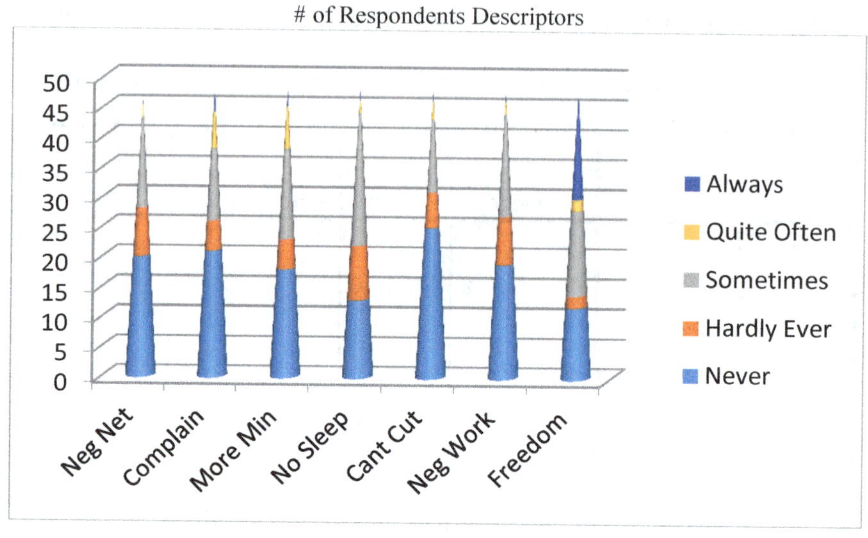

Figure 17.4: Problems Encountered with Internet Usage (n=48)

The descriptors receiving the greatest frequency of response were "never" except for No Sleep and Freedom. No Sleep meant that the respondent lost sleep because of late-night internet usage. It had the greatest frequency of "sometimes" for a response by 44% of the surveys. Freedom meant that the respondent had more freedom in relationship development because of electronic communication as a whole. It had the greatest frequency of "sometimes" for a response by 28% of the surveys. This was near the frequency of Freedom's response cohort of "never" at 24% and "always" at 20% of the surveys.

None of the negative responses for problematic usage were above the 2-4% range for Ugandan respondents; therefore, freedom in using the internet was rarely a problem for others complaining about personal usage, for more minutes used, for attempts to cut usage, or for negative implications for work tasks. Overall, respondents reported greater communication freedom produced by internet and other electronic usage, with little social or psychological hindrances.

Mechanical hindrances of slow, weak connections in Uganda were aptly compared to US usage in cost and accessibility by the open access question, Neg Net. Few respondents filled in the question asking why they did not like the internet. Most echoed, "There are no disadvantages to the internet." Those answering negatively, or semi-negatively were:

4 responses: "It gives answers to all my questions, but it takes time."
4 responses: "It is costly and you get addicted to it,"
4 responses: "It is hard for me to access it if I am away from the university and library."
"Not everyone has access to it. Some of my family members do not have internet so they cannot email."
"I have to use it at the library. No privacy."

Frustrations and fears were noted by some for their internet usage:

"Confusing to understand. Not many people check email on a frequent basis."
"Difficult to use. Too much material. Hard to find information."
"You have to be careful who you talk to since there are dangerous people on the internet. It is hard to always understand a person's intentions."
"The dangers involving women and children."

Two hypotheses were also tested.

H1: Time spent living or studying outside Uganda will be positively associated with electronic communication usage.

Table 17.12 Correlation of Living or Studying Outside Uganda and Electronic Communication Usage (n=50, 49)

	Outside Living	Outside Study	Internet Usage	Cellular Usage
Outside Living Sig. N	1 50	.503* .000 50	111 .443 50	.077 .597 49
Outside Study Sig. N	.503* .000 50	1 50	-.034 .815 50	-.169 .245 49
Internet Usage Sig. N	.111 .443 50	-.034 .815 50	1 50	.722* .000 49
Cellular Usage Sig. N	.077 .597 49	-.169 .245 49	.722* .000 49	1 49

*Pearson correlations are significant at the 0.01 level (2-tailed).

No significant correlations were found for living or studying outside Uganda with either internet usage or cellular usage. Hypothesis 1 was not supported.

H2: Age will be positively correlated to cellular and internet usage, and number of media choices.

Table 17.13 Correlation of Age, Cellular and Internet Usage, and Number of Media Choices (n=50, 49)

	Age	# Media Choices	Internet Usage	Cellular Usage
Age Sig. N	1 50	-.107 .457 50	.228 .111 50	.319** .026 49
Media Choices Sig. N	-.107 .457 50	1 50	.368* .008 50	.307 .032 49
Internet Usage Sig. N	.228 .111 50	.368* .008 50	1 50	.722* .000 49
Cellular Usage Sig. n	.319** .026 49	.307** .032 49	.722* .000 49	1 49

*Pearson correlations are significant at the 0.01 level (2-tailed).

** Pearson correlations are significant at the 0.05 level (2-tailed).

Significance was found for age and cellular usage (.319**), media choices and internet usage (.368*), internet and cellular usages (.722*), and media choices and cellular usage (.307**). All correlations were positive except age and number of media choices. Hypothesis 2 was supported for the other correlations.

SUMMARY

This mixed study of qualitative and quantitative methods was undertaken over a two-year period extending from 2013-2015. Extensive data from 50 Ugandans were found to answer the six research questions, and test two hypotheses.

Participant age was predominantly within the 18-25 year cohort with 66% of respondents. The sample contained slightly more females than males; 54% and

46% respectively. The largest categories of education completed were for preparatory or secondary school (40%), and institute, college or university (38%). No school completion, and graduate school (MA, MS, PhD, MD, LD) completion were noted for 8% of respondents each. Six per cent reported completing primary school only. The gender and educational demographics were balanced, representing all categories. The trend of cellular and internet usage by the newer generations of 18-25 is corroborated by Ugandan respondents.

Little variety in types of electronic communication are used in Uganda. Four percent had used no electronic devices at all, and the greatest percentages were for one or two devices used. These devices were most frequently reported as the radio (80%) and cellular telephone (72%); television (52%) and internet (48%) were reported next in frequency of use.

Cellular telephone usage in Uganda compares to that in neighboring countries of Oman, China, Thailand, Pakistan, India, and South Sudan. Uganda has one of the longest usage periods of frequency for five years or more for 34% of this study's respondents. India, South Sudan, and Thailand were comparative for the longer period of five years or more, but their percentages were much higher at India 86%, South Sudan 46%, and Thailand 57%.

Internet usage in Uganda is quite similar. Uganda's longest usage period of frequency was again for five years or more, but 46% were recorded as in this period. Uganda also compares to India, South Sudan, and Thailand at 93%, 46% and 61% respectively. Ugandans' media usage compares to South Sudan for number of electronic devices and modes of delivery. Both countries had a high percentage of respondents who have used one to two devices at least once. Uganda's cohort is smaller at 36% than South Sudan's at 62%. No other cohort is smaller by percentage than Uganda. The largest cohort percentage of developing countries studied is for Oman (61%) and South Sudan (62%). Six to nine items used were recorded for Oman and India in their highest frequency cohorts.

Approximately 16% of the Ugandan sample had not ever used a cellular telephone, and 4.1% had used one less than a year. Nearly one fourth of the sample had used a cellular telephone between two and three years. The highest percentage (36%) had used a cellular telephone for five years or more in Uganda.

Ten percent of those surveyed reported no internet usage, and 8% reported less than a year of usage. Twenty-four percent reported usage between two and three years. The highest percentage (46%) had used the internet for five years or more in Uganda. The main reasons for using the internet in Uganda netted a 90% response rate for this open-ended question. Main reasons were communication, research, entertainment and utilitarian tasks. Over 50% mentioned the top reason:

Staying in touch with friends and family. Respondents catalogued facebooking, chatting, email and Skype as the means. Research, reading and writing (news and homework) were mentioned by 49% of the respondents. Internet usage was not used by 10% of Ugandan respondents. Third in importance was entertainment (movies, YouTube videos, music, games, photos) as cited by 31%. Shopping and looking at web sites were also mentioned (12%). Least mentioned were work and utilitarian tasks (discussion, storage, record keeping) due to less expense and necessity (2%).

Internet usage for chatting duration was never a problem for 12% of Ugandan respondents. Another 10% reported that it was hardly ever a problem. The highest percentage (38%) reported that chatting was always a problem in internet usage. Collapsed responses for sometimes, quite often, and always (noticeable problematic chat usage) totaled 74%, nearly three quarters of the sampled responses. Internet usage was not used by 10% of Ugandan respondents. Another 16% reported less than one hour usage per week. The highest percentage (30%) reported that they used the internet from one to five hours per week. Collapsed responses for 10 or more hours of usage per week totaled 32%, nearly one third of the sampled responses.

Problematic usage was recorded by several internally reliable means. The descriptors receiving the greatest frequency of response were "never" except for No Sleep and Freedom. No Sleep meant that the respondent lost sleep because of late-night internet usage. It had the greatest frequency of "sometimes" for a response by 44% of the surveys. Freedom meant that the respondent had more freedom in relationship development because of electronic communication as a whole. It had the greatest frequency of "sometimes" for a response by 28% of the surveys. This was near the frequency of Freedom's response cohort of "never" at 24% and "always" at 20% of the surveys which clouded the results for this subject pool.

None of the negative responses for problematic usage were above the 2-4% range for Ugandan respondents; therefore, freedom in using the internet was rarely a problem for others complaining about personal usage, for more minutes used, for attempts to cut usage, or for negative implications for work tasks. Overall, respondents reported greater communication freedom produced by internet and other electronic usage, with little social or psychological hindrances.

Mechanical hindrances of slow, weak connections in Uganda were aptly compared to US usage in cost and accessibility by the open access question, Neg Net. Few respondents filled in the question asking why they did not like the internet. Most echoed, "There are no disadvantages to the internet." Frustrations and fears were noted by four respondents for their internet usage.

The first hypothesis, "Time spent living or studying outside Uganda will be positively associated with electronic communication usage," had no significant correlations, and was not supported. Living or studying outside Uganda was not a factor in adopting or continuing internet usage or cellular usage.

The second hypothesis, "Age will be positively correlated to cellular and internet usage, and number of media choices," had significance for age and cellular usage (.319), media choices and internet usage (.368), internet and cellular usages (.722), and media choices and cellular usage (.307). All correlations were positive except age and number of media choices. Hypothesis 2 was supported.

Strengths and Limitations

The disciplines of communication studies, development and sociology; and social work can readily access current documentation of electronic usage for the emerging, developing economy of Uganda. Diffusion of Innovation theory was reinforced as an explanation of internet usage not dependent upon education or living arrangements outside Uganda. This study truly examines Ugandan users of electronic communication in many levels.

Strengths of the study include the near equal representation of male and female response, and the illustration of many electronic communication devices and modes of delivery. Cellular and internet usage were confirmed to be far less problematic than helpful to Ugandans. Another strength of the study is the general breadth of many survey questions allowing comparison to other developing countries of the area, and allowing comparison over the second decade of research.

The study has the limitation of not collecting data on social networking as defined by Facebook and other international venues readily accessible in the 21st century. This was beyond the scope of the original project, and would not have been compared to earlier studies. Social networking was used as a research tool, however. It aided the faster collection of surveys and snowball method of contacting Ugandan nationals, a weighty problem for distribution over a specific population. Another limitation would be the limited respondents available for the age cohort of 36-45 (10%), 46-55 (2%), and 56-65 (0%) age cohort. Those electronic users age 35 and younger appear to represent Ugandan usage, which may not be accurate. This could have been offset by purposively collecting data on more business owners, educators and government service workers.

Further Directions

Ugandans are incorporating electronic and mobile communication platforms into their everyday lifestyles at a pace that excels growth in ICT infiltration in developed countries. Africa, as a whole, is the world's fastest growing market for ICT usage, and Ugandans are the fastest growing sector of users. Several catalysts are responsible for these indicators including charities that donate refurbished computers and phones; special functions and applications for low budget mobile phones such as flashlights, radio, and African language options; outside investment in ICT infrastructure and development in sub-Saharan Africa; and expanded energy access and alternative energy sources, such as charging stations and solar power.

While increases in the usage of digital and electronic technology platforms are often identified as a positive step for human development and opportunity, more studies need to identify the consequences of growth in ICT infiltration in countries that are considered as undeveloped and underdeveloped. Particular focus should be on cyber security and identify theft where fewer safeguards currently exist.

Crime is high in Uganda. In 2013, the International Criminal Police Organization (INTERPOL) reported surges in human, drug, and arms traffickers, and in 2008, counterfeit medical products and fake antimalarial drugs for children. Last year INTERPOL reported rises in organized crime groups from outside Uganda selling counterfeit electrical cables that failed to meet safety standards, and increases in cybercrimes such as mobile money, transaction, and banking fraud; hacking (unauthorized access); and intellectual property crime. Therefore, future study should address the challenges in countering cybercrime activity with the following questions in mind:

1. Are training and tools (anti-spyware software) readily available to safeguard identity and personal information?
2. Does existing legislation and enforcement adequately address cybercrime?
3. Are governmental organization monitoring cybersecurity incidents and collaborating with service providers and organizations on cybersecurity-related issues?

While ICT access and development remain limited in Uganda when compared with developed countries, the Ugandan desire for electronic and digital platforms can be seen in long-range planning reports such as municipal comprehensive plans, retail and industry budgets, banking, integration into individual quality-of-life development, and a growing usage by young adult demographics. Particularly evident since 2000, are growths in ICT usage in education and healthcare. Uganda's first phase for introducing free education began in 1997 with primary schools, and now includes secondary and post-secondary education. The 2000 Ebola outbreak initiated telemedicine practices, such as video conferencing, radio, fax, and television, and tele-pharmacy, referring to pharmaceutical care provisions. Smart phone technology now offers apps and plug-in devices for remote healthcare monitoring, therapy, blood pressure readings, ear infection diagnosis, and monitor mental health. Although telemedicine addresses cost and access concerns, practicing medicine in a virtual environment should raise concerns about hacking personal and confidential information, and breaches to physician and patient confidentiality.

REFERENCES

Akpan-Obong, Patience, Carlos A. Thomas, Kilbily D. Samake, Moses Niwe, and Victor W. Mbarika. "An African Pioneer Comes of Age: Evolution of Information and Communication Technologies in Uganda." *Journal of Information, Information Technology, and Organizations* 4 (2009): 148-169.

Anderson, T. "Mobile Phones Reach Uganda's Villages." *BBC News,* 2007. http://news.bbc.co.uk/go/pr/fr/-/2/hi/business/7071636.stm

Blauw, Sanne L., and Philip H. Franses. "The Impact of Mobile Telephone Use On Economic Development of Households in Uganda." Discussion paper / Tinbergen Institute, (2011): 1-48. http://repub.eur.nl/res/aut/2/

Burrell, Jenna. "Evaluating Shared Access: Social Equality and the Circulation of Mobile Phones in Rural Uganda." *Journal of Computer-mediated Communication* 15, no. 2 (2010): 230–250. doi: 10.1111/j.1083-6101.2010.01518.x

Buskins, Ineke, and Anne Webb. *African Women & ICTs: Investigating Technology, Gender and Empowerment.* London: Zed Books, 2009.

Chang, Larry W., Andrew Mwanika, Dan Kaye, Wilson W. Muhwezi, Rose Nabirye, Scovia Mbalinda ... and Robert C. Bollinger. "Information and Communication Technology and Community-based Health Sciences

Training in Uganda: Perceptions and Experiences of Educators and Students." *Information and Communication Technology and Community-based Healthcare* 37, no. 1 (2012): 1-11.

Contreras, Felix. "Young Latinos, Blacks Answer Call of Mobile Devices." National Public Radio, 2009 September. www.npr.org/templates/story/story.php?storyId=120852934

Ferris, Shaun, Engoru, Patrick, and Kaganzi, Elly. "Making Market Information Services Work Better for the Poor in Uganda." *Collective Action and Property Rights (CAPRi), CAPRi Working Papers* no. 77 (2008): May.

International Food Policy Research Institute (IFPRI), www.capri.cgiar.org/pdf/capriwp77.pdf.

Horst, Heather, and Daniel Miller. *The Cell Phone: An Anthropology of Communication.* London: Berg, 2006.

International Criminal Police Organization. *Uganda.* "2014 Trafficking in Persons

Report." US Department of State.

_ _ _ . "Operation Mamba (IMPACT). Targeting Counterfeit Medicines in Tanzania and Uganda." October 29, 2008. http://www.interpol.int/News-and-media/News/2008/N20081029

John, Sydelle. "Internet & Telephone Usage in Africa." Kirkland, WA: Demand Media, Inc., 2012. http://www.ehow.com/facts_6807828_internet-telephone-usage-africa.html#ixzz24DJY5xJn

Johnson, David S., Rajinder P. Goel, Paul Birtwistle, and Phil Hirst. "Transferring Medical Images on the World Wide Web for Emergency Clinical Management: A Case Report." *British Medical Journal* 316 (1998,

March): 988. doi: 10.1136/bmj.316.7136.988

Katahoire, Anne R., Grace Baguma, and Florence Etta. "Schoolnet Uganda-curriculumnet." International Development Research Center, *Information & Communication Technologies for Development in Africa* 3 (2001): 215-249.

Katungi, Enid, Svetlana Edmeades, and Melinda Smale. "Gender, Social Capital and Information Exchange in Rural Uganda." *Collective Action and Property Rights (CAPRi), Working Papers* no. 59 (2006): 1-32. International Food Policy Research Institute (IFPRI), www.capri.cgiar.org

Luscombe, Belinda. "TIME's Mobile Tech Issue: Tracking Disease One Text at a Time." *TIME Healthland* (2012): August. *Healthland.time.com/2012/08/15/disease-cant-hide*

Madanda, Aramanzan N. "Uganda: Is technology a blessing or a curse?" Association for Progressive Communications post on Makerere University seminar, *APC Take Back the Tech! To End Violence Against Women*, May 21, 2010.

Madanda, Aramanzan N., Berna Noglobe, and Goretti Z. Amuriat. "Uganda: Violence Against Women and Information and Communication Technologies." Association for Progressive Communications (APC). Department of Women and Gender Studies, Makerere University (2009): September. www.womenstudies.mak.ac.ug www.apc.org/en/system/files/APCWNSP_MDG3_2011_EN%20pdf.pdf

MapSwitch Uganda. "Driving Financial Access through Technology in Uganda, Transforming Financial Services for Under-served Markets." 2009. www.mapinternational.net

May, Julian, and Kathleen Diga. "Mobile Cell Phones and Poverty Reduction: Technology Spending Patterns and Poverty Level Change among Households in Uganda." *M.A. Thesis*, University of KwaZulu; Natal, Durban, 2007. http://hdl.handle.net/10413/2073

McKay, Betsy, Nicholas Bariyo, and Drew Hinshaw. "A Tale of Two Africas." *Wall Street Journal*. 23-24 August, 2014.

Minges, M. "Mobile Internet for Developing Countries." 2007, October. https://www.isoc.org/isoc/conferences/inet/01/CD_proceedings/G53/

Miniwatts Marketing Group. "Internet Usage and Population Statistics for Africa." 2012. www.internetworldstats.com

Mitchell, Kimberly J., Sheana Bull, Julius Kiwanuka, and Michelle L. Ybarra. "Cell Phone Usage among Adolescents in Uganda: Acceptability for Relaying Health Information." *Health Education Research* 26, no. 5 (2011): 770-781. doi: 10.1093/her/cyr022

Mobile Internet for Developing Countries. *Conference Proceedings* 23 (2007): 26. https://www.isoc.org/isoc/conferences/inet/01/CD_proceedings/G53/

Mutambi, Joshua, Joseph K. Byaruhanga, B. Kariko Buhwezi, Lena Trojer, and Peter Okidi-Lating. "Transferring Best Practices for Uganda Technological Innovation and Sustainable Growth." Second

International Conference on Advances in Engineering and Technology, New Delhi, India, August 25-30 (2011): 256-262.

Muto, Megumi. "The Impacts of Mobile Phones and Personal Networks on Rural-to-Urban Migration: Evidence from Uganda." *Journal of African Economies* (2012): April 16. doi: 10.1093/jae/ejs009

Muto, M., and Yamano T. Takashi. "The Impact of Mobile Phone Coverage Expansion on Market Participation: Panel Data Evidence from Uganda." *World Development* 37, no. 12 (2009, December): 1887–1896. http://dx.doi.org/10.1016/j.worlddev.2009.05.004

Nalubega, F. "Using the Mobile Phone to Boost Agri-business." *Daily Monitor* (2012): June 5.

Ndawula, Stephen, David H. Ngobi, Deborah Namugenyi, and Rose C. Nakawuki. "A Study of End-users' Attitudes Towards Digital Media Approach: The Experience of a Public University in Uganda." *International Journal of Higher Education* 1, no. 2 (2012): 150-156.

Reynolds, Laura, and Daniel Ninsiima. "Mobile Phone Technology Improves Farmers' Fortunes in Uganda." *iConnecT-Online* (2012): August. http://blogs.worldwatch.org/nourishingtheplanet/mobile-phone-technology

Republic of Uganda. *National Development Plan, 2010-2014.* Ministry of Finance, Planning and Economic Development, 2010.

Siedner, Mark J., Jessica E. Haberer, Mwebesa B. Bwana, Norma C. Ware, and David R. Bangsberg. "High Acceptability for Cell Phone Text Messages to Improve Communication of Laboratory Results with HIV-infected Patients in Rural Uganda: A Crosssectional Survey Study." *BMC Medical Informatics and Decision Making* 12, no. 56 (2012): additional file. doi: 10.1186.1472-6947-12-56

Swendeman, Dallas, and Mary J. Rotheram-Borus. "Innovation in Sexually Transmitted Disease and HIV Prevention: Internet and Mobile Phone Delivery Vehicles for Global Diffusion." *Current Opinion Psychiatry* 23, no. 2 (2010): 139–144. doi: 10.1097/YCO.0b013e328336656a

Uganda Digital Education Resource Bank. *Ministry of Education and Sports.* Schoolnet Uganda. (2007). http://www.schoolnetuganda.sc.ug/projects/on-going-projects/uderb.htm

Waiswa, Robert, and Constant Okello-Obura. "To What Extent Have ICTs

Contributed to E-Governance in Uganda?" *Library Philosophy and Practice.* 2014, July. University of Nebraska-Lincoln.

Wang, Le, Edward Mutafungwa, Yeswanth Puvvala, and Jukka Manner. "Strategies for Energy Efficient Mobile Web Access: An East African Case Study." E-infrastructure and E-services for Developing Countries [lecture notes]. *Institute for Computer Sciences, Social Informatics and Telecommunications Engineering* 92 part 1 (2012): 74-83. doi: 10.1007/987-3-642-29093-0_7

World Economic Forum. "Network Readiness Index." (2014). http://www3.weforum.org/docs/GITR/2014/GITR_OverallRanking_201 4.pdf

Ybarra, Michele L., Ruth Biringi, Tonya Prescott, and Sheana S. Bull. "Usability and Navigability of an HIV/AIDS Internet Intervention for Adolescents in a Resource-limited Setting." *Computers, Informatics and Nursing Journal,* 2012. doi: 10.1097/NXN.obo13e318266cb0e

Ybarra, Michele L., Julius Kiwanuka, Nneka Emenyonu, and David R. Bangsberg. "Internet Use Among Ugandan Adolescents: Implications for HIV Intervention." *PLoS Med 3,* no. 11(2006): e433. doi:10.1371/journal.pmed.0030433

CHAPTER 18

Internet and Mobile Phone Usage in Education in United Arab Emirates

Zahrah Aljaber

The United Arab Emirates (UAE) is located in Southwest Asia, bordering the southeast end of the Arabian Gulf. It is between Oman and Saudi Arabia; in a strategic location along southern approaches to the Strait of Hormuz which is a transit point for world oil. UAE's population is 9.2 million. The country was established in December 1971, and has grown to be the second largest economy in the Arab world between 2010 and 2013.

Education in the United Arab Emirates has developed since 2011. Spending on education technology in the UAE is one of the highest in the world. UAE is seeking to use the best methods of digital education. The government provides online education (Tele-education) such as the Open University system, under the supervision of a professor in a foreign university, and mobile phone education. The UAE is constantly looking to better the education system.

This study focuses on the use of internet technology and mobile phone usage in education in United Arab Emirates, and addresses two points: First, Robotel smart class multimedia helps to create a generation of known information between people and develop students' skills. Second, mobile phone education is seeking to access free education without paper using a mobile phone. Structuration theory is the essential theory of the study which focuses on working together as a group from different places by using different technology to effect the learning audience. The data is collected by questionnaires that United Arab Emirates citizens have completed. Five research questions will be analyzed and reported within this chapter.

LITERATURE REVIEW

The United Arab Emirates is a federation of seven emirates. Each emirate is governed by a hereditary emir. One of the emirs is selected as the President of the

United Arab Emirates. The constituent emirates are Abu Dhabi, Ajman, Dubai, Fujairah, Rasal-Khaimah, Sharjah, and Umm al-Quwain. Abu Dhabi contains the capital of UAE. Islam is the religion of the UAE, and Arabic is the official language. Islamic law is a significant source of the UAE's legislation, and it influences current educational practices. According to Al-Kilefi (2012), in 2011, the economy in UAE has developed into a modern, albeit developing, state. UAE has the second largest economy in the Arab world. UAE is ranked as the 14th best nation in the world for doing business based on its economy and regulatory environment. Petroleum and natural gas exports play an important role in the economy.

Education in UAE is free for citizens. It consists of primary schools, middle schools and high schools, and the curriculum is created to enrich United Arab Emirates development's goals and values. The Arabic language is used in education with English as a second language. According to the QS Rankings (2007) the UAE has kept improving education and research that includes the establishment of the CERT Research Centers, the Masdar Institute of Science and Technology, and Institute for Enterprise Development. According to Etahad news (2013) the volume of spending on education technology in the UAE comes within the highest in the world. The UAE government is seeking through the initiative to reach a paper free learning environment. This is what has already been accomplished via a mobile phone education initiative in the federal higher education, which was launched in September of 2012.

Theory

Structuration theory is a general theory of social action. This theory is tested through ordinary practice. As communication acts strategically to achieve goals, participants do not realize that they are simultaneously creating forces that return to affect future actions. Structuration theory highlights ways of communication within large groups and individually. Interpersonal relationships are very important within the culture, the persons, and what they care about. According to Giddens (1984) using concepts from objectivist and subjectivist social theories, he discarded objectivism's focus on detached structures, which lacked regard for humanist elements and subjectivism's exclusive attention to individual or group agency without consideration for socio-structural context.

DeSanctis and Poole (1994) adapted and augmented this theory by researching interest in the relationship between technology and social structures such as information technology in organizations. DeSanctis and Poole proposed an adaptive Structuration theory with respect to the emergence and use of group decision support systems. In particular, they chose Giddens' notion of modalities to consider how technology is used with respect to its spirit. Appropriations are the immediate, visible actions that reveal deeper structuration processes and are enacted with moves. Appropriations may be faithful or unfaithful, and become instrumental when used with various attitudes.

Giddens defines three dimensions of structuration that explain how people create a right way to use social structures. They act accordingly for social structures to emerge from these actions. The first is knowledge, which uses technology according to moral codes, and people can use it in a community, so they can reach the goals. The second is the use of a virtual learning environment. That means that some structuration ideas on how technology works can influence social action. According Loureiro-Koechlin and Allan (2010), Giddens introduces the concept of time–space distanciation as a key feature of modernity. This refers to the separation of time and space, which in traditional societies is linked through place, and their recombination in forms which permit the precise time-space zoning of social life (Walsham 1998). For example, the time and space for a taught module are linked to the classroom. The use of a visual learning experience (VLE) changes this system so the interactions can take place at different times and from different places (2009). Third, the most well-known uses of structuration in the internet technology (IT) field is Orlikowski's enactment of technologies-in-practice. Orlikowski (2000) separates two aspects of technology: technology as artifact which means using tools of technology, and technology-in-practice which means practicing the technology to understand the interaction between people and social action.

Electronic Communication

Today, there are so many electronic devices that have developed and can be used in different ways, especially in communicating traditional ways. People were using computers for counting, but now they are using this technology for different purposes such as shopping, learning, and contacting others in different places. People found that new tools of technology have a powerful ability and speed to send or receive messages between people. For example, e-mail, mailing lists, the World Wide Web, and access to information in the world helps academics, researchers, and students contribute to empirical knowledge. Researchers found that students like to use modern electronic communication to better learn from traditional ways. Powell (2015) says it has been argued that online discussions provide opportunities for students to learn from each other through a collaborative process of sharing experiences and views, creating discussions that involve an exchange of multiple perspectives. The asynchronous nature of online discussions is believed to allow students to unpick and analyze themes in a way which face-to-face communication cannot. It reveals a preference by students to use the discussion board to share opinions and experiences, rather than to learn about a subject. Learning is identified as arising from the act of reading and thinking.

Education in United Arab Emirates

United Arab Emirates is seeking to use the best methods of digital education in the world. In UAE they are using technological education to support educational practices in the transfer of teaching methods to the modern digital pattern. Also, students are qualifying for dealing with information and communication technology, including raising their competitiveness in the world. According to Etahad news (2013), the volume of spending on education technology in the UAE comes within the highest in the world. The UAE government is seeking through the initiative to reach a paper-free classroom. This is what has already been accomplished via a mobile phone education initiative in the federal higher education, which was launched in September of 2012.

Communication technologies contribute to raising the level of education and scientific research. They help move the invention of records on the internet sites, raise the level of education and training, develop teaching methods, and facilitate remote transfer of education and training services to remote areas. Also, they provide online education (tele-education) such as the Open University system, the supervision of a professor in a foreign university, and the use of the most important elements of information technology: audio and video. Studies show that the use of these methods may make the student learning process easier, and lift the motivation of defaulting students. However, the results of the use of communication and information technology in teaching depends on the teacher for the organization of the education process. The technology alone does not achieve significant change in raising the level of education and learning what is not employed in innovative ways by the teacher.

Problems with Education Technology in United Arab Emirates

Education in the Emirates depends on memorization of information and transfer of this information to students without encouraging them to use their potential to help them find solutions, develop a spirit of research, innovation, and self-learning. Students are expected to listen and remember the information, but never give an opinion. As educational institutions have not been able to achieve a balance between the demands of the labor market, the skills, and the available supply of the educational system of various training centers, there is widespread unemployment among graduates, especially graduates with university degrees, in addition to the inflation of academic education on technological and research skills development. Therefore, UAE seeks to provide educational solutions and applications systems (Smart Class Multimedia) in public schools in order to raise the quality and viability of the development level of learning, especially in the grades that support multimedia applications. This initiative aims to improve the UAE region through the latest computers and communications technologies in the classroom.

Role of Education Technology

Communication technology in education has the potential to be extended in many innovative ways. Robotel Smart Class Multimedia helps teachers and trainers to use interactive tools to present the curriculum, share information, monitor student comprehension, and to overcome the challenges facing the education process. Mohammed Saeed (2011) asserts that programs create a generation knowing information technology applications, and enhance educational experiences using innovative teaching methods. Students gain necessary skills to succeed in their future careers. The Emirates aims are to provide a better education system, and to provide international students with the knowledge, understanding, skills and values to ensure their success in personal and national leveling the UAE now provides.

Mobile Phone Education

United Arab Emirates is seeking to access paper-free education by using a mobile phone. According to Michael Folan (2010), this technology impacts in strengthening and deepening the things that students learn. It is also at a time when the world is witnessing a digital revolution which will help this reference to the correct choice in the field of digital education. This technology is available and easy to use. It helps teachers choose teaching methods and install the most feasible practices, noting that the process of identifying the gap in the digital education helps digital educational software developers to create more effective educational solutions.

Groves, Meisenbach, and Scott-Cawiezell (2011) discuss when many people died in USA hospitals because there was no safety culture; they also discuss how to improve safety culture in healthcare organizations and how to keep patients safe. However, they suggest that using Structuration theory helped them understand the problem. Structuration theory is like a system which includes both individual actions and group structures to work and find solutions. It uses nursing, safety, and organizational culture to make all these work together to improve safety culture. Also, everybody in healthcare organizations is responsible for patient safety; especially nurses, because they are in constant attendance and nearest to patients. Structuration theory helps them to understand their culture system by studying their formal and informal messages. Everybody works together as a structure. However, there are some limitations in this theory such as complicated concepts which can be difficult to understand. Furthermore, it requires more of explanation and development by testing it in the theory form. These limitations may make communication difficult, and limit improvement in the safety culture system.

Chan, Deave, and Greenhalgh (2010) use Structuration theory to look at childhood obesity in transition zones. They notice that many children are obese especially in the fastest growing countries. However, they think that Structuration theory may help with the interaction between population and individual causes of obesity. They did research in overweight preschool children in Hong Kong to find reasons. They did find different reasons for different children. There are micro-level and macro-level reasons. Some of the reasons are social structures in the family, caregiving, environment, or disjointed problems. Some of the solutions are in the structure of security, caregivers, controls for family stress, and mealtime routines that can limit child obesity. They find that when they use Structuration theory to organize children's routine lives, they will help to limit childhood obesity.

Using Structuration theory differently, Loureiro-Koechlin, and Allan (2010) encourage women to progress into employment. They organize learning activities in the form of a lecture, face-to-face, and online. Then, they used Structuration theory to focus on the relationships between human agency as big, and social structures as small. Structuration theory is used as an analytical tool to help understand what happened within the project. They are interested in implementing e-learning processes as a part of a project; however, they used Structuration theory to explain how the e-learning and e-mentoring structures emerged and from which sources. They used Structuration theory to design a module to enable students to develop the necessary knowledge and skills to allow them to enter LaSCI at a management level. Structuration theory is a system of relationships that determines the actions of individuals, and then these actions of individuals lead to the creation of a successful in society. By using learning structures, students have received automated emails with reminders and updates to organize activities and to facilitate learning. However, there were two factors that influenced the project and their learning activities: the issue of absence and presence online, and the issue of time frame changes for online users. The theory helps to find solutions to this project.

Fulk (1993) talks about technology in an article with empirical evidence of communication technology patterns. The article found these patterns in a study of electronic mail. They used a group of scientists and engineers. They found that people like to work in large groups more than work as individuals to find when people work in a group, they will interact more, and find deeper analysis and information by using Structuration theory to explain the patterns which are found in communication technology. In addition, communication technologies are both a cause and consequence of work as a structure. Structuration theory worked in this article because they found a good result of working together.

Moreover, Flynn (2001) used Structuration theory to explain information systems development, and used this information in a UK public National Health Service (NHS) organization. Flynn wanted to use Structuration theory to understand the social aspects of this process. Also, the article helps to explain the complex organizational processes in development and use. The theory changed a UK public National Health Service (NHS) organization. Also, they felt that using the internet will help to change the organizational culture, worker values, and what they believe. The study shows that Structuration theory is a powerful framework for social shaping of behavior, so it helps to organize the work in the UK Public National Health Service (NHS) organization between people and the network to work together to reach their goals.

Andersson and Hatakka (2010) describe how distance education in developing countries is growing with information and communication technologies. They study distance education in Bangladesh and Sri Lanka because these countries use different technologies for implementing interactivity; for example, the internet, computers, video, and mobile phones. However, they want to use these technologies differently in learning practices to change teachers' and students' beliefs and values to provide everyone more possibilities for interaction together. They think that e-learning involves a shift in the educational structure from traditional transmission to interaction. Then, they test everything on Structuration theory and compare the traditional transmission and development interaction. They found that, when people work together, collaboration and the use of self-assessment tools make students take more ownership of their learning.

Structuration theory will be useful in United Arab Emirates because people work as a structure in large groups from different cities. Also, they use different communication technology to reach their goals. I like this theory. It is beneficial to understand a developing country. The condition of the internet and mobile devices usage will be shown by research and examining the following research questions.

- RQ1: What age, gender, group of people use technology communication in United Arab Emirates?
- RQ2: What kinds of technology tools are used in United Arab Emirates?
- RQ3: Is technology communication used correlated to religion, moral codes, and economics?
- RQ4: Do people prefer to use electronic communication to communicate and form a relationship in United Arab Emirates?
- RQ5: Have individual lives developed by using different technology in United Arab Emirates?

METHOD

The study discuses internet and mobile phone usage in United Arab Emirates and how these technologies effect UAE's education. Therefore, all participants who complete the questionnaire should be United Arab Emirates citizens or expatriates. There are two students at Fort Hays State University who I met face to face and five friends who are from United Arab Emirates that I sent the research questions. They sent it to their families, friends, and people who work at United Arab Emirates' schools. The subject pool is from different age brackets from 18-25 years old, 25-35 years old, 35-45 years old, and 45- 55 years old.

Instrumentation and Procedures

Since this study is going to discuss two kinds of technologies that are used in United Arab Emirates, the method which is used in this study combines quantitative and qualitative research. The data collected is primarily from United Arab Emirates citizens. Forty questions in the research surveys are in English. Most students in United Arab Emirates know English because they study English in the school as a second language, but there are translated forms for older people who do not know English. How people use technology and the internet in their daily lives and why they choose and prefer to communicate with others by using different electronic devices were discussed in the survey questions. In addition, there is an online survey, Survey Monkey which is a data representation survey program that will help to reach the goal collecting surveys in United Arab Emirates.

RESULTS

The survey data was collected in November, 2014 and April, 2015. There were 31 respondents from United Arab Emirates who participated in the survey. Most were returned in English, but two of them were in Arabic. The Survey Monkey link was sent to people from United Arab Emirates, and some people who have studied abroad. 31 people responded, and 40 questions were answered for each respondent.

RQ1: What age, gender, and group of people use technology communication in United Arab Emirates?

Survey results show that all respondents use the internet and cellular telephones more than any other technology. The largest percentage of internet and cellular telephone usage is in the 25-35 age group, which was 17 respondents with 54.83%. The second largest group was 10 respondents from the 18-25 age group with 32.25%. Only 3 respondents with 9.08% from 35-45 year olds made this group the smallest proportion. From the 45-65 age group there was 1 respondent with 3.23%. See Figure 18.1.

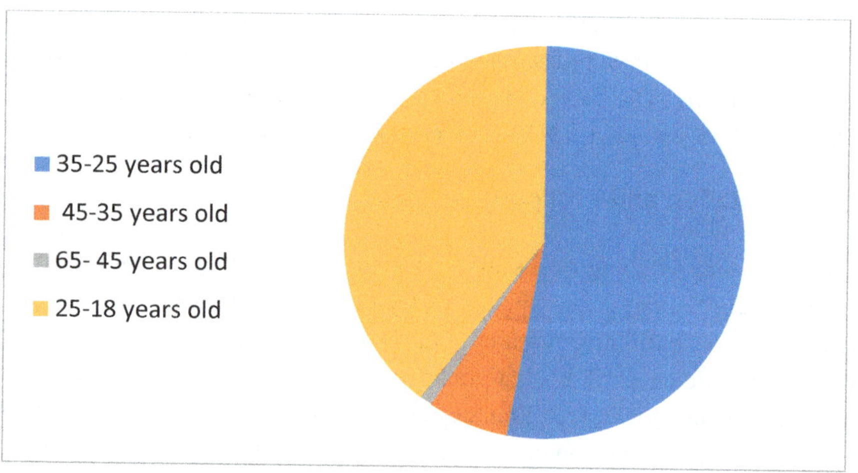

Figure 18.1: Respondent Age Demographics of Emirates (n=31)

Figure 18.2 shows all the participants; 20 respondents are male (64.52%) while only 11 respondents are female (35.48%); therefore, more people who participated were male.

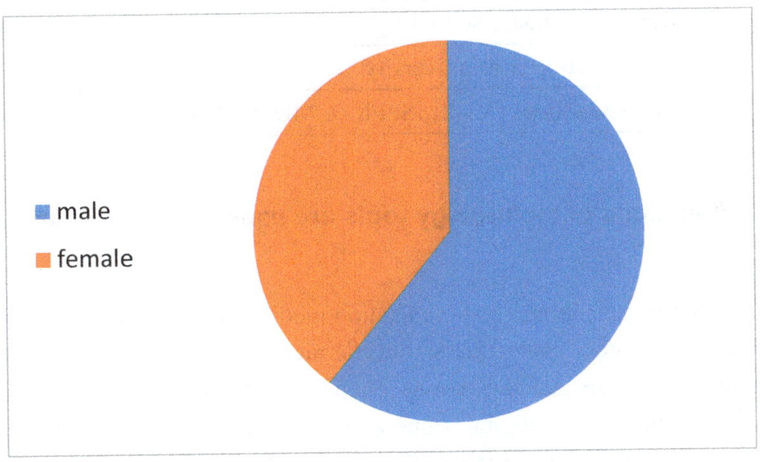

Figure 18.2: Respondents Gender Demographics of Emirates (n=31)

Most respondents have completed institute, college or university that constituted 64.52%. Education level completion was 29.3% for graduate school (MA, MS, Ph.D, MD, LD), and 6.45% for high or secondary school. See Figure 18.3.

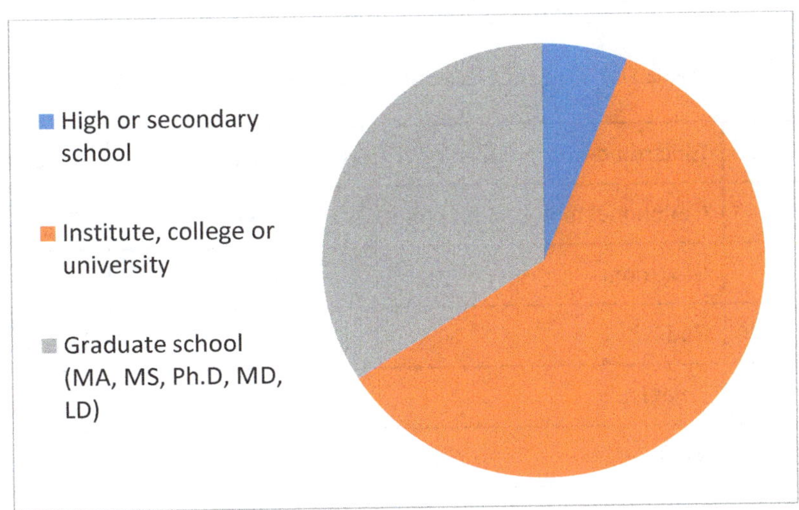

Figure 18.3: Education Level Completion (n=31)

Legend for Pie Chart

Education Level Completion	Respondents
Primary school	0
High or secondary school	2
Institute, college or university	20
Graduate school (MA, MS, Ph.D, MD, LD)	9

RQ 2: What kinds of technology tools are used in United Arab Emirates?

There are fourteen electronic communication media that were recorded as used in United Arab Emirates. Table 18.1 shows the media use of respondents in United Arab Emirates. These include: mobile phone, email account, online chat room, television, satellite dish, digital camera, radio, e-commerce website, banking kiosk or money card, the internet listserv, VCR, the internet bulletin board, voicemail, and proxima are included as communication media. The cell or mobile phone with 100% of 31 respondents and the television with 96.77% of 29 respondents were used commonly in United Arab Emirates. The least percentage of the electronic communication usage was satellite dish, Voicemail and proxima (or other presentation device) have the least percentage of the electronic communication usage with 6.12%.

Table 18.1 Types of Electronic Communication Used in UAE

Number	Electronic Communication Types	Respondents	Percentage
1	Cellphone, moto, mobile phone	31	100.0%
2	Chatroom	10	32.25%
3	Radio	27	87.09%
4	Television	30	96.77%
5	VCR	6	19.35%
6	Satellite dish	5	16.12%
7	Digital camera	20	64.51%

8	E-commerce website	7	22.58%
9	Banking kiosk or money card	10	32.25%
10	Internet listserv	7	22.85%
11	Internet bulletin board	4	12.9%
12	Email account	8	25.80%
13	Voicemail	8	25.80%
14	Proxima (or other presentation device)	5	16.12%

RQ3: Is technology communication used correlated to religion, moral codes, and economics?

The electronic communication usage in United Arab Emirates relates to religion and moral codes because there is a strict boundary between male and female responsibility such us dress, movement, eye contact or speech. Men and women can communicate and use electronic communication together in accordance to Islamic roles. Moreover, the electronic communication usage does not correlate with the economic status in United Arab Emirates. Figure 18.4 shows that electronic device users have no reliable income in United Arab Emirates with 54.83% and have enough to provide for their own and others' needs on a regular basis with 19.35%. These people were students and teachers who live in United Arab Emirates or study overseas.

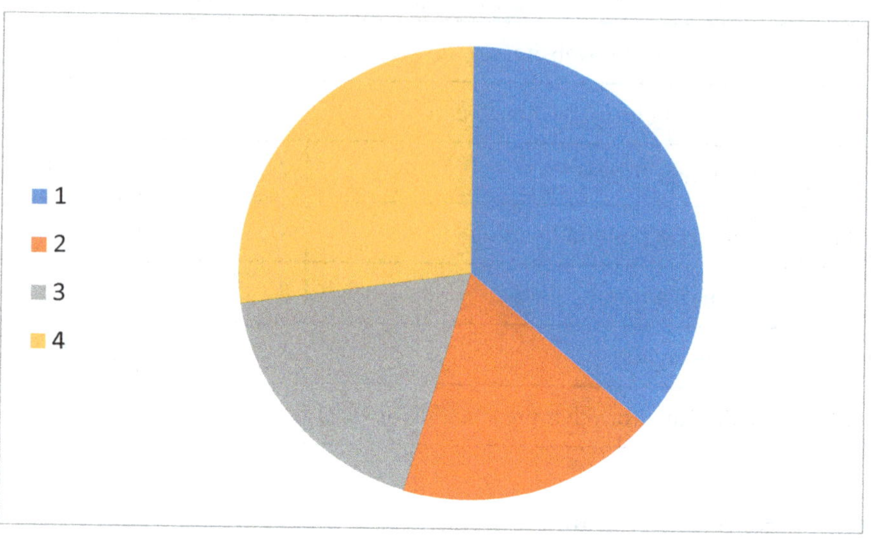

Figure 18.4: Economic Status of United Arab Emirates Electronic Device Users

Legend for Pie Chart

Number	Economic Status	Respondents	Percentage
1	I have no reliable income.	16	54.83%
2	I have a fixed income/pension less than 1,000 USD per month.	4	12.90%
3	I have a job or business providing housing, food, clothing and life necessities.	4	12.90%
4	I have enough to provide for myself and others' needs on a regular basis.	6	19.35%

RQ 4: Do people prefer to use electronic communication to communicate and form a relationship in United Arab Emirates?

Figure 18.5 shows that people in United Arab Emirates prefer to communicate personally more then use electronic communication. There were 26 respondents who preferred to communicate in person with 83.87%. There were 4 respondents who preferred to communicate by telephone with 12.9%. However, most respondents like to form and create new relationships online (72.73%), and most of them had hardly ever met new electronic friends face-to-face (54.45%). See figure 18.6 and table 18.2.

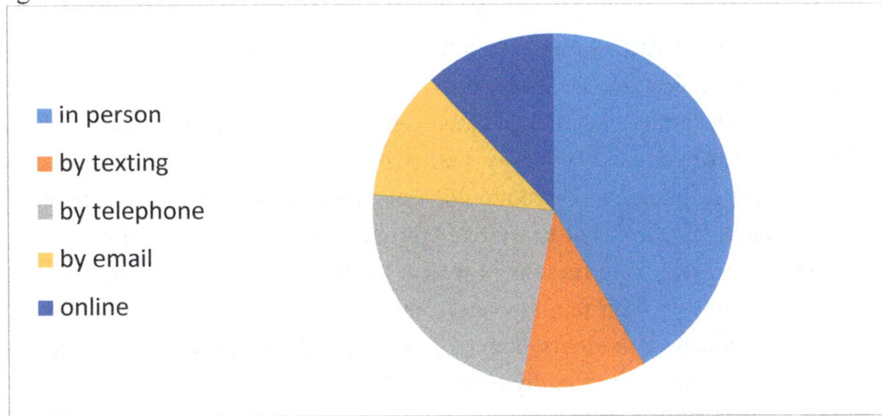

Figure 18.5: Communication Choice (n=31)

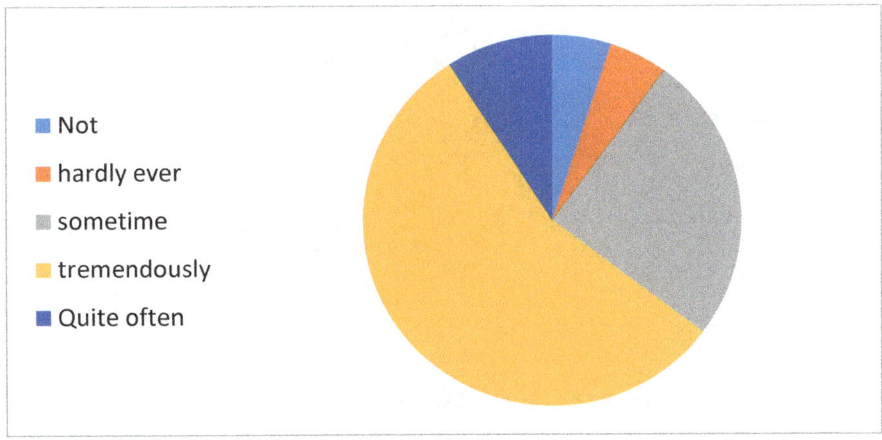

Figure 18.6: Form New Relationships Online (n=31)

Table 18.2 Meet New Relationships Online (n=31)

Meet New Electronic Friends Face-to-face Descriptors	Percentage
Never	27.27%
Hardly ever	45.45%
Sometimes	18.18%
Always	9.9%

RQ5: Have individual lives developed by using different technology in United Arab Emirates?

Recently in United Arab Emirates, individuals' lives have developed by using electronic communication. The traditional way, which is visiting and spending the whole time with family and friends, was used in United Arab Emirates between individuals. However, now the way that people communicate has changed by using different technology. There are 26 respondents who chose "tremendously" in questionnaires which I sent to United Arab Emirates people with 83.87% of all participants. The two main reasons of their answer were first, it was amazing and easy to communicate with families and friends, and second, it saves time and is fast to communicate with others. Three of the respondents replied "sometimes" at 9.67%, 1 respondent chose "hardly ever" at 3.22%, and 1 respondent replied "not" at 3.22%. See Figure 18.7.

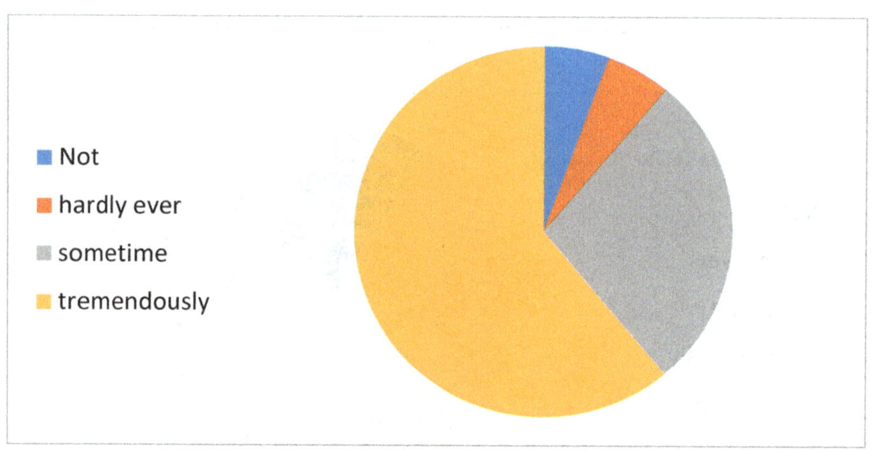

Figure 18.7: Ways of Communication (n=31)

Some respondents wrote:

> "Because I can reach friends who are in different countries."
> "Because it speeds a way to get the information."
> "Because it helps to find more about the word."
> "Because it is easy to use to communicate and search for a lot of things."
> "Because it makes the communication easier."

DISCUSSION

United Arab Emirates people are used to using communication technology to communicate and develop their country. Five research questions were answered.

Summary

RQ 1: What age, gender and education levels use electronic devices in United Arab Emirates?

The results of the study showed that all respondents of United Arab Emirates people use electronic devices. People who are 25-35 years old are more likely to use the internet and cellular phone than any other technology, which was shown with 17 respondents in this age group with 54.83%. Ten respondents from the 18-25 age group use technology. Only 3 respondents with 9.08% of 35-45 years old reported using technology. There were 20 respondents who were male (64.52%) while only 11 respondents were female (35.48%). Moreover, most respondents had completed institute, college or university school that constituted 64.52%. Education level completion was 29.3% for graduate school (MA, MS, Ph. D, MD, LD), and 6.45% for high or secondary school.

RQ 2: What kinds of technology tools are used in United Arab Emirates?

Fourteen types of electronic communication media were recorded as used in United Arab Emirates. The study showed which was the most frequently used media by respondents in United Arab Emirates. These technologies include: mobile phone, email account, online chat room, television, satellite dish, digital camera, radio, e-commerce website, banking kiosk or money card, the internet listserv, VCR, the internet bulletin board, voicemail, and proxima. In United Arab Emirates 100% of 31 respondents use a mobile phone and 96.77%, or 30 respondents use the television. The least percentage of the electronic communication usage was satellite dish,Voicemail and proxima (or other presentation device) with 6.12%. The results showed that the main reasons for using electronic devices in United Arab Emirates were communicating with other people and an easy way to reach information about the world.

RQ3: Is technology communication used correlated to religion, moral codes, and economic status?

The result showed that the electronic communication usage in United Arab Emirates is directly related to religion and moral codes because there is a strict boundary between male and female responsibility such us dress, movement, eye contact, or speech. Men and women can communicate and use electronic communication together in accordance to Islamic roles. Moreover, the electronic communication usage is not correlated to economic status in United Arab Emirates. The electronic devices users have no reliable income in United Arab

Emirates with 54.83% and have enough to provide for their needs as well as others on a regular basis with 19.35%. These people were students and teachers who live in United Arab Emirates or study abroad.

RQ 4: Do people prefer to use electronic communication to communicate and form a relationship in United Arab Emirates?

People in United Arab Emirates prefer to communicate personally more than communicate by using electronic communication. The study showed that 26 respondents prefer to communicate in person with 83.87%. There were 4 respondents who prefer to communicate by telephone with 12.90%. Most respondents like to form and create new relationships online (72.73%), and some of them hardly ever meet new electronic friends face-to-face (54.45%). However, the result showed that the traditional way to meet or create relationships, which is face-to-face interpersonal communication, is still important in United Arab Emirates.

RQ5: Have individual lives developed by using different technology in United Arab Emirates?

Results showed that individual lives have developed by using electronic communication. However, the traditional way, which is visiting and spending the whole time with family and friends, was used in United Arab Emirates between individuals. The way that people communicate has changed by using different technology. There were 26 respondents who chose "tremendously" changed lives in questionnaires which I sent to United Arab Emirates people with 83.87% of all participants responding in this manner. The two main reasons of their answer were first, "amazing and easy to communicate with families and friends," and second, "saves time and is fast to communicate with others." Moreover, 3 respondents replied "sometimes" changed lives for 9.67%, 1 respondent chose "hardly ever" 3.22%, and 1 respondent replied "not" changed life at 3.22%. The qualitative responses included:

- Because it speeds the way to get the information
- Because it helps to find more about the word.
- Because it easiest way to use to communicate and search for a lot of things.
- Because it make the communication easier. I can reach a friend who is in a different country.

Limitations

There were three main limitations of this study. First, the research was extremely lacking of rural resident respondents. All participants of the survey are urban residents in areas which have more development of technology; therefore, the study cannot show the difference between people who live in cities and people who live in the country. Second, the participation of men and women was not similar due to culture and equality in numbers returned. There were 31 people who participated in this study; twenty were male respondents (64.52%) while 11 respondents were female (35.48%). Also, men and women in United Arab Emirates use different technology as a result of their religion. What type do women use for communication? It is a question that needed to be answered. A second limitation was that some of the questions in the survey elicited general answers including: "I prefer to communicate by email, in person, etc.," or "I meet new electronic friends face-to-face." The results showed that respondents chose what type of communication they preferred, and if they meet new electronic friends face-to-face without any specific personal reasons. These two limitations, and the small sample size mean the study does not show the main reasons males and females as a group use electronic communication in United Arab Emirates.

Suggestions for Future Research

In United Arab Emirates, electronic communication users are people who are young adults with a higher degree. They use these electronic communications to access social networks which help them to communicate with their family and friends, find information, and go shopping; especially people who study abroad. Specifically, this study does not show the difference between men and women when using electronic communication. In a future study, we may find more reasons for the particular national usage.

This research and study have been aided by Structuration theory which is a general theory of social action. Structuration theory highlights ways of communication within large groups and individually. Interpersonal relationships are very important within the culture of UAE, its people both male and female, and what they care about personally and as a nation. Structuration theory is useful in United Arab Emirates because its people work as a structure in large groups from different cities. Also, they use different communication technology to reach their goals. Structuration theory is beneficial to researchers in understanding a developing country. In contrast, people in United Arab Emirates need to learn more about electronic communication usage because these electronic communications help to develop technology, economy, and education.

REFERENCES

Alghazo, Iman M. "Student Attitudes toward Web-Enhanced Instruction in an Educational Technology Course." *College Student Journal* 40, no. 3 (Sep. 2006): 620-630.

Alsawaie, Othman N., and Iman M. Alghazo. "Enhancing Mathematics Beliefs and Teaching Efficacy of Prospective Teachers through an Online Video Club." *International Journal of Applied Educational Studies* 8, no. 1 (Aug. 2010): 44-56.

Andersson, Annika, and Mathias Hatakka. "Increasing Interactivity in Distance Educations: Case Studies Bangladesh and Sri Lanka." *Information Technology for Development* 16, no. 1 (Oct, 2010): 16-33.

Chan, Christine, Toity Deave, and Trisha Greenhalgh. "Childhood Obesity in Transition Zones: An Analysis Using Structuration Theory." *Sociology of Health & Illness* 32, no. 5 (Jul. 2010):711-729.

Clarke, Matthew. "The Discursive Construction of Interpersonal Relations in an Online Community of Practice." *Journal of Pragmatics* 41, no. 11 (Nov. 2009): 2333-2344.

Flynn, Donal. "Using Structuration Theory to Explain Information Systems Development and Using a Public Health Organization." *Global Co-Operation in the New Millennium* 36, no. 6 (June 2001): 1- 12.

Fulk, Janet. "Social Construction of Communication Technology." *The Academy of Management Journal* 36, no. 5 (Oct. 1993): 921-950.

Groves, Patricia S., Rebecca J. Meisenbach, and Jill Scott-Cawiezell. "Keeping Patients Safe in Health Care Organizations: A Structuration Theory of Safety Culture." *Journal of Advanced Nursing* 67, no. 8 (Aug. 2011): 1846-1855.

Hargis, Jace, Cathy Cavanaugh and Tayeb Kamali. "A Federal Higher Education iPad Mobile Learning Initiative: Triangulation of Data to Determine Early Effectiveness." *Innovative Higher Education* 39, no. 1 (Feb. 2014): 45-57.

Loureiro-Koechlin, Cecilia., and Barbara Allan. "Time, Space and Structure in an E-Learning and E-mentoring Project That Develops Women

Community." *British Journal of Educational Technology* 41, no. 5 (Sep. 2010): 721- 735.

Powell, Alison B. (2015) "Open Culture and Innovation: Integrating Knowledge Across Boundaries." *Media, Culture and Society* 37, no. 3 (2015): 376-393.

Schmid, Richard F., Robert M. Bernard, and Eugene Borokhovski. "The Effects of Technology Use in Postsecondary Education: A Meta-analysis of Classroom Applications." *Computers & Education* 72, no. 4 (May 2014): 271-291.

Shana, Zuhrieh. "Learning with Technology: Using Discussion Forums to Augment a Traditional-Style Class." *Journal of Educational Technology & Society* 12, no. 3 (Jul. 2009): 214-228.

Appendix A

Country Comparisons

Comparative Collection Data for *ECDC*, Volumes 1 and 2
24 Developing Countries

COUNTRY *ECDC* Volume	Number of Respondents	GENDER		AGE COHORT of Largest %	
		Male %	Female %		
Argentina 2	61	39	61	18-25	84
Brazil 1	7 case studies	NA	NA	8-75	100
China 1	425	42	57	18-25	41
Guinea 2	33	34	66	18-25	58
India 1	14	21	79	45-65	43
Indonesia 2	36	60	40	18-25	54
Jamaica 2	45	48	52	18-25	48
Kenya 2	30	53	47	25-35	37
Kyrgyzstan 2	41	43	57	18-25	48
Laos 2	43	47	53	18-25	63
Malaysia 2	60	84	16	25-35	48
Maldives 2	25	48	52	18-25	28
Mexico 2	28	74	26	18-25	74
Nepal 2	19	60	34	18-35	59
Oman 1	211	57	36	15-20	35
Pakistan 1	287	31	69	18-25	67
Paraguay 2	43	43	50	25-35	46
Philippines 2	28	44	56	25-35	50
Russia 2	41	51	49	18-25	37
South Africa 1	22	100	0	25-35	68
South Sudan 1	37	39	56	25-35	42
Thailand 1	250	46	54	18-25	66
Uganda 2	50	65	35	25-35	55
UAE* 2	31	52	47	18-25	34
			Double Modes		
TOTAL	1868		14x	18-25	
			8x	25-35	

Note: Only complete, valid surveys were recorded as individual respondents. Brazilian subjects (approximately 600) were observed only; seven case studies were recorded for the sociological study, and gender was hidden for anonymity.

*United Arab Emirates is arguably a developed country.

Appendix B

Maps and Communication Facts

Argentina

BOLIVIA
BRAZIL
PARAGUAY
Rio Paraguay
Rio Paraná
Salta
San Miguel de Tucumán
Corrientes
Posadas
Rio Paraná
ANDES
San Juan
Córdoba
Santa Fe
Concepción del Uruguay
Cerro Aconcagua
San Lorenzo
San Martin
Rosario
URUGUAY
SOUTH PACIFIC OCEAN
Mendoza
BUENOS AIRES
La Plata
Rio de la Plata
PAMPAS
Bahia Blanca
Mar del Plata
CHILE
San Carlos de Bariloche
Viedma
Punta Colorada
PATAGONIA
ANDES
SOUTH ATLANTIC OCEAN
Comodoro Rivadavia
Laguna del Carbón
Falkland Islands
(Islas Malvinas)
(administered by U.K., claimed by ARGENTINA)
Río Gallegos
Strait of Magellan
Ushuaia
Scotia Sea
Beagle Channel
Drake Passage
0   200   400 km
0   200   400 mi

Telephones - main lines in use:
10 million (2012)
country comparison to the world: 22

Telephones - mobile cellular:
58.6 million (2012)
country comparison to the world: 23

Telephone system:
general assessment: in 1998 Argentina opened its telecommunications market to competition and foreign investment encouraging the growth of modern telecommunications technology; fiber-optic cable trunk lines are being installed between all major cities; major networks are entirely digital and the availability of telephone service is improving domestic: microwave radio relay, fiber-optic cable, and a domestic satellite system with 40 earth stations serve the trunk network; fixed-line teledensity is increasing gradually and mobile-cellular subscribership is increasing rapidly; broadband Internet services are gaining ground international: country code - 54; landing point for the Atlantis-2, UNISUR, South America-1, and South American Crossing/Latin American Nautilus submarine cable systems that provide links to Europe, Africa, South and Central America, and US; satellite earth stations - 112; 2 international gateways near Buenos Aires (2011)

Broadcast media:
government owns a TV station and a radio network; more than 2 dozen TV stations and hundreds of privately owned radio stations; high rate of cable TV subscription usage (2007)

Internet country code:
.ar

Internet hosts:
11.232 million (2012)
country comparison to the world: 13

Internet users:
13.694 million (2009)
country comparison to the world: 28

GUINEA

Telephones - main lines in use:

18,000 (2012)

country comparison to the world: 192

Telephones - mobile cellular:

4.781 million (2012)

country comparison to the world: 115

Telephone system:

general assessment: inadequate system of open-wire lines, small radiotelephone communication stations, and new microwave radio relay system domestic: Conakry reasonably well-served; coverage elsewhere remains inadequate and large companies tend to rely on their own systems for nationwide links; fixed-line teledensity less than 1 per 100 persons; mobile-cellular subscribership is expanding and exceeds 40 per 100 persons international: country code - 224; satellite earth station - 1

Intelsat (Atlantic Ocean) (2011)

Broadcast media:

government maintains marginal control over broadcast media; single state-run TV station; state-run radio broadcast station also operates several stations in rural areas; a steadily increasing number of privately owned radio stations, nearly all in Conakry, and about a dozen community radio stations; foreign TV programming available via satellite and cable subscription services (2011)

Internet country code:

.gn

Internet hosts:

15 (2012)

country comparison to the world: 223

Internet users:

95,000 (2009)

country comparison to the world: 161

INDONESIA

Telephones - main lines in use:

37.983 million (2012)

country comparison to the world: 8

Telephones - mobile cellular:

281.96 million (2012)

country comparison to the world: 4

Telephone system:

general assessment: domestic service includes an interisland microwave system, an HF radio police net, and a domestic satellite communications system; international service good domestic: coverage provided by existing network has been expanded by use of over 200,000 telephone kiosks many located in remote areas; mobile-cellular subscribership growing rapidly international: country code - 62; landing point for both the SEA-ME-WE-3 and SEA-ME-WE-4 submarine cable networks that provide links throughout Asia, the Middle East, and Europe; satellite earth stations - 2 Intelsat (1 Indian Ocean and 1 Pacific Ocean) (2011)

Broadcast media:

mixture of about a dozen national TV networks - 2 public broadcasters, the remainder private broadcasters - each with multiple transmitters; more than 100 local TV stations; widespread use of satellite and cable TV systems; public radio broadcaster operates 6 national networks as well as regional and local stations; overall, more than 700 radio stations with more than 650 privately operated (2008)

Internet country code:

.id

Internet hosts:

1.344 million (2012)

country comparison to the world: 42

Internet users:

20 million (2009)

country comparison to the world: 22

JAMAICA

Telephones - main lines in use:

265,000 (2011)

country comparison to the world: 123

Telephones - mobile cellular:

2.665 million (2012)

country comparison to the world: 135

Telephone system:

general assessment: fully automatic domestic telephone network domestic: the 1999 agreement to open the market for telecommunications services resulted in rapid growth in mobile-cellular telephone usage while the number of fixed-lines in use has declined; combined mobile-cellular teledensity exceeded 110 per 100 persons in 2011 international: country code - 1-876; the Fibralink submarine cable network provides enhanced delivery of business and broadband traffic and is linked to the Americas Region Caribbean Ring System (ARCOS-1) submarine cable in the Dominican Republic; the link to ARCOS-1 provides seamless connectivity to US, parts of the Caribbean, Central America, and South America; the ALBA-1 fiber-optic submarine cable links Jamaica, Cuba, and Venezuela; satellite earth stations - 2 Intelsat (Atlantic Ocean) (2010)

Broadcast media:

3 free-to-air TV stations, subscription cable services, and roughly 30 radio stations (2013)

Internet country code:

.jm

Internet hosts:

3,906 (2012)

country comparison to the world: 149

Internet users:

1.581 million (2009)

country comparison to the world: 80

KENYA

Telephones - main lines in use:

251,600 (2012)

country comparison to the world: 124

Telephones - mobile cellular:

30.732 million (2012)

country comparison to the world: 33

Telephone system:

general assessment: inadequate; fixed-line telephone system is small and inefficient; trunks are primarily microwave radio relay; business data commonly transferred by a very small aperture terminal (VSAT) system domestic: sole fixed-line provider, Telkom Kenya, is slated for privatization; multiple providers in the mobile-cellular segment of the market fostering a boom in mobile-cellular telephone usage with teledensity reaching 65 per 100 persons in 2011 international: country code - 254; landing point for the EASSy, TEAMS and SEACOM fiber-optic submarine cable systems; satellite earth stations - 4 Intelsat (2011)

Broadcast media:

about a half-dozen large-scale privately owned media companies with TV and radio stations as well as a state-owned TV broadcaster provide service nation-wide; satellite and cable TV subscription services available; state-owned radio broadcaster operates 2 national radio channels and provides regional and local radio services in multiple languages; a large number of private radio stations broadcast on a national level along with over 100 private and non-profit provincial stations broadcasting in local languages; transmissions of several international broadcasters available (2014)

Internet country code:

.ke

Internet hosts:

71,018 (2012)
country comparison to the world: 88

Internet users:

3.996 million (2009)
country comparison to the world: 59

LAOS

Telephones - main lines in use:

112,000 (2012)

country comparison to the world: 143

Telephones - mobile cellular:

6.492 million (2012)

country comparison to the world: 99

Telephone system:

general assessment: service to general public is improving; the government relies on a radiotelephone network to communicate with remote areas domestic: 4 service providers with mobile cellular usage growing very rapidly international: country code - 856; satellite earth station - 1 Intersputnik (Indian Ocean region) and a second to be developed by China (2012)

Broadcast media:

6 TV stations operating out of Vientiane - 3 government-operated and the others commercial; 17 provincial stations operating with nearly all programming relayed via satellite from the government-operated stations in Vientiane; Chinese and Vietnamese programming relayed via satellite from Lao National TV; broadcasts available from stations in Thailand and Vietnam in border areas; multi-channel satellite and cable TV systems provide access to a wide range of foreign stations; state-controlled radio with state-operated Lao National Radio (LNR) broadcasting on 5 frequencies - 1 AM, 1 SW, and 3 FM; LNR's AM and FM programs are relayed via satellite constituting a large part of the programming schedules of the provincial radio stations; Thai radio broadcasts available in border areas and transmissions of multiple international broadcasters are also accessible (2012)

Internet country code:

.la

Internet hosts:

1,532 (2012)

country comparison to the world: 166

Internet users:

300,000 (2009)

country comparison to the world: 130

MALAYSIA

Telephones - main lines in use:
4.589 million (2012)
country comparison to the world: 34

Telephones - mobile cellular:
41.325 million (2012)
country comparison to the world: 30

Telephone system:
general assessment: modern system featuring good intercity service on Peninsular Malaysia provided mainly by microwave radio relay and an adequate intercity microwave radio relay network between Sabah and Sarawak via Brunei; international service excellent domestic: domestic satellite system with 2 earth stations; combined fixed-line and mobile-cellular teledensity roughly 140 per 100 persons international: country code - 60; landing point for several major international submarine cable networks that provide connectivity to Asia, Middle East, and Europe; satellite earth stations - 2 Intelsat (1 Indian Ocean, 1 Pacific Ocean) (2011)

Broadcast media:
state-owned TV broadcaster operates 2 TV networks with relay throughout the country, and the leading private commercial media grou

operates 4 TV stations with numerous relays throughout the country satellite TV subscription service is available; state-owned radi broadcaster operates multiple national networks as well as regional an local stations; many private commercial radio broadcasters and som subscription satellite radio services are available; about 55 radio station overall (2012)

Internet country code:
.my

Internet hosts:
422,470 (2012)
country comparison to the world: 53

Internet users:
15.355 million (2009)
country comparison to the world: 26

MALDIVES

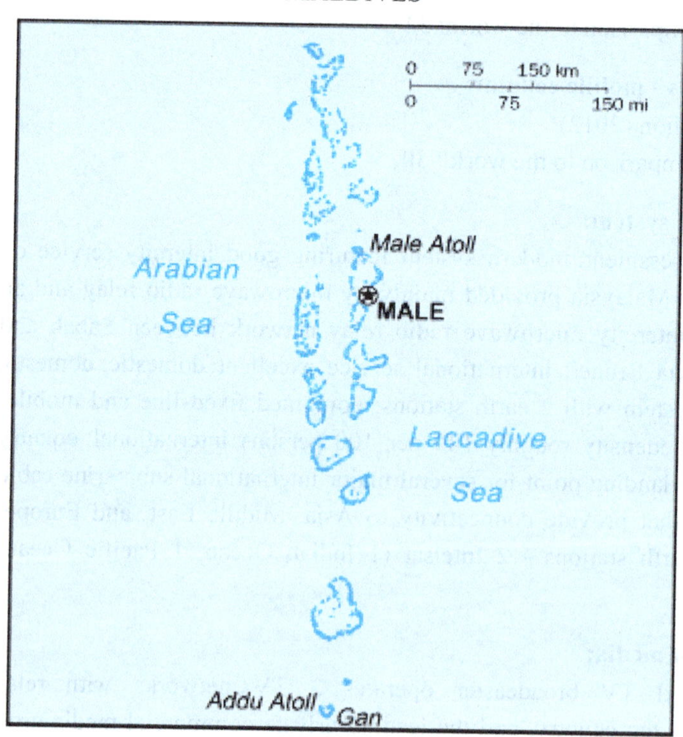

Telephones - main lines in use:

23,140 (2012)

country comparison to the world: 183

Telephones - mobile cellular:

560,000 (2012)

country comparison to the world: 166

Telephone system:

general assessment: telephone services have improved; inter-atoll communication through microwave links; all inhabited islands and resorts are connected with telephone and fax service domestic: each island now has at least 1 public telephone, and there are mobile-cellular networks with a rapidly expanding subscribership that has reached 135 per 100 persons international: country code - 960; linked to international submarine cable Fiber-Optic Link Around the Globe (FLAG); satellite earth station - 3 Intelsat (Indian Ocean) (2011)

Broadcast media:

state-owned radio and TV monopoly until recently; state-owned TV operates 2 channels; 3 privately owned TV stations; state owns Voice of Maldives and operates both an entertainment and a music-based station; 5 privately owned radio stations (2012)

Internet country code:

.mv

Internet hosts:

3,296 (2012)

country comparison to the world: 153

Internet users:

86,400 (2009)

country comparison to the world: 164

Mexico

Telephones - main lines in use:

20.22 million (2012)

country comparison to the world: 14

Telephones - mobile cellular:

100.786 million (2012)

country comparison to the world: 13

Telephone system:

general assessment: adequate telephone service for business and government; improving quality and increasing mobile cellular availability, with mobile subscribers far outnumbering fixed-line subscribers; domestic satellite system with 120 earth stations; extensive microwave radio relay network; considerable use of fiber-optic cable and coaxial cable domestic: despite the opening to competition in January 1997, Telmex remains dominant; Fixed-line teledensity is less than 20 per 100 persons; mobile-cellular teledensity is about 80 per 100 persons international: country code - 52; Columbus-2 fiber-optic submarine cable with access to the US, Virgin Islands, Canary Islands, Spain, and Italy; the Americas Region Caribbean Ring System (ARCOS-1) and the MAYA-1 submarine cable system together provide access to Central America, parts of South America and the Caribbean, and the US; satellite earth stations - 120 (32 Intelsat, 2 Solidaridad (giving Mexico improved access to South America, Central America, and much of the US as well as enhancing domestic communications), 1 Panamsat, numerous Inmarsat mobile earth stations); linked to Central American Microwave System of trunk connections (2011)

Broadcast media:

many TV stations and more than 1,400 radio stations with most privately owned; the Televisa group once had a virtual monopoly in TV broadcasting, but new broadcasting groups and foreign satellite and cable operators are now available (2012)

Internet country code:

.mx

Internet hosts:

16.233 million (2012)

country comparison to the world: 9

Internet users:

31.02 million (2009)

country comparison to the world: 12

NEPAL

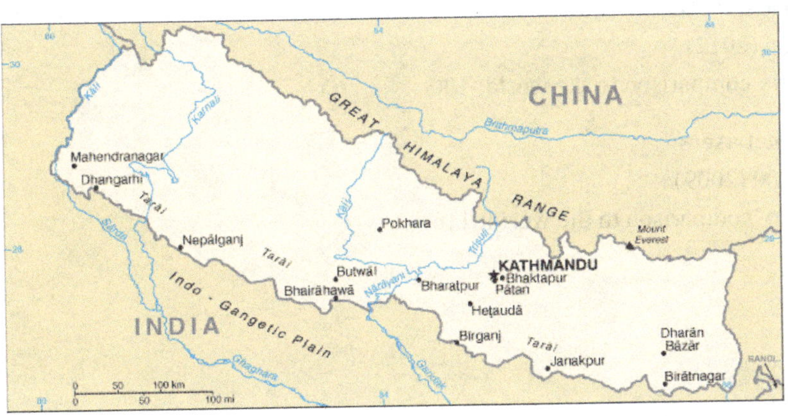

Telephones - main lines in use:

834,000 (2013)

country comparison to the world: 83

Telephones - mobile cellular:

18.138 million (2013)

country comparison to the world: 54

Telephone system:

general assessment: poor telephone and telegraph service; fair radiotelephone communication service and mobile-cellular telephone network domestic: mobile-cellular telephone subscribership base is increasing with roughly 90% of the population living in areas covered by mobile carriers international: country code - 977; radiotelephone communications; microwave and fiber landlines to India; satellite earth station - 1 Intelsat (Indian Ocean) (2011)

Broadcast media:

state operates 2 TV stations as well as national and regional radio stations; roughly 30 independent TV channels are registered with only about half in regular operation; nearly 400 FM radio stations are licensed with roughly 300 operational (2007)

Internet country code:

.np

Internet hosts:

41,256 (2012)

country comparison to the world: 100

Internet users:

577,800 (2009)

country comparison to the world: 116

PARAGUAY

Telephones - main lines in use:

376,000 (2012)

country comparison to the world: 107

Telephones - mobile cellular:

6.79 million (2012)

country comparison to the world: 95

Telephone system:

general assessment: the fixed-line market is a state monopoly and fixed-line telephone service is meager; principal switching center is in Asuncion domestic: deficiencies in provision of fixed-line service have resulted in a rapid expansion of mobile-cellular services fostered by competition among multiple providers international: country code - 595; satellite earth station - 1 Intelsat (Atlantic Ocean) (2010)

Broadcast media:

6 privately owned TV stations; about 75 commercial and community radio

stations; 1 state-owned radio network (2010)

Internet country code:

.py

Internet hosts:

280,658 (2012)

country comparison to the world: 65

Internet users:

1.105 million (2009)

country comparison to the world: 94

PHILIPPINES

Telephones - main lines in use:
3.939 million (2012)
country comparison to the world: 43

Telephones - mobile cellular:
103 million (2012)
country comparison to the world: 12

Telephone system:
general assessment: good international radiotelephone and submarine cable services; domestic and interisland service adequate domestic: telecommunications infrastructure includes the following platforms: fixed-line, mobile cellular, cable TV, over-the-air TV, radio and Very Small Aperture Terminal (VSAT), fiber-optic cable, and satellite; mobile-cellular communications now dominate the industry international: country code - 63; a series of submarine cables together provide connectivity to Asia, US, the Middle East, and Europe; multiple international gateways (2011)

Broadcast media:
multiple national private TV and radio networks; multi-channel satellite and cable TV systems available; more than 350 TV stations - 4 major TV networks operating nationwide with 1 being government-owned; some 1100 cable TV providers and some 1,200 radio stations broadcasting; the Philippines is scheduled to complete the switch from analog to digital broadcasting by the end of 2015 (2012)

Internet country code:
.ph

Internet hosts:
425,812 (2012)
country comparison to the world: 52

Internet users:
8.278 million (2009)
country comparison to the world: 34

RUSSIA

Telephones - main lines in use:

42.9 million (2012)

country comparison to the world: 6

Telephones - mobile cellular:

261.9 million (2012)

country comparison to the world: 5

Telephone system:

general assessment: the telephone system is experiencing significant changes; there are more than 1,000 companies licensed to offer communication services; access to digital lines has improved, particularly in urban centers; Internet and e-mail services are improving; Russia has made progress toward building the telecommunications infrastructure necessary for a market economy; the estimated number of mobile subscribers jumped from fewer than 1 million in 1998 to more than 235 million in 2011; fixed line service has improved but a large demand remains domestic: cross-country digital trunk lines run from Saint Petersburg to Khabarovsk, and from Moscow to Novorossiysk; the telephone systems in 60 regional capitals have modern digital infrastructures; cellular services, both analog and digital, are available in many areas; in rural areas, the telephone services are still outdated, inadequate, and low density international: country code - 7; Russia is connected internationally by undersea fiber optic cables; satellite earth stations provide access to Intelsat, Intersputnik, Eutelsat, Inmarsat, and Orbita systems (2011)

Broadcast media:

6 national TV stations with the federal government owning 1 and holding a controlling interest in a second; state-owned Gazprom maintains a controlling interest in a third national channel; government-affiliated Bank Rossiya owns controlling interest in a fourth and fifth, while the sixth national channel is owned by the Moscow city administration; roughly 3,300 national, regional, and local TV stations with over two-thirds completely or partially controlled by the federal or local governments; satellite TV services are available; 2 state-run national radio networks with a third majority-owned by Gazprom; roughly 2,400 public and commercial radio stations (2007)

Internet country code:

.ru; note - Russia also has responsibility for a legacy domain ".su" that was allocated to the Soviet Union and is being phased out

Internet hosts:

14.865 million (2012)

country comparison to the world: 10

Internet users:

40.853 million (2009)

country comparison to the world: 10

UGANDA

Telephones - main lines in use:

315,000 (2012)

country comparison to the world: 113

Telephones - mobile cellular:

16.355 million (2012)

country comparison to the world: 58

Telephone system:

general assessment: mobile cellular service is increasing rapidly, but the number of main lines is still deficient; work underway on a national backbone information and communications technology infrastructure; international phone networks and Internet connectivity provided through satellite and VSAT applications domestic: intercity traffic by wire, microwave radio relay, and radiotelephone communication stations, fixed and mobile-cellular systems for short-range traffic; mobile-cellular teledensity about 50 per 100 persons in 2010 international: country code - 256; satellite earth stations - 1 Intelsat (Atlantic Ocean) and 1 Inmarsat;

analog links to Kenya and Tanzania (2011)

Broadcast media:

public broadcaster, Uganda Broadcasting Corporation (UBC), operates radio and TV networks; Uganda first began licensing privately owned stations in the 1990s; by 2007 there were nearly 150 radio and 35 TV stations, mostly based in and around Kampala; transmissions of multiple international broadcasters are available in Kampala (2007)

Internet country code:

.ug

Internet hosts:
32,683 (2012)
country comparison to the world: 106

Internet users:
3.2 million (2009)
country comparison to the world: 66

UNITED ARAB EMIRATES (UAE)

Telephones - main lines in use:
1.967 million (2012)
country comparison to the world: 59

Telephones - mobile cellular:
13.775 million (2012)

country comparison to the world: 61

Telephone system:

general assessment: modern fiber-optic integrated services; digital network with rapidly growing use of mobile-cellular telephones; key centers are Abu Dhabi and Dubai domestic: microwave radio relay, fiber optic and coaxial cable international: country code - 971; linked to the international submarine cable FLAG (Fiber-Optic Link Around the Globe); landing point for both the SEA-ME-WE-3 and SEA-ME-WE-4 submarine cable networks; satellite earth stations - 3 Intelsat (1 Atlantic Ocean and 2 Indian Ocean) and 1 Arabsat; tropospheric scatter to Bahrain; microwave radio relay to Saudi Arabia (2011)

Broadcast media:

except for the many organizations now operating in Dubai's Media Free Zone, most TV and radio stations remain government-owned; widespread use of satellite dishes provides access to pan-Arab and other international broadcasts (2007)

Internet country code:

.ae

Internet hosts:

337,804 (2012)

country comparison to the world: 61

Internet users:

3.449 million (2009)

country comparison to the world: 61

All maps and communication data were obtained from:
https://www.cia.gov/library/publications/the-world-factbook/geos/br.html

CPSIA information can be obtained
at www.ICGtesting.com
Printed in the USA
FSOW03n0837040116
15343FS